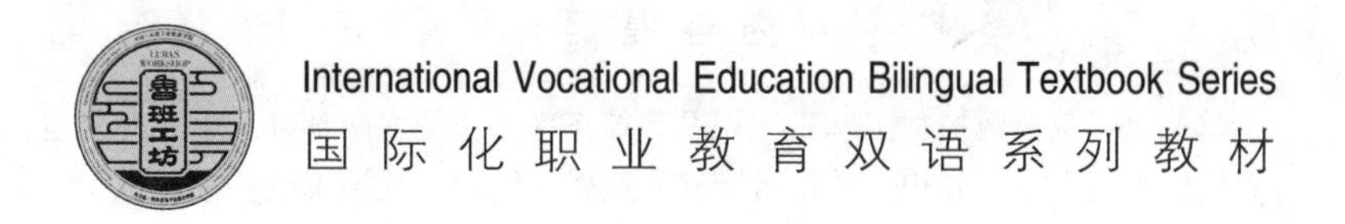

CNC Machining Technology
数控加工技术

Wang Xiaoxia
王晓霞　主　编
Wang Qinglong　Liu Sheng　Dai Dongchen
王庆龙　刘　晟　戴冬晨　副主编

Beijing
Metallurgical Industry Press
2020

内 容 提 要

本书主要包括数控加工基本知识及数控车削加工技术两部分。第一部分包含数控加工技术基本知识、数控机床基本操作与维护及数控编程基本知识；第二部分包含台阶轴的数控加工、圆锥面的数控加工、切槽与切断圆弧面的数控加工、螺纹零件的数控加工、套类零件的数控加工。

本书既可作为职业院校机电相关专业的国际化教学用书，也可作为相关专业企业员工的培训教材和有关专业人员的参考书。

图书在版编目(CIP)数据

数控加工技术 = CNC Machining Technology：汉、英/王晓霞主编．—北京：冶金工业出版社，2020. 8

国际化职业教育双语系列教材

ISBN 978-7-5024-8588-7

Ⅰ.①数…　Ⅱ.①王…　Ⅲ.①数控机床—加工—高等职业教育—双语教学—教材—汉、英　Ⅳ.①TG659

中国版本图书馆 CIP 数据核字(2020)第 152643 号

出 版 人　苏长永

地　　址　北京市东城区嵩祝院北巷 39 号　邮编　100009　电话　(010)64027926

网　　址　www. cnmip. com. cn　电子信箱　yjcbs@ cnmip. com. cn

责任编辑　俞跃春　耿亦直　美术编辑　彭子赫　版式设计　孙跃红　禹　蕊

责任校对　郑　娟　责任印制　李玉山

ISBN 978-7-5024-8588-7

冶金工业出版社出版发行；各地新华书店经销；三河市双峰印刷装订有限公司印刷

2020 年 8 月第 1 版，2020 年 8 月第 1 次印刷

787mm×1092mm　1/16；20. 5 印张；493 千字；307 页

59. 00 元

冶金工业出版社　投稿电话　(010)64027932　投稿信箱　tougao@cnmip. com. cn

冶金工业出版社营销中心　电话　(010)64044283　传真　(010)64027893

冶金工业出版社天猫旗舰店　yjgycbs. tmall. com

Editorial Board of International Vocational Education Bilingual Textbook Series

国际化职业教育双语系列教材

编 委 会

Foreword

With the proposal of the ‘Belt and Road Initiative’, the Ministry of Education of China issued *Promoting Education Action for Building the Belt and Road Initiative* in 2016, proposing cooperation in education, including ‘cooperation in human resources training’. At the Forum on China-Africa Cooperation (FOCAC) in 2018, President Xi proposed to focus on the implementation of the ‘Eight Actions’, which put forward the plan to establish 10 Luban Workshops to provide skills training to African youth. Draw lessons from foreign advanced experience of vocational education mode, China's vocational education has continuously explored and formed the new mode of vocational education with Chinese characteristics. Tianjin, as a demonstration zone for reform and innovation of modern vocational education in China, has started the construction of ‘Luban Workshop’ along the ‘Belt and Road Initiative’, to export high-quality vocational education achievements.

The compilation of these series of textbooks is in response to the times and it's also the beginning of Tianjin Polytechnic College to explore the internationalization of higher vocational education. It's a new model of vocational education internationalization by Tianjin, response to the ‘Belt and Road Initiative’ and the ‘Going Out’ of Chinese enterprises. Tianjin Polytechnic College and Uganda Technical College-Elgon reached a cooperation intention to establish the Luban Workshop to carry out vocational education cooperation on mechatronics technology and ferrous metallurgy technology major in 2019. The establishment of Luban Workshop is conducive to strengthen the cooperation between China and Uganda in vocational education, promote the export of high-quality higher vocational education resources, and serve Chinese enterprises in Uganda and Ugandan local enterprises. Exploring and standardizing the overseas operation of Chinese colleges, the expansion of international influences of China's higher vocational education is also one of the purposes.

The construction of 'Luban Workshop' in Uganda is mainly based on the EPIP (Engineering, Practice, Innovation, Project) project, and is committed to cultivating high-quality talents with innovative spirit, creative ability and entrepreneurial spirit. To meet the learning needs of local teachers and students accurately, the compilation of these international vocational skills bilingual textbooks is based on the talent demand of Uganda and the specialty and characteristics of Tianjin Polytechnic.

These textbooks are supporting teaching material, referring to Chinese national professional standards and developing international professional teaching standards. The internationalization of the curriculums takes into account the technical skills and cognitive characteristics of local students, to promote students' communication and learning ability. At the same time, these textbooks focus on the enhancement of vocational ability, rely on professional standards, and integrate the teaching concept of equal emphasis on skills and quality. These textbooks also adopted project-based, modular, task-driven teaching model and followed the requirements of enterprise posts for employees.

In the process of writing the series of textbooks, Wang Xiaoxia, Li Rui, Wang Zhixue, Ge Huijie, Yu Wansong, Wang Lei, Li Xiujuan, Gong Na, Zhang Zhichao, Jia Yanlu, Chen Baoling and other chief teachers, professional teams, English teaching and research office have made great efforts, receiving strong support from leaders of Tianjin Polytechnic College. During the compilation, the series of textbooks referred to a large number of research findings of scholars in the field, and we would like to thank them for their contributions.

Finally, we sincerely hope that the series of textbooks can contribute to the internationalization of China's higher vocational education, especially to the development of higher vocational education in Africa.

Principal of Tianjin Polytechnic College Kong Weijun

May, 2020

序

随着“一带一路”倡议的提出，2016年中华人民共和国教育部发布了《推进共建“一带一路”教育行动》，提出了包括“开展人才培养培训合作”在内的教育合作。2018年习近平主席在中非合作论坛上提出，要重点实施“八大行动”，明确要求在非洲设立10个鲁班工坊，向非洲青年提供技能培训。中国职业教育在吸收和借鉴发达国家先进职教发展模式的基础上，不断探索和形成了中国特色职业教育办学模式。天津市作为中国现代职业教育改革创新示范区，开启了“鲁班工坊”建设工作，在“一带一路”沿线国家搭建“鲁班工坊”平台，致力于把优秀职业教育成果输出国门与世界分享。

本系列教材的编写，契合时代大背景，是天津工业职业学院探索高职教育国际化的开端。“鲁班工坊”是由天津率先探索和构建的一种职业教育国际化发展新模式，是响应国家“一带一路”倡议和中国企业“走出去”，创建职业教育国际合作交流的新窗口。2019年天津工业职业学院与乌干达埃尔贡技术学院达成合作意向，共同建立“鲁班工坊”，就机电一体化技术专业、黑色冶金技术专业开展职业教育合作。此举旨在加强中乌职业教育交流与合作，推动中国优质高等职业教育资源“走出去”，服务在乌中资企业和乌干达当地企业，探索和规范我国职业院校“鲁班工坊”建设和境外办学，扩大中国高等职业教育的国际影响力。

中乌“鲁班工坊”的建设主要以工程实践创新项目（EPIP：Engineering，Practice，Innovation，Project）为载体，致力于培养具有创新精神、创造能力和创业精神的“三创”复合型高素质技能人才。国际化职业教育双语系列教材的编写，立足于乌干达人才需求和天津工业职业学院专业特色，是为了更好满足当地师生学习需求。

本系列教材采用中英双语相结合的方式，主要参照中国专业标准，开发国际化专业教学标准，课程内容国际化是在专业课程设置上，结合本地学生的技术能力水平与认知特点，合理设置双语教学环节，加强学生的学习与交流能力。同时，教材以提升职业能力为核心，以职业标准为依托，体现技能与质量并重的教学理念，主要采用项目化、模块化、任务驱动的教学模式，并结合企业岗位对员工的要求来撰写。

本系列教材在撰写过程中，王晓霞、李蕊、王治学、葛慧杰、于万松、王磊、李秀娟、宫娜、张志超、贾燕璐、陈宝玲等主编老师、专业团队、英语教研室付出了辛勤劳动，并得到了学院各级领导的大力支持，同时本系列教材借鉴和参考了业界有关学者的研究成果，在此一并致谢！

最后，衷心希望本系列教材能为我国高等职业教育国际化，尤其是高等职业教育走进非洲、支援非洲高等职业教育发展尽绵薄之力。

天津工业职业学院书记、院长　孔维军

2020年5月

Preface

Tianjin Polytechnic College and Uganda Technical College−Elgon reached a cooperation intention to establish the Luban workshop to carry out vocational education cooperation on mechatronics technology and ferrous metallurgy technology major in 2019. In order to strengthen the cooperation between China and Uganda in vocational education, the two colleges plan to compile a series of international vocational skills bilingual textbooks.

This book is one of the international vocational skills bilingual textbooks.

This book is a supporting teaching textbook for the construction project of Uganda Luban Workshop of Tianjin Polytechnic College. According to China's national professional standards of turning workers, the main assessment contents of intermediate turning workers are turning machining of medium complex shafts, sleeves, and ordinary threaded parts, and the major assessment contents of senior turning workers are turning machining of complex curved shafts, sleeves, and multi − thread threaded parts. This book refers to China's national professional standards for CNC technology, follows the requirements of corporate positions for employees. It is guided by the cultivation of engineering consciousness of students, and it is based on professional standards, focuses on improving professional abilities and machining training. Using a modular, task−driven teaching mode, combined with the actual use of equipments.

This book practices the teaching philosophy of equal emphasis on skills and quality, and adopts a hierarchical teaching system, implements training content by category, and promotes the training process step by step. It mainly includes two basic projects of CNC machining basic knowledge and CNC turning processing technology, a total of nine sub−modules. Each sub−module consists of task description, competences, related knowledge, task implementation, assessment and exercise. The task implementation includes the content of national occupation standard assessment requirements such as part drawing reading, processing technology analysis, workpiece processing, and precision inspection. Relevant knowledge is aimed at specific tasks, introducing its process knowledge, programming knowledge, processing and dimensional measurement related knowledge. This book incorporates advanced technology and equipments into the teaching contents in a timely manner, focuses on the combination of traditional manufacturing technology and advanced manufacturing technology, and combines the training of basic engineering skills with the cultivation of engineering innovation ability.

This book mainly introduces CNC turning programming and machining of typical parts such as inner and outer cylindrical surface, conical surface, arc surface, end face, thread, grooving, boring, etc. For students majoring in mechatronics technology, machinery manufacturing technology, numerical control technology, etc., they can carry out teaching on CNC machining cognition, basic operation of CNC machine tools, manual programming of CNC machining, and maintenance of CNC machine tools. It is an important carrier for cultivating composite technical and technical talents.

The book's specific lesson allocation suggestions are as follows:

No.	Contents		Hours
Project 1	Basic knowledge of CNC machining	Task 1. 1 Basic knowledge of CNC machining technology	4
		Task 1. 2 Basic operation and maintenance of CNC machine tools	6
		Task 1. 3 Basic knowledge of CNC programming	6
Project 2	CNC turning technology	Task 2. 1 CNC machining of stepped shaft	6
		Task 2. 2 CNC turning of conical	8
		Task 2. 3 Grooving and cutting of parts	6
		Task 2. 4 CNC turning of arc surface	8
		Task 2. 5 CNC turning of threaded parts	8
		Task 2. 6 CNC machining of sleeve parts	8
Total			60

This book is completed by the CNC programming teaching team of Tianjin Polytechnic College. This book is written by Wang Xiaoxia as editor in chief, and written by Wang Qinglong、Liu Sheng、Dai Dongchen as deputy editor in chief. The specific division of labor is as follows: Wang Xiaoxia is responsible for project 1 task 1. 1, task 1. 3, project 2 task 2. 1. Liang Guoyong is responsible for project 1 task 1. 2, Feng Yanhong is responsible for project 2 task 2. 2. Li Yan is responsible for project 2 task 2. 3. Wang Qinglong is responsible for project 2 task 2. 4. Liu Sheng is responsible for project 2 task 2. 5. Dai Dongchen is responsible for project 2 task 2. 6. This book is unified by Wang Xiaoxia. Principal Kong Weijun and Director Li Meihong put forward many valuable opinions to this book, thank you very much!

Due to the limited level of editors and the hasty time, there are unavoidable omissions and deficiencies in the book, and We hope that readers can provide us valuable comments and suggestions.

The editor
May, 2020

前　言

2019年天津工业职业学院与乌干达埃尔贡技术学院达成合作意向，共同建立“鲁班工坊”，就机电一体化技术专业、黑色冶金技术专业开展职业教育合作，双方计划编撰国际化职业教育双语系列教材。

本书是国际化职业教育双语系列教材之一。

本书是天津工业职业学院乌干达鲁班工坊建设项目配套教材。按照我国车工国家职业标准，中级车工主要考核中等复杂轴套类及普通螺纹零件的车削加工，高级车工主要考核复杂曲面轴套类、多线螺纹零件的车削加工。本书参照我国车工国家职业标准，遵循企业岗位对就业人员的要求，以培养学生的工程意识为导向、职业标准为依托、提升职业能力为核心、加工过程训练为重点，采用模块化、任务驱动的教学模式，结合实际使用设备情况而编写。

本书践行技能与素质并重的教学理念，采用层次化教学体系，实训内容分类实施、实训过程梯次推进，主要包括数控加工基本知识和数控车削加工技术两个项目，共9个任务。每个任务由任务描述、任务目标、相关知识、任务实施、任务评价和同步思考与训练六部分组成。任务实施部分包括零件图识读、加工工艺分析、工件加工、精度检验等国家职业标准考核的内容，相关知识是针对具体任务，介绍其工艺知识、编程知识、加工与尺寸测量相关知识，将先进的技术及设备适时融入教学内容中，注重传统制造技术与先进制造技术相结合，工程基本能力训练与工程创新能力培养相结合。

本书主要介绍了内外圆柱面、圆锥面、圆弧面、端面、螺纹、切槽、镗孔等典型零件的数控车削编程与加工，可面向机电一体化技术、机械制造技术、数控技术等专业的学生开展数控加工认知、数控机床基本操作、数控手工编程、数控机床维护保养等教学，是培养复合型技术技能人才的重要载体。

本书的具体学时分配建议如下表：

序号	课程内容		课时
项目 1	数控加工基本知识	任务 1.1　数控加工技术基本知识	4
		任务 1.2　数控机床基本操作与维护	6
		任务 1.3　数控编程基本知识	6
项目 2	数控车削加工技术	任务 2.1　台阶轴的数控加工	6
		任务 2.2　圆锥面的数控加工	8
		任务 2.3　切槽与切断	6
		任务 2.4　圆弧面的数控加工	8
		任务 2.5　螺纹零件的数控加工	8
		任务 2.6　套类零件的数控加工	8
合　计			60

本书由天津工业职业学院数控编程教学团队完成。由王晓霞担任主编，王庆龙、刘晟、戴冬晨担任副主编。具体分工如下：王晓霞负责项目 1 任务 1.1、任务 1.3，项目 2 任务 2.1；梁国勇负责项目 1 任务 1.2；冯艳宏负责项目 2 任务 2.2；李焱负责项目 2 任务 2.3；王庆龙负责项目 2 任务 2.4；刘晟负责项目 2 任务 2.5；戴冬晨负责项目 2 任务 2.6。本书由王晓霞统稿。天津工业职业学院孔维军院长、机械工程系李梅红主任对本书提出了很多宝贵意见，非常感谢！

由于编者水平有限，书中难免存在疏漏和不足之处，希望读者提出宝贵意见和建议。

编者

2020 年 5 月

Contents

目　录

Project 1 Basic Knowledge of CNC Machining

CNC machining refers to a kind of process method of machining parts on CNC machine tools, which is basically the same as the traditional machine tools. The difference is that CNC machine tools use digital and alphabetical forms to indicate the shape and size of the workpiece and other technical requirements and processing technology. It is required that the control system sends instructions to control the tools for processing. CNC machining is an effective way to solve the problems of changing parts variety, complex shape, small batch size, high precision etc. , and thus to achieve the goal of efficient and automated machining.

Task 1.1 Basic Knowledge of CNC Machining Technology

Introduction

CNC machining technology is the basis of modern manufacturing technology. Its widespread applications have replaced ordinary machinery with CNC machinery, which has fundamentally changed the global manufacturing industry. The level, ownership and popularity of CNC machining technology have became one of the important indicators to measure a country's comprehensive national strength and industrial modernization level. This task mainly solves the following issues:

(1) What parts can the CK6140 CNC lathe complete?

(2) What are the characteristics of CK6140 CNC lathe?

(3) What are the main components of the CK6140 CNC lathe?

(4) Which type does CK6140 CNC lathe belong to according to the different classification methods?

Competences

(1) Knowledge:

1) Understand the process of the production and development of CNC machine tools.

2) Familiar with CNC lathe structure.

3) Recognize the development trend of CNC machine tools.

(2) Skill:

1) Understand digital control, numerical control technology, numerical control system, computer numerical control system.

2) Have the ability to explain CNC machining process, characteristics and applications.

3) Have the ability to classify CNC machine tools in different forms.

(3) Quality:

1) Cultivate the ability to consult information and teamwork.

2) Develop a good habit of learning attitude and fluent expression.

Relevant Knowledge

1. 1. 1 The Concept, Production and Development of Numerical Control

1. 1. 1. 1 Basic Concepts

(1) Numerical Control. Numerical control is abbreviated as NC, which is an automatic method of programming and controlling a certain working process by means of numbers, characters or other symbols.

(2) Numerical Control Technology. Numerical control technology is a technology that uses digital information to control mechanical movements and working processes. It usually controls mechanical quantities such as position, angle, and speed, and the switching quantities related to the flow of mechanical quantities. It has become the basic technology for manufacturing industry to realize automatic, flexible, and integrated production.

(3) Numerical Control System. Numerical control system is a special computer system that performs some or all numerical control functions according to the control program stored in computer memory, and is equipped with interface circuit and servo drive device.

(4) Computer Numerical Control. The computer numerical control system is abbreviated as CNC. It is a numerical control system that uses a special computer which can store programs to perform some or all basic numerical control functions.

1. 1. 1. 2 The Generation and Development of CNC Machine Tools

Numerical control machine tools refer to a type of machine tool that use numerical control technology to control the machining process automatically. The fifth Technical Committee of the International Federation of Information Processing Union (IFIP) defines a CNC machine tool as follows: CNC machine tool is a machine tool equipped with a program control system, which can logically process the program with number or other symbol coded instructions. The program control system mentioned in the definition is the numerical control system.

In 1952, the first CNC machine tool (three-axis vertical CNC milling machine) came out in the United States, which became an epoch-making event in the history of the world's machinery industry. China has been developing CNC machine tools since 1958, and successfully produced the first CNC machine tools in 1966.

Along with the development of electronic technology and computer technology, CNC machine tools have experienced the following stages:

From 1952 to 1959, for the first generation of CNC machine tools, electronic components were electronic tubes;

From 1959 to 1964, for the second generation of CNC machine tool, electronic components were transistors;

From 1965 to 1970, for the third generation of CNC machine tools, and the electronic components were small and medium-sized integrated circuits;

From 1971 to 1974, for the fourth generation of CNC machine tools, electronic components were small-scale general-purpose electronic computer controll systems (CNC) with large-scale integrated circuits;

After 1974, for the fifth generation of CNC machine tools, electronic components were micro electronic computer controlled systems (MNC).

The first three generations were NC hardwired technology, and the latter two generations were CNC softwired technology. At present, the sixth generation CNC machine tools are being developed.

1.1.2 CNC Machining Process, Characteristics and Applications

1.1.2.1 CNC Machining Process

When a CNC machine tool processes a part, it must first record the process information required for processing the part in the form of a program, store it on some carriers, enter it into the CNC system, and the CNC device processes the program to send a control signal to instruct the servo system to drive the servo motor, coordinate the movement of the machine tool, make it produce a series of machine tools movements of the main movement and the feed movement, and complete the processing of the parts. The process of CNC machined parts is shown in Figure 1-1.

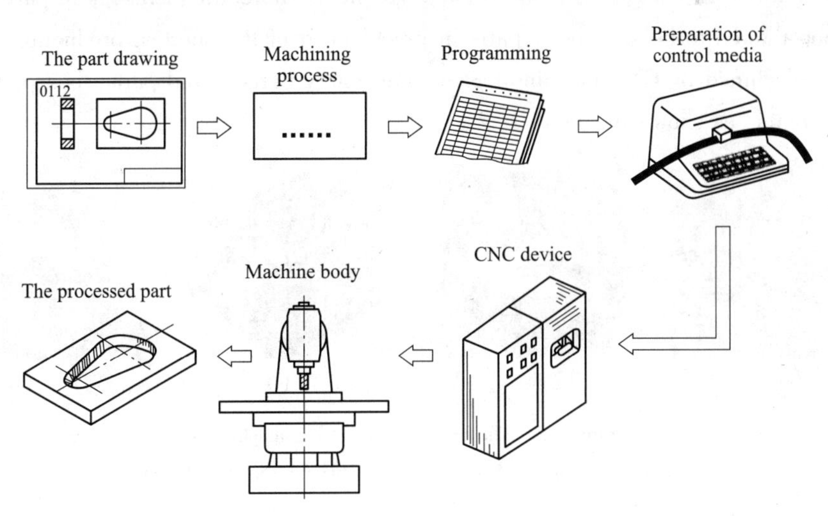

Figure 1-1 The process of CNC machining

1.1.2.2 CNC Machining Characteristics

CNC machining has the following advantages:

(1) Significantly reduce the number of tooling, and no complicated tooling is required to machine

complex shaped parts. If you want to change the shape and size of the part, you only need to modify the part programming, which is suitable for the development and modification of new products.

(2) Stable machining quality, high machining precision and high repetition precision.

(3) In the case of multi-variety and small-batch production, the production efficiency is higher, which can reduce the time of production preparation, machine tool adjustment and process inspection.

(4) It can machine complex surface which is difficult to be machined by conventional methods, and even machine some position which cannot be observed.

The disadvantage of CNC machining is that the machine tools is more expensive. It must be programmed, installed, operated and maintained by senior technicians, which requires the high level of repair personnel.

1.1.2.3 Application of CNC Machining

With the development of electronic technology and computer technology, CNC machine tools are constantly updated and developed rapidly. Today, CNC machine tools have been widely used in almost all manufacturing industries, such as spaceship, ship, machine tool, automobile, rail transit, energy (thermal power, hydropower, thermal power, wind power, etc.), metallurgy, light industry, textile, electronics, medical devices, general machinery, engineering machinery, etc.

As a highly mechatronic product, CNC machine tool has high technical content and high cost. From the most economical aspect, CNC machine tool is suitable for processing a variety of small batch parts, parts with complex structure and high precision requirements, key parts with high price and not allowed to be scrapped, parts in urgent need of the shortest production cycle. The typical parts machined by CNC machine tool include rotary parts, mold parts, box parts and special-shaped parts, as shown in Figure 1-2.

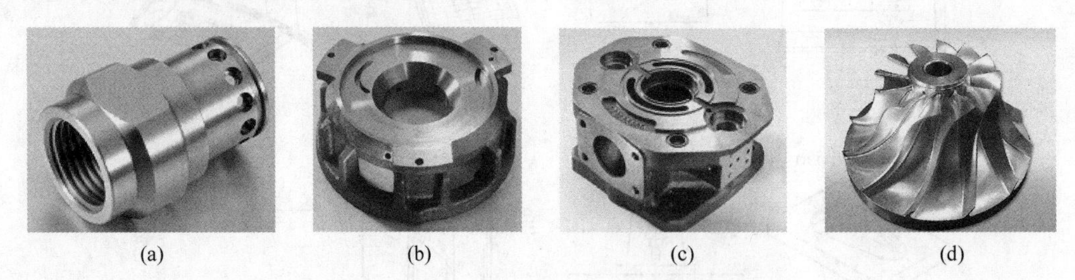

(a) (b) (c) (d)

Figure 1-2 The application of CNC machining

(a) Rotary part; (b) Mold part; (c) Box part; (d) Special-shaped part

1.1.3 Classification of CNC Machine Tool

1.1.3.1 Classification by Control Movement Mode

According to the tool movement mode, CNC machine tools are classified as point control, linear control and contour control, as shown in Figure 1-3.

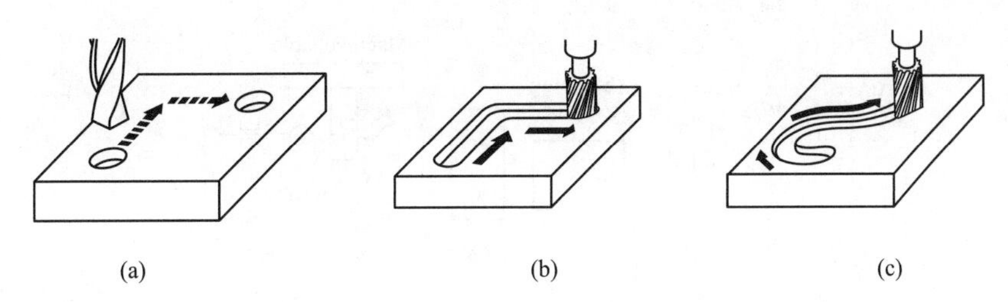

(a) (b) (c)

Figure 1-3 Classification of CNC machine tools according to control motion

(a) Point control; (b) Linear control; (c) Contour control

(1) Point control CNC machine tool. The point control system moves the tool from one point to another. CNC drilling machine is the most typical point control machine tool. In the process of the drill moving from one point to the next, the path and feed is not important, and it does not control the movement trajectory from the start to the end.

(2) Linear control CNC machine tool. The characteristic of linear control CNC machine tool is not only to control the precise position of the target point of the moving parts of the machine tool, but also to realize the cutting feed movement along the straight line or oblique line between the starting point and the target point. Typical linear control CNC machine tools include early CNC lathe, CNC milling machine and CNC boring machine.

(3) Contour control CNC machine tool. The contour control system can simultaneously adjust the feed of the table (or spindle) on at least two coordinate axes. This control method requires complex control and drive systems. CNC lathes, CNC milling machines, machining centers, and other CNC machine tools commonly use contour control motions.

1.1.3.2 Classification by the Control Mode of Servo System

CNC machine tools can be divided into open-loop control and closed-loop control according to the presence or absence of the feedback device for the servo drive controlled quantity. Depending on where the detection device is installed, closed-loop control is divided into two types: full-closed control and semi-closed-loop control, as shown in Figure 1-4.

(1) Open-loop control CNC Machine Tool. The open-loop control system does not provide position feedback information to the control unit. The advantages of the open-loop control system are low cost and the disadvantage is that it is difficult to detect position errors. Economical CNC machines generally use open-loop control.

(2) Full closed-loop control CNC machine tool. The full closed-loop control system has high precision, and usually uses AC, DC or hydraulic servo motors. CNC machine tools such as CNC boring and milling machines, ultra-precision CNC lathes and CNC machining centers that require high accuracy generally use full closed-loop control.

(3) Semi-closed loop control CNC machine tool. The semi-closed-loop control CNC machine tool installs the position detection device at the end of the drive motor or transmission screw to indirectly measure the actual position or displacement of the actuator. Its accuracy is lower than that

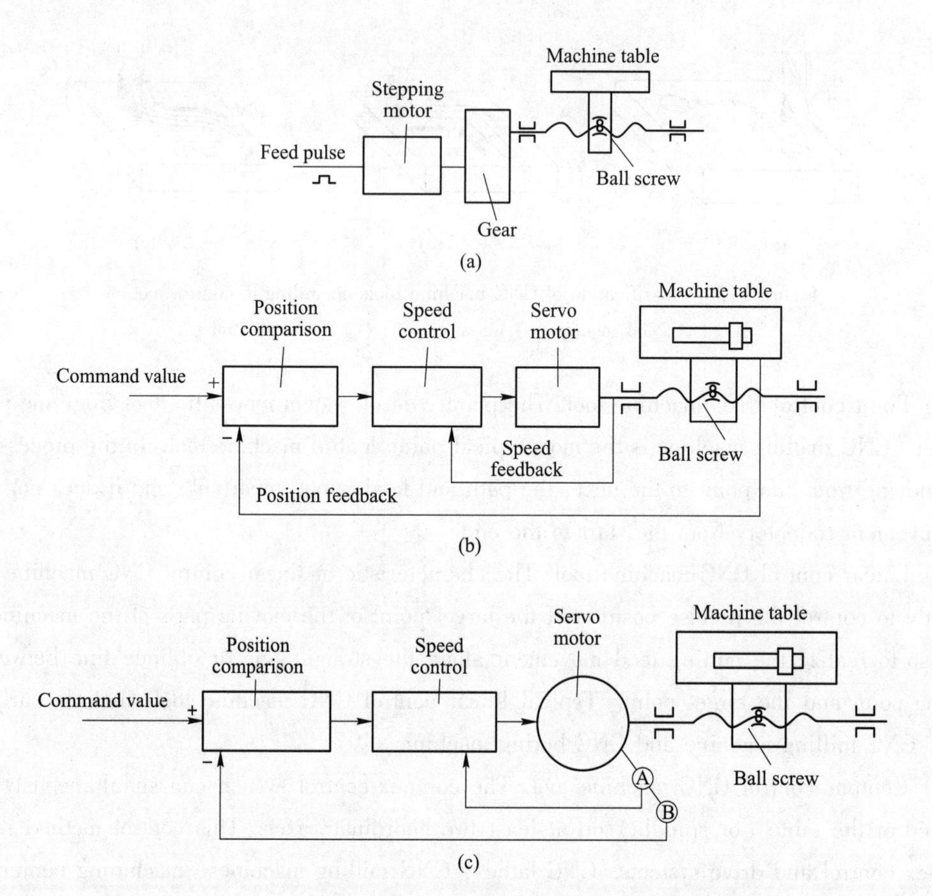

Figure 1-4 Classification of CNC machine tools according to the control mode of servo system

(a) Open loop control CNC machine tool; (b) Full closed-loop control CNC machine tool;

(c) Semi closed loop control CNC machine tool

of the closed-loop system, but the measurement device has a simple structure and easy to install and debug. Mid-range CNC machine tools generally use semi-closed loop control.

1.1.3.3 Classification by Process Use

(1) Cutting CNC Machine Tools.

Cutting CNC machine tool refers to the CNC machine tool with cutting functions.

1) General CNC machine tools. The most commonly used general CNC machine tools are CNC lathes, CNC milling machines, CNC boring machines, CNC drilling machines, CNC grinding machines, CNC gear processing machines, etc., and there are many varieties in each type, as shown in Figure 1-5.

2) Machining centers. The machining center is equipped with a tool magazine and an automatic tool change device on the general CNC machine tool. After the workpiece is clamped once, it can be machined in multiple processes, reducing the number of auxiliary times such as workpiece loading and unloading and tool replacement, and the production efficiency of the machine tool is high. Common machining centers are shown in Figure 1-6.

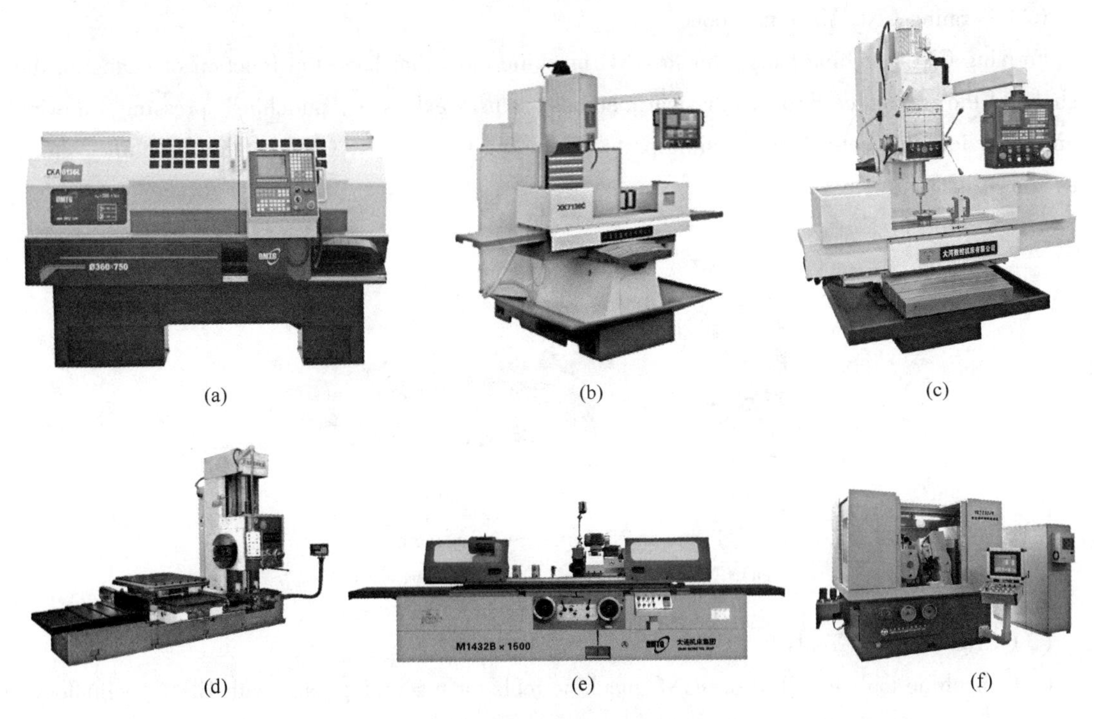

Figure 1-5 General CNC machine tools

(a) CNC lathe; (b) CNC milling machine; (c) CNC drilling machine; (d) CNC boring machine; (e) CNC grinding machine; (f) CNC gear processing machine

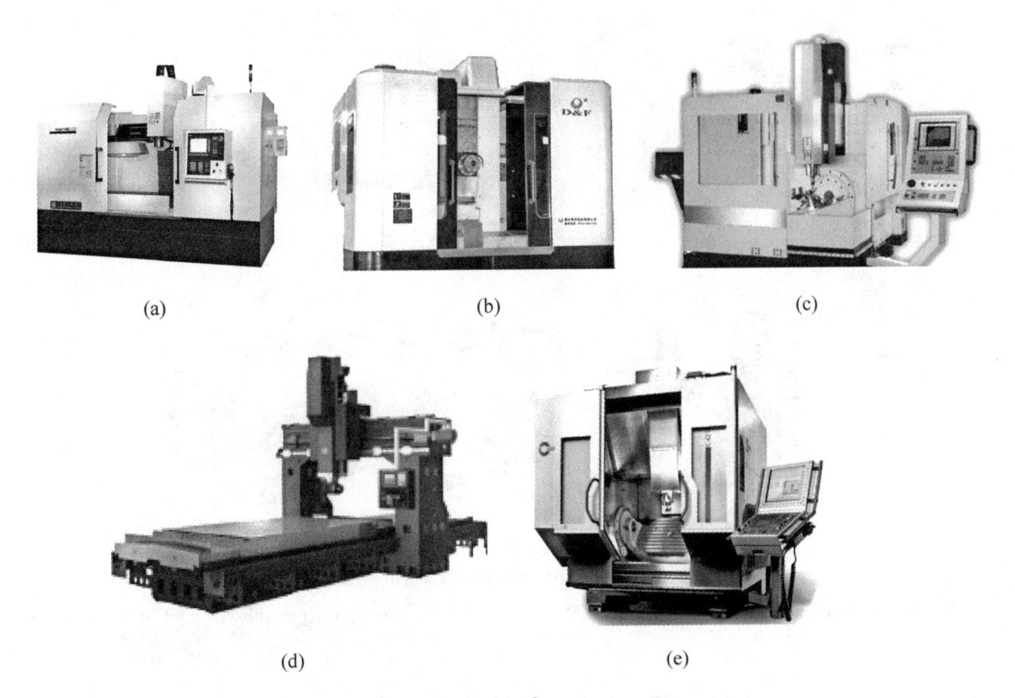

Figure 1-6 Machining centers

(a) Vertical machining center; (b) Horizontal machining center; (c) Universal machining center; (d) Gantry machining center; (e) Five axis machining center

(2) Forming CNC Machine Tools.

Forming CNC machine tools refer to CNC machine tools that have the function of changing the shape of the workpiece through physical methods. It uses extrusion, punching, pressing, drawing and other forming methods to machine parts, including CNC press, CNC bending machine, etc., as shown in Figure 1-7.

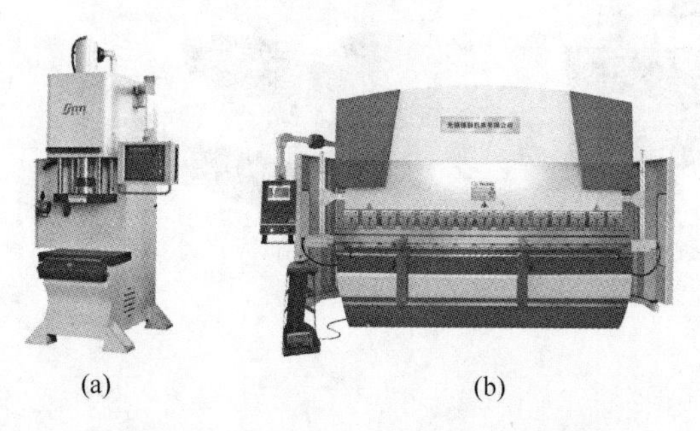

(a) (b)

Figure 1-7 Forming CNC machines

(a) CNC press machine; (b) CNC bending machine

(3) CNC Machine Tools for EDM.

CNC machine tools for EDM are CNC machine tools for machining parts with EDM Technology. The common CNC EDM machine tools are CNC sinker EDM machine, CNC WEDM machine, CNC flame cutting machine, CNC laser beam machining machine, etc., as shown in Figure 1-8.

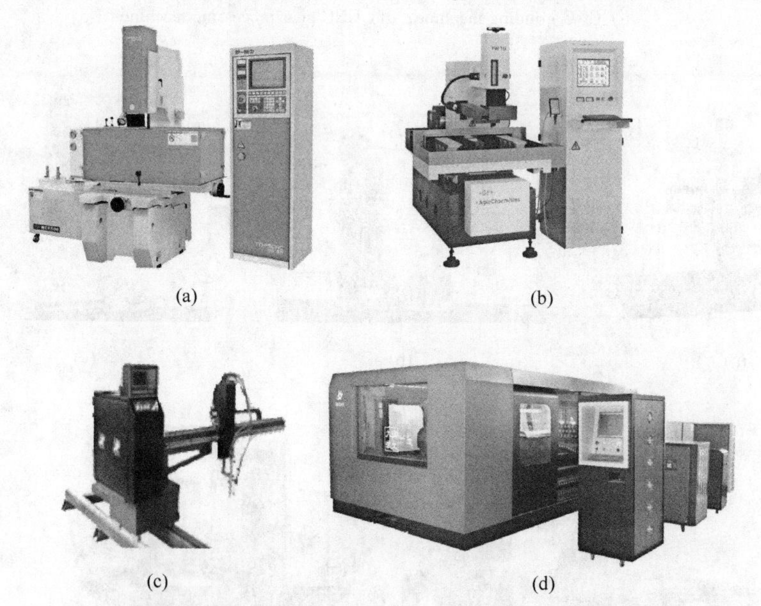

(a) (b)

(c) (d)

Figure 1-8 CNC machine tools for EDM

(a) CNC Sinker EDM Machine; (b) CNC WEDM Machine; (c) CNC flame cutting machine; (d) CNC LBM machine

1.1.4 Composition and Classification of CNC Lathes

The CNC lathe is a computer numerical control lathe. The most machining operations on the CNC lathe are end face turning, cylindrical turning, grooving, cutting, drilling, boring, threading and other processing operations.

1.1.4.1 CNC Lathe's Components

The main components of a CNC lathe are headstock, bed, guide rail, turret, tailstock, feeding mechanism, operator panel and auxiliary systems etc. As shown in Figure 1-9.

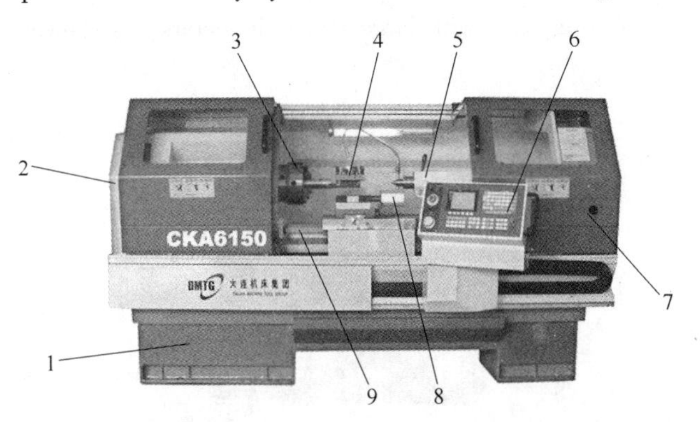

Figure 1-9 Components of the CNC lathe

1—Bed; 2—Headstock; 3—Three jaw chuck; 4—Turret; 5—Tailstock; 6—Operation panel; 7—Protective shield; 8—Turret motor; 9—Guide rail

The bed is fixed on the base, on which the main components of the lathe are installed, and they maintain an accurate relative position during work.

The headstock is fixed on the far left of the bed, and mainly used to support and drive the main shaft to realize the main movement of the machine tool.

The blade slide consists of a longitudinal (Z-direction) slide and a horizontal (X-direction) slide. Longitudinal slides are installed on the bed rails to achieve longitudinal (Z-direction) movement along the bed. Horizontal slides are installed on the longitudinal slides, and lateral (X-direction) movements are achieved along the rails on the longitudinal slides. The tool-holder slide enables longitudinal and lateral feed movements of the tools mounted on it.

The turret tool post is mounted on the slide to clamp various cutting tools, which can be automatically selected as required for certain machining purpose.

The tailstock is installed on bed rail and can be moved longitudinally along the rail to adjust the position. It is mainly used to install the tip and assist the workpiece during machining. It can also be equipped with drills, reamer and other tools for hole machining.

The shield is mounted on the base to protect operators and the workshop from chips.

Hydraulic transmission provides some auxiliary movements, mainly, spindle gear shift, tailstock moving, and automatic clamping.

The electric control system consists of a CNC system (including a CNC device, a servo system and a PLC) and an electrical control system. It realizes automatic machine tool control.

1. 1. 4. 2 CNC Lathe's Classification

(1) Classification by the Spindle's Position. According to the spindle position, CNC lathes are divided into horizontal CNC lathes and vertical CNC lathes. Horizontal CNC lathes are used to process the rotating surfaces of various shaft, sleeve and disk parts, such as internal and external cylindrical surfaces, conical surfaces, and threaded surfaces, as shown in Figure 1-9. Vertical CNC lathes are used for turning parts with large rotary diameters, as shown in Figure 1-10.

Figure 1-10 Vertical CNC lathe

(2) Classification by the Features of the Control System. According to the control system, CNC lathes are classified as economy CNC lathes, full-function CNC lathes and turning centers.

1) Economy CNC lathes. Economical CNC lathes are generally designed on the basis of ordinary lathes, equipped with an open-loop servo system, and the control device is a single board machine. This type of lathes has a simple structure and low price. Compared with other CNC lathes, there is no automatic compensation of the tool nose arc radius and constant linear speed cutting, as shown in Figure 1-11.

Figure 1-11 Economy CNC lathe

2) Full-function CNC lathes. Full-function CNC lathes are often referred to as CNC lathes for short, as well as standard CNC lathes, as shown in Figure 1-9. The full-function CNC lathe has a standard system with a high-resolution CRT display, with graphics simulation, tool compensation, communication or network interface, multi-axis linkage and other functions. The full-function CNC lathe adopts a closed-loop or semi-closed-loop control servo system, which has the characteristics of high rigidity, high precision and high efficiency.

(3) Turning centers. As shown in Figure 1-12, the turning center is a full-function CNC lathe. It is equipped with parts such as a tool magazine, an automatic tool changer, an indexing device, a milling device, and a manipulator to achieve multi-step composite processing. After the workpiece is clamped at one time, the turning center can complete the turning, milling, drilling, reaming, tapping and other machining operations. Its efficiency and degree of automation are high, but the price is more expensive.

Figure 1-12 Turning center

1.1.5 Development Trend of CNC Machine Tools

With the development of science and technology, today's CNC machine tools are constantly adopting the latest technological achievements, and are developing in the direction of high speed, high precision, multi-function, intelligence, automation and maximum reliability, which are specifically manifested in the following aspects.

1.1.5.1 High Speed and High Precision

(1) The numerical control machine tool system uses a processor with a higher number of bits and a higher frequency to increase the basic operation speed of the system, making high-speed operation, modularization, and multi-axis group control systems possible.

(2) Based on the full digital servo system, the linear servo motor is used to directly drive the "Zero drive" linear servo feed mode of the machine tool table.

(3) The non-linear compensation control technology of static and dynamic friction of machine tool is adopted.

(4) High speed and high power motorized spindle is applied.

(5) It is equipped with high-speed and powerful built-in PLC.

1.1.5.2 Multifunction

(1) CNC machine tools have multiple functions, which can maximize the utilization of the equipment.

(2) Adopt the technology of "foreground machining and background editing".

(3) CNC equipment networking.

1.1.5.3 Intellectualization

(1) Pursuing intelligence in machining efficiency and machining quality, such as adaptive control (AC) of machining process, automatic generation of process parameters.

(2) Improving the drive performance and using the intelligent connections, such as feedforward control, adaptive calculation of motor parameters, automatic load identification, and automatic model selection.

(3) Simplification of programming and intelligent operation, such as intelligent automatic programming, intelligent man-machine interface, etc.

(4) The content of intelligent diagnosis and intelligent monitoring is convenient for system diagnosis and maintenance.

1.1.5.4 Automation

(1) CAD/CAM graphic interactive automatic programming.

(2) CAD/CAPP/CAM integrated fully automatic programming.

1.1.5.5 Reliability Maximization

(1) The CNC system will use a more integrated circuit chip.

(2) Through automatic running diagnostics, online diagnostics, offline diagnostics and other diagnostic programs, fault diagnosis and alarming of hardware, software and various external devices in the system are realized.

Implemention

Visit the CNC machining training center, observe the CK6140 CNC Lathe according to the learning, and complete Table 1-1.

Table 1-1 The task implementation worksheet

Machine name		CK6140 CNC lathe	Fill in the record sheet according to the observation results
No.	Items		Record
1	Machining contents	Typical parts machined by CNC machine tools	
		Parts suitable for machining by CNC machine	
		Surfaces of CNC machined parts	
2	Characteristic	Advantages of CNC machining	
		Disadvantages of CNC machining	
3	Structure composition	Basic compositions	
		The role of each component	

Continued Table 1-1

No.	Items		Record
4	Category	Classification by control movement mode	
		Classification by servo system control mode	
		Classification by process use	
		Classification by spindle position	
		Classification by control system characteristics	

Assessment

The assessment table is shown in Table 1-2.

Table 1-2 The assessment table

Item	No.	Standard	Partition	Score
Machining contents (15%)	1	Typical parts machined by CNC machine tools	5	
	2	Parts suitable for machining by CNC machine	5	
	3	Surfaces of CNC machined parts	5	
Characteristics (15%)	4	Advantages of CNC machining	12	
	5	Disadvantages of CNC machining	3	
Structure compositions (30%)	6	Basic compositions	10	
	7	The role of each component	20	
Category (20%)	8	Classification by control movement mode	4	
	9	Classification by servo system control mode	4	
	10	Classification by process use	4	
	11	Classification by spindle position	4	
	12	Classification by control system characteristics	4	
Vocational ability (20%)	13	Learning ability	4	
	14	Communication skills	4	
	15	Team cooperation	4	
	16	Safe operation and civilized production	8	
Total				

Exercises

(1) Answer the following questions

1) Briefly describe characteristics of CNC machining.

2) Which kind of parts are suitable for CNC machining?

3) Which types of CNC machine tools can be divided into according to process use?

(2) Synchronous training

1) Watch the process of CK6140 CNC lathe and explain the process of CNC machining.

2) Observe the pictures of CKA6150 CNC lathe, and explain the structure and category of the lathe.

Task 1.2 Basic Operation and Maintenance of CNC Machine Tools

Introduction

For those new to CNC machine tools, it is very important to master certain CNC machine tools operation skills. On the one hand, it can avoid the occurrence of machine tool collision accidents, leading to machine tool damage. On the other hand, it can quickly improve the operator's skills in a short period of time, and is qualified for his own job. For training equipment, this task requires operations to be completed:

(1) Start CK6140 CNC lathe.

(2) Realize the forward rotation of the spindle in manual mode and MDI mode.

(3) Achieve tool feed in manual mode and manual mode.

(4) Complete any tool change operation.

(5) Complete routine maintenance of CK6140 CNC lathe.

Competences

(1) Knowledge:

1) Know the mechanical structure of CNC lathes.

2) Identify the CNC lathe system panel and operation panel.

3) Familiar with the safety operation rules of CNC lathes.

(2) Skill:

1) Learn how to start and close CNC lathes in accordance with operating procedures.

2) Skilled in manual and MDI operation of CNC lathes.

3) Have the ability to complete the routine maintenance of CNC lathes as required.

(3) Quality:

1) Train students to think independently and pay attention to details.

2) Cultivate students not to give up lightly and have a sense of responsibility.

Relevant Knowledge

Computer Numerical Control lathes, also known as CNC lathes, are digitally controlled lathes with a computer. It has the characteristics of flexible processing, high versatility, and can adapt to frequent changes in product varieties and specifications. It can meet the requirements of new product development and multiple varieties, small batches, and production automation. Therefore, they are widely used in machinery manufacturing, such as automobiles manufacturers, engine manufacturers, etc.

1.2.1 Mechanical Structure of CK6140 CNC Lathe

As shown in Figure 1-13, the CK6140 CNC lathe is a 2-axis CNC lathe that works under the control of a computer. It is equipped with a 4-station tool post. The spindle speed is adjusted through frequency changes. The main drive motor realizes stepless speed regulation under the control of the inverter. It is suitable for turning inner and outer cylindrical surfaces, end faces, arbitrary tapered surfaces, arc surfaces and various threads. It can also complete turning, cutting, grooving, drilling, cutting, reaming and other processes. It has high processing efficiency, high precision, convenient use and safe and reliable operation.

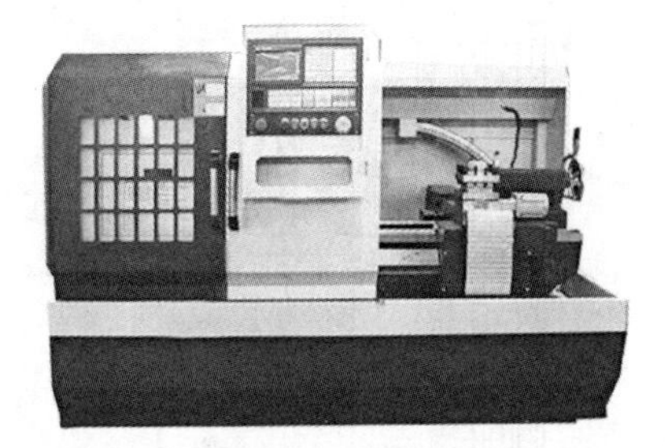

Figure 1-13 CK6140 Horizontal CNC lathe

1.2.1.1 Lathe Bed

Adopt flat bed structure, resin sand molding, high quality cast iron casting. The guide rail adopts medium frequency quenching grinding and sticking plastic technology, which has good wear resistance and accuracy retention.

1.2.1.2 Spindle Box

The main shaft structure is shown in Figure 1-14. It adopts a typical structure of two-point support at the front and rear ends, with high rigidity and rotation accuracy. The taper of the main shaft hole is MT6.

CNC lathe spindle adopts V-belt drive. V-belt drive can adjust belt tension to obtain the best transmission performance. Its advantage is no noise when the CNC lathe changes speed, as shown in Figure 1-15.

1.2.1.3 Feed Motion

The CNC lathe feed system is divided into longitudinal (*X* axis) and transverse (*Z* axis). The feed motion is driven by the ball motor of the servo motor to achieve rapid movement and feed motion, as shown in Figure 1-16.

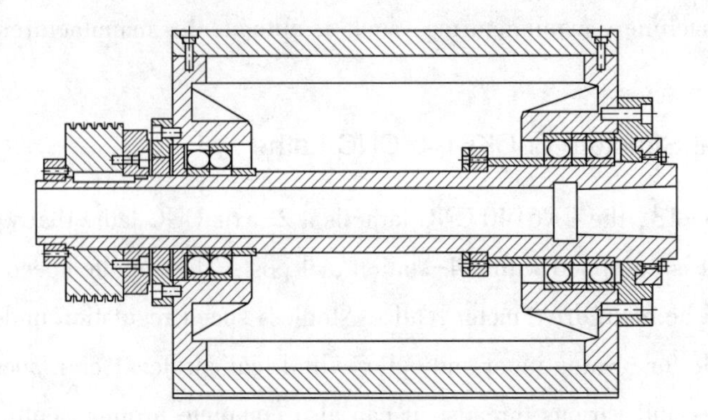

Figure 1-14 Main shaft structure

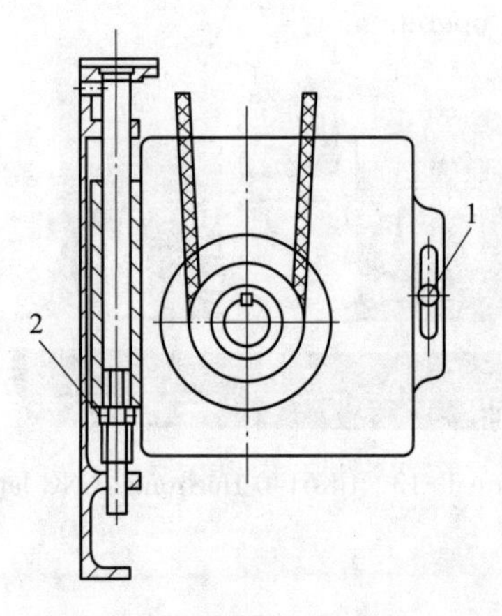

Figure 1-15 Schematic diagram of spindle motor belt adjustment

1—Locking bolt; 2—Main shaft belt adjusting bolt

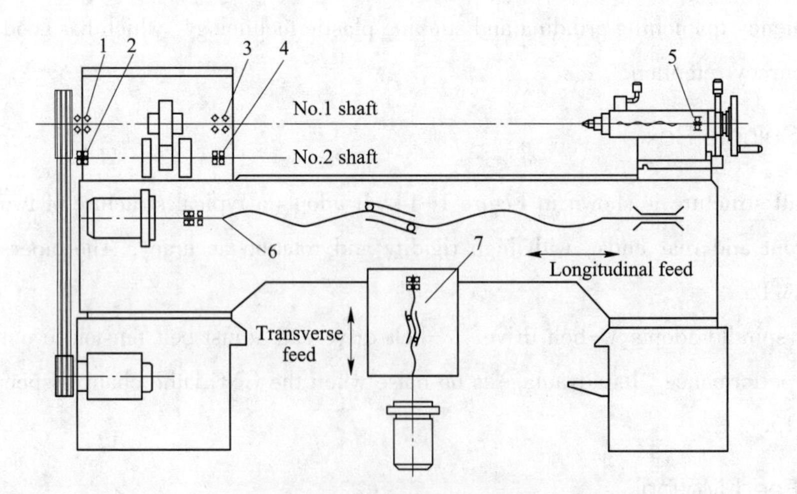

Figure 1-16 Feed motion diagram

1~5—Bearing; 6—Longitudinal feed lead screw; 7—Transverse feed lead screw

Longitudinal feed motion (X axis): The X axis is radial and parallel to the horizontal slide.

Transverse feed motion (Z axis): The main axis transmitting the main cutting force is the Z axis and is perpendicular to the workpiece clamping surface. If there are multiple spindles, choose a spindle that is perpendicular to the workpiece clamping surface as the Z axis.

1.2.1.4 Main Technical Parameters of CNC Lathe

The main technical parameters of CK6140 CNC lathe are shown in Table 1-3.

Table 1-3 The main technical parameters of CK6140-750 CNC lathe

Items	Parameters
Maximum turning diameter on the machine	400mm
Maximum swing bracket	210mm
Maximum length of workpiece	750mm
Spindle speed range	80~750/400~1800r/min or 80~1800r/min
Spindle hole	ϕ52/ϕ82mm
Taper of spindle hole	MT6
Tool rest position	4
Maximum size of tool holder	20×20 (4 stations)
Taper of tailstock sleeve	MT4
Maximum range of tailstock sleeve	140mm
Main motor power	5.5kW
(Z) Cut off the motor (power supply)	6Nm
(X) Cut off the motor (power supply)	4Nm
Package size	2300mm×1380mm×1720mm
Net weight	14000kg

1.2.2 CK6140 CNC Lathe Panel

The CNC machine tool operation panel is an important component of CNC machine tool. It is a tool for operators to interact with CNC machine tool. It mainly consists of display device, NC keyboard, MCP, status light and other parts. There are many types of CNC lathes and CNC systems. The operation panels designed by various manufacturers are not the same, but the basic functions and use methods of various knobs, buttons and keyboards in the operation panel are basically the same.

CK6140 CNC lathe panel is composed of system panel and operation panel, as shown in Figure 1-17 and Figure 1-18. The names of the keys on the CNC lathe system panel are shown in Table 1-4, and the functions of the keys on the CNC lathe operation panel are shown in Table 1-5.

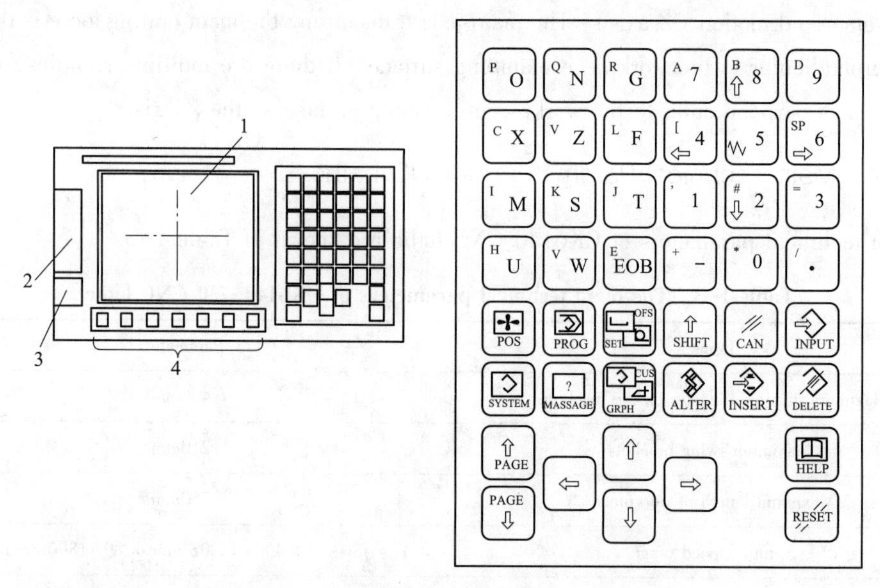

Figure 1-17 FANUC series 0i model F system panel

1—LCD; 2—PCMIA; 3—USB; 4—Soft key;

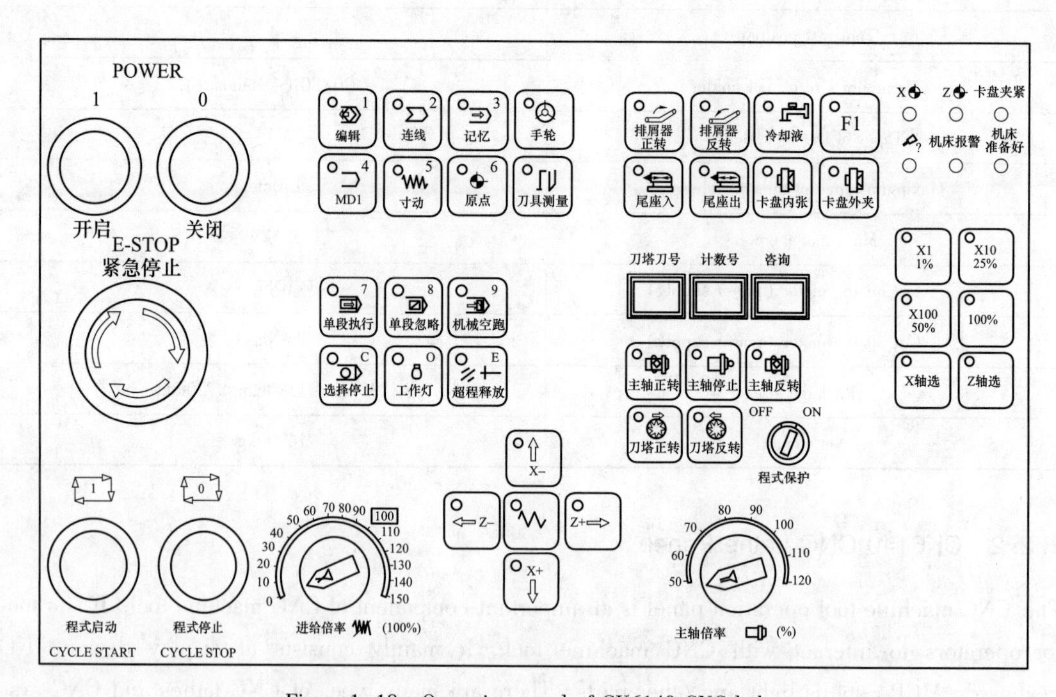

Figure 1-18 Operation panel of CK6140 CNC lathe

Table 1-4 Main key names of CNC system panel

Key	Name	Key	Name
◄□□□□□□►	Soft key	P O	Address and number keys

Continued Table 1-4

Key	Name	Key	Name
POS	Position key	PROG	Program key
OFS SET	OFS/SET key	⇧ SHIFT	SHIFT key
CAN	Cancel key	INPUT	Input key
SYSTEM	System key	? MASSAGE	Massage key
CUS GRPH	Graphics key	ALTER	Alter key
INSERT	Insert key	DELETE	Delete key
⇧ PAGE PAGE ⇩	Page key	⇧ ⇦ ⇨ ⇩	Cursor key
HELP	Help key	RESET	Reset key
E EOB	End of block key		

Table 1-5 Key functions of CNC lathe operation panel

Name	Key	Function
Power switch	1 开启	Power on
	0 关闭	Power off
Emergency stop	E-STOP 紧急停止	It is red. If there is or will be an accident, press this button to cut off the power and stop the servo control. If you want to restore the power supply, turn this button slightly clockwise, and then reset the origin again

Continued Table 1-5

Name	Key	Function
Operation mode	1 编辑	Edit mode: edit programs, etc
	2 连线	Connection mode: on the one hand, the machine can execute program transmission with personal computer, on the other hand, it can machine at the same time
	3 记忆	Memory mode: the computer executes memory storage programs
	4 MDI	MDI mode: this key is used for manual data input. It is used for operations that are not memorized during machining. It is usually limited to input and execute a single section of instruction
	5 寸动	Inching mode: use the manual movement of each axis of the machine tool to select the axial movement direction key and movement speed, and select this button to control
	6 原点	Origin mode: return to origin in each axis. (Before the *Z* axis performs home position return, the *X* axis home position must be performed first)
	手轮	Handwheel mode: when using handwheel operation mode, select this button
Spindle function	主轴正转	Spindle forward rotation: (limited to manual mode) press this key to perform forward rotation, and the indicator light will be on
	主轴反转	Spindle reverse: (limited to manual mode) press this key to perform the reverse operation of the spindle, and the indicator light is on
	主轴停止	Spindle stop: press this key to stop the spindle immediately
	50 60 70 80 90 100 110 120 主轴倍率 (%)	Spindle rotation speed adjustment: During automatic operation, turning this button can adjust the spindle rotation speed, but it is invalid when cutting threads. Its adjustment range is 50%~120%
Automatic operation	1 程序启动 CYCLE START	CYCLE START: This key is used to start the automatic operation cycle. When the automatic operation is performed, this light will stay on until the execution is completed

Continued Table 1-5

Name	Key	Function
Automatic operation	程序停止 CYCLE STOP	CYCLE STOP: Press this button in automatic operation to stop the tool feed immediately. When this button is pressed again, the unfinished program instruction can be resumed
	7 单段执行	Single block execution: Each time the "program start" key is pressed, the program executes one block
	8 单段忽略	Single block ignore: When the program is executed to the block with "/" at the beginning, it will not execute this block and continue to the next block
	9 机械空跑	Mechanical dry run: The cutting rate of the axis in the program is controlled by the feed rate selection switch on the panel
	C 选择停止	Select stop: This button must cooperate with the M01 instruction in the program. When this instruction is executed, press this button. When the program reaches M01, the machine will automatically stop processing; if this button is not pressed, M01 The instruction will be ignored without execution stop
	0 工作灯	Work light: When this key is pressed, the key indicator light will be on immediately, and the work light will be on immediately. If the key is pressed again, the key indicator will disappear and the work light will go out
	E 超程释放	Overtravel release: When this key is held down, the overtravel state or emergency stop state can be released
Fast movement rate	X1 1%	When the machine tool is manual, this button can control the movement speed of *X/Z* axis, each time the axis/position is moved by 0.001mm; When the program is executed automatically, this button can control the rapid displacement (G00) rate, which is reduced to 1%; When using the handwheel, each turn of the slide moves 0.001mm
	X10 25%	When the machine tool is manual, this button can control the *X/Z* axis movement rate, each time the axis/position is moved by 0.01mm; When the program is executed automatically, this button can control the rapid displacement (G00) rate, which is reduced to 25%; When using the handwheel, each turn of the slide moves 0.01mm
	*X*100 50%	When the machine tool is manual, this button can control the *X/Z* axis movement rate, and the axis/position moves 0.1mm each time; When the program is executed automatically, this button can control the rapid displacement (G00) rate, which is reduced to 50%; When using the handwheel, each turn of the slide moves 0.1mm

Continued Table 1-5

Name	Key	Function
Fast movement rate	100%	When the machine tool is manual, this button can control the X/Z axis movement rate, and the axis/position moves 1.0mm each time; When the machine tool executes automatically, if 100% is selected, the movement speed of the machine tool is consistent with the programmed G00 speed. When using the handwheel, each slide rotates 1.0mm
Feed rate	0 10 20 30 40 50 60 70 80 90 100 110 120 130 140 150 进给倍率 (100%)	Feed rate adjustment: In the automatic state, the feed speed specified by the F code can be adjusted with this switch, the adjustment range is 0% to 150%; adjustment is not allowed when turning threads
Axis selection	X轴选	Select the X axis
	Z轴选	Select the Z axis
Rotary handwheel		Rotating in the "-" direction (counterclockwise) indicates feeding in the negative direction of the axis, and rotating in the "+" direction (clockwise) indicates feeding in the positive direction of the axis
Axis/position	X-	Moving in the negative direction of the X axis, the tool approaches the workpiece in the lateral direction
	X+	Moving in the positive direction of the X axis, the tool moves away from the workpiece in the lateral direction
	Z-	Moving in the negative direction of the Z axis, the tool approaches the workpiece in the longitudinal direction
	Z+	Moving in the positive direction of the Z axis, the tool moves away from the workpiece in the longitudinal direction
		Rapid movement: rapid movement along the selected axis
Auxiliary functions	冷却液	Manual coolant: When this key is pressed, the key indicator lights up immediately and the cutting fluid will be ejected immediately; if you press it again, the light will go out and the cutting fluid will stop ejecting at the same time
	OFF ON 程式保护	Program protection switch: This switch is used to prevent programs stored in memory from being modified by mistake. When modifying the program, turn the program protection switch to the " I " position

Continued Table 1-5

Name	Key	Function
Turret operation	刀塔正转	Turret forward rotation: Press this button once to perform a forward rotation of the turret. Press and hold this button to continue forward rotation (only in manual mode)
	刀塔反转	Turret reversal: Press this key once to perform a reversal of the turret. Press and hold this key to continue the reversal (only in manual mode)
Indicator lights	机床准备好	The system is powered on and the power indicator is on
	X	The X-direction reference point is completed, and the X-zero indicator is on
	Z	The Z direction reference point is completed, and the Z-zero indicator is on
	机床报警	Mechanical failure warning light
	?	Warning light for skid oil shortage, it is used to lubricate the ball screw and bed. When the light is on, it must be refueled

1.2.3 Basic Operation of CNC Lathe

1.2.3.1 Machine Power on and Power off

(1) Power on operation sequence.

1) Turn on the external main power switch.

2) Switch the main power switch of the electrical box of the machine tool to the "ON" state, and turn on the main power of the machine tool.

3) Press the switch (1 开启) on the operation panel to turn on the NC power.

4) Release emergency stop (E-STOP 紧急停止) switch.

(2) Power off Operation Sequence.

1) Press emergency stop (E-STOP 紧急停止), to make the machine enter the emergency stop state.

2) Press the switch on the operation panel to turn off the NC power.

3) Switch the main power switch of the electrical box of the machine tool to the "OFF" state, and turn off the main power of the machine tool.

4) Turn off the external main power switch.

1. 2. 3. 2 Emergency Stop Switch

When the machine tool is in a critical situation, press the emergency stop button 

immediately, and the machine tool will stop all controls immediately. When released, turn the button clockwise to resume the standby state. After the emergency stop is released, each axis must return to the machine reference point before continuing operation.

1. 2. 4 Manual Reference Point Return

With an incremental measuring system, the reference point return operation must be performed before the machine tool works. Once the machine tool has a power failure, emergency stop or overtravel alarm signal, the CNC system loses the memory of the reference point coordinates, and the operator must perform the operation of returning to the reference point after troubleshooting. It is not necessary to return to the reference point with an absolute measuring system. The operation steps for manual reference point return are as follows:

(1) Press Key.

(2) Press the key and key, the tool returns to the reference point quickly, the zero return indicator lights up, and check whether the mechanical coordinate value on the CRT display is zero.

Note: The order of returning to the reference point of the machine tool is X axis first, then Z axis, to prevent the tool post from colliding with the tailstock. In addition, when the stop on the slide is less than 30mm from the reference point, use the key to move the slide to the negative direction of the reference point before returning to the machine reference point.

1. 2. 5 Manually Operated Machine Tools

1. 2. 5. 1 Manual Feed of Skateboard

The fast moving function can move each axis for a long distance and fast. The steps are as follows:

(1) Select the Key.

(2) Adjust the fast moving speed percentage to the desired position.

(3) Press the axis/position selection switch to move each axis in the required direction.

1.2.5.2 Manual Control of Spindle Rotation

(1) Spindle Start.

1) Press the key [O 4 MDI].

2) Press the key [PROG], the program screen under MDI appears on LCD display.

3) Input "M03" or "M04", input "S××", such as "M03 S500", press the key [E EOB]、[INSERT].

4) Start by cycle [1 程式启动 CYCLE START] Key, the spindle rotates at the set speed.

(2) Spindle Stop.

1) Input "M05" in MDI screen and press the key [E EOB]、[INSERT].

2) Press the key [1 程式启动 CYCLE START], spindle stop.

The above method is used for the first spindle rotation after the machine is turned on. Later operations can be performed in manual mode to directly rotate the spindle forward, reverse or stop.

1.2.5.3 Manual Operation of Tool Holder Rotation

(1) Press the key [O 4 MDI].

(2) Press the key [PROG].

(3) Input T××, such as "T01", press the key [E EOB]、[INSERT].

(4) Press the key [1 程式启动 CYCLE START], No. 1 Tool is turned to the working position.

1.2.6 Daily Maintenance of CNC Lathes

After using the CNC lathe for a period of time, the parts that are in contact with each other will wear out, and their working performance will be gradually affected. At this time, some parts of the CNC lathe should be properly adjusted and maintained to restore the machine tool to normal technology status. The daily maintenance and primary maintenance of the CNC lathe are completed by the operator, and the secondary maintenance is jointly performed by the operator and the maintenance staff. The maintenance cycle and the time of each maintenance should also be reasonably determined according to the different conditions of the CNC lathe structure, roughing and finishing, and often adjusted according to the actual situation.

1.2.6.1 Operators of CNC Lathe Shall Carry out Corresponding Maintenance Every Day

(1) Check whether the oil volume of the lubricating oil tank is normal. Turn on the machine to

check whether the lubrication system can work normally.

(2) Check whether the oil volume of the hydraulic oil tank is normal. Turn on the machine to check whether the hydraulic system can work normally.

(3) Check whether the amount of cutting fluid in the cooling tank is normal. Turn on the machine to check whether the cutting fluid system can work normally.

(4) Turn on the machine to check whether the CNC system is normal, whether each fan chip remover is running normally, and whether the pressure gauge values are normal.

(5) Remove the iron filings and sundries from each guide rail in the CNC lathe, and check whether the surface is scratched.

(6) After the machine is turned on, it must be idling for a certain period of time to check whether it is running normally. After completing the one day machining task, turn off the CNC system of the CNC lathe, turn off the power, and clean the work place.

1.2.6.2 Primary Maintenance of CNC Lathe

(1) Wipe the exterior of the CNC lathe, including covers, covers, and accessories, to ensure that it is free of oil, no rust, corrosion, no iron filings and debris, and that it is clean inside and out.

(2) Clean the iron filings on the rotary tool post and tailstock and check that they are operating normally.

(3) Check whether there are loose screws on the lathe, and whether the connections of the fuel tanks, pipes and oil standards are stable. If so, there is any repair in time.

(4) Check the abrasion and tension of the spindle drive belt, X-axis drive belt, and Z-axis drive belt.

(5) Check whether the hydraulic system and the hydraulic pump, hydraulic motor and other components under the system are operating normally, and whether there are abnormal noises during operation.

(6) Check whether the cooling system and the cooling pump, cooling motor and other components under the system operate normally, and whether there are abnormal noises during operation.

(7) Check whether the chip removal system and the chip remover, chip removal motor and other components under the system are operating normally, and whether there are abnormal noises during operation.

(8) Check whether the lubrication system is operating normally, ensure the oil circuits are unblocked. Check and replace the filters as needed; check whether the oil outlets are normal to lubricate the machine tool (rails, tailstock).

(9) Check all the oil tanks and cooling water tanks, and replenish them as needed. Fill the oil injection points of the machine as needed to ensure that the oil tanks, oil gauges and oil windows are bright. Add rust inhibitor to cutting fluid according to requirements in summer.

(10) Check whether the CNC system is normal after startup, and whether the axes of the CNC lathe and the rotary tool post are moving normally.

(11) Wipe the electrical box to ensure that the inside and outside are clean, check whether each line is leaking, whether the contacts are in good contact, and check whether the limit device and ground are safe and reliable.

1.2.6.3 Secondary Maintenance of CNC Lathe

(1) Routine inspections. After the machine has been running for 500 hours, regular inspection and maintenance are required. Generally, the operator should be the main operator and the maintenance personnel should cooperate to perform the inspection. The power must be cut off before the inspection. The inspection items are shown in Table 1-6.

Table 1-6 Routine inspection items

No.	Position	Inspection items
1	Electrical system	Whether the emergency stop button is sensitive and reliable. Whether the motor is running normally, and whether there is abnormal heating. Whether the wires and cables are damaged. Whether the position switch and button function normally and the operation is reliable
2	CNC system	Whether the CNC system starts and shuts down normally and reliably
3	Cooling system lubricating system	Whether cutting fluid and lubricant meet requirements. Whether the liquid level of the oil tank and cutting fluid tank meets the required requirements. Whether each lubrication point is reasonably lubricated. Whether the cutting fluid is obviously contaminated and whether the quality of the lubricant is acceptable. Whether the scraper is damaged
4	Motor device	Whether the tension of V-belt is appropriate and whether there is crack on the surface. Whether the pulley operates normally
5	Protection cover	Check whether the protective cover is dirty, resulting in the decrease of visibility

(2) Regular inspections. After the machine tool has been running for a certain period of time, the parts in contact with each other will wear out. Its working performance is gradually affected, and the machine tool should be checked and adjusted regularly. Regular inspection is generally conducted by the operator under the guidance of maintenance personnel. As shown in Table 1-7.

Table 1-7 Regular inspection items

No.	Inspection site	Inspection and maintenance	Interval
1	Electrical installations	Check and tighten all wiring screws. Check the grounding device	6 Months
2	Operating system	Check whether the CNC system starts and shuts down normally and reliably	Check per shift

Continued Table 1-7

No.	Inspection site	Inspection and maintenance	Interval
3	Cooling system	Clean the chip tray	Timely conduct
		Change the chip liquid	2 Months (8 hours per day)
		Clean the filter screen and water tank	6 Months
4	Lubrication system	Check the lubricating pump and oil separator. Check whether the oil circuit is smooth. Check the oil quality	1 year
5	Safeguard	Check whether the safety device is reliable	6 Months
6	Triangle belt	Appearance inspection and tightness inspection. Clean the pulley	6 Months
7	Protection cover	Check whether the protective cover is dirty and the visibility is reduced. Wipe the stains gently with a soft cloth and detergent. Then wipe it with a clean cloth	1 Months

Note: unless otherwise specified, the interval is determined on the premise of two shift system.

CNC lathe is a kind of advanced processing equipment with high degree of automation, complex structure and high cost, which integrates electricity, machine and liquid, and has the characteristics of technology intensive and knowledge intensive. In order to give full play to its benefits and reduce the occurrence of faults, maintenance work must be done well. Therefore, it is required that the CNC lathe maintenance personnel not only has the knowledge of machinery, processing technology and hydraulic and pneumatic aspects, but also has the knowledge of electronic computer, automatic control, drive and measurement technology, so as to fully understand and control the CNC lathe and timely do a good job in maintenance work.

1.2.6.4 Working Environment of Machine Tools

(1) Ambient temperature: within the range of 5~40℃, and the average temperature within 24 hours shall not exceed 35℃.

(2) Relative humidity: within the range of 30% ~ 95%, and the principle of humidity change is not to cause condensation.

(3) Altitude: below 1000m.

(4) Atmosphere: no excessive dust, acid gas, corrosive gas and salt.

(5) Avoid changes in ambient temperature caused by direct sunlight or thermal radiation of the machine.

(6) Install away from vibration source.

(7) The installation position should be far away from inflammable and explosive materials.

1.2.6.5 Accuracy of Machine Tools

(1) Precision of processed workpiece: IT6 ~ IT7.

(2) Surface roughness of processed workpiece: *Ra*1.6μm.

(3) Positioning accuracy (*X/Z*): 0.03/0.04mm.

(4) Repeated positioning accuracy (*X/Z*): 0.012/0.016mm.

1.2.7 Safety Operation Regulations for CNC Lathe

1.2.7.1 Student Rules

(1) Protective equipments must be worn before starting work. Gloves are not allowed during machining. Female students must wear work hats. Hair is not allowed to stay outside. High heels are not allowed. Accessories is not allowed.

(2) Do maintenance work on equipment before using CNC machine tools.

(3) The blanks and measuring tools should be placed in a fixed position, and the drawings or instructions should be placed in a convenient place.

(4) The work piece must be fastened, and the tool must be tightened to prevent it from being injured by loosening and throwing. Before turning on the machine, check whether the chuck wrench is removed from the chuck.

(5) The machine's protective cover should be closed during processing, and it is forbidden to touch the rotating workpiece by hand.

(6) When loading and unloading the workpiece, measuring the machining surface, and manually changing the speed, the machine must stop first.

(7) When the lathe is operating abnormally or malfunctions during machining, stop the machine tool immediately and report to the instructor to avoid danger.

(8) Carefully clean the machine after the daily practice to ensure that the bed surface and guide rails are clean and lubricated.

(9) Organize tools, gages and workpieces.

(10) Observe the rules and regulations of the internship management. For students who violate discipline and rules and regulations, the instructor shall give the necessary criticism and education. In serious cases, the instructor has the right to stop his internship.

1.2.7.2 Safe Operating Procedures

(1) Students must operate CNC machine tools under the guidance of teachers.

(2) It is forbidden for multiple people to operate at the same time.

(3) When returning to the reference point manually, the position of each axis of the machine tool should be more than 30mm from the reference point.

(4) When using the handwheel or rapid movement to move each axis, be sure to look at the "+" and "−" directions of each axis before moving. When moving, moveing slowly and then fast.

(5) If the student encounters a problem, report it to the instructor immediately. Trying operation is prohibited.

(6) Before running the program, check the position of the cursor in the program, the position of the function buttons of the machine tool, and whether there are any debris or tools on the guide rail.

(7) When starting the program, be sure to press the start button with one hand and the other hand on the emergency stop button. The hand cannot leave the emergency stop button while the program is running. Press the emergency stop button immediately in case of emergency.

(8) When the machine is running, close the protective door to prevent chips and lubricant from flying out.

(9) There is a pause instruction in the program. When the workpiece size needs to be measured, the standby bed is completely stopped and the spindle is stopped for measurement. At this time, do not touch the start button to avoid personal accidents.

(10) Be careful not to get your hands, body, and clothing close to the workpiece or machine parts that are rotating.

(11) During high speed cutting, do not remove chips directly by hand and use special hooks.

Task Implemention

(1) Start CK6140 CNC lathe.

(2) Realize the forward rotation of the spindle in manual mode and MDI mode.

(3) The tool feed is realized in manual mode and hand-crank mode, respectively.

(4) Complete the tool change operation for any tool number.

(5) Complete the daily maintenance of CK6140 CNC lathe.

Assessment

The assessment table is shown in Table 1-8.

Table 1-8 The assessment table

Item	No.	Standard		Partition	Score
Machine tool basic operation (90%)	1	Machine start, shut down	6 points will be deducted for each step error	12	
	2	Spindle forward rotation in manual mode	2 points will be deducted for each step error	14	
	3	Spindle forward rotation in MDI mode	4 points will be deducted for each step error	12	
	4	Tool feed in manual mode	4 points will be deducted for each step error	12	
	5	Tool feed in manual mode	4 points will be deducted for each step error	10	

Continued Table 1-8

Item	No.	Standard		Partition	Score
Machine tool basic operation (90%)	6	Tool changing operation	5 points will be deducted for each step error	10	
	7	Routine maintenance	5 points will be deducted for each step error	20	
Occupation Capacity (10%)	8	Learning ability		2	
	9	Communication skills		2	
	10	Team work		2	
	11	Safe operation and civilized production		4	
Total					

Exercises

(1) Answer the following questions:

1) What are the parts of the CK6140 CNC lathe mechanical structure?

2) Briefly describe the role of the emergency stop button.

3) Briefly describe the process of turning off the CNC lathe.

4) Describe the process of manual return of the CNC machine tool to the machine reference point.

5) What is the daily maintenance and maintenance of CNC lathes?

(2) Synchronous training:

1) In MDI mode, realize the forward rotation of the spindle at the speed 400r/min.

2) No. 01 tool is in the working position now, change No. 04 tool to the working position.

Task 1.3 Basic Knowledge of CNC Programming

Introduction

The part drawing is shown in Figure 1-19, and the finishing program is shown in Table 1-9. Task requirements: input the program on the CNC machine and simulate.

Competences

(1) Knowledge:

1) Understand the contents and methods of CNC programming.

2) Explain the structure and format of CNC program.

3) Describe the coordinate system of CNC turning.

(2) Skill:

1) Have the ability to explain the meaning of NC program function words.

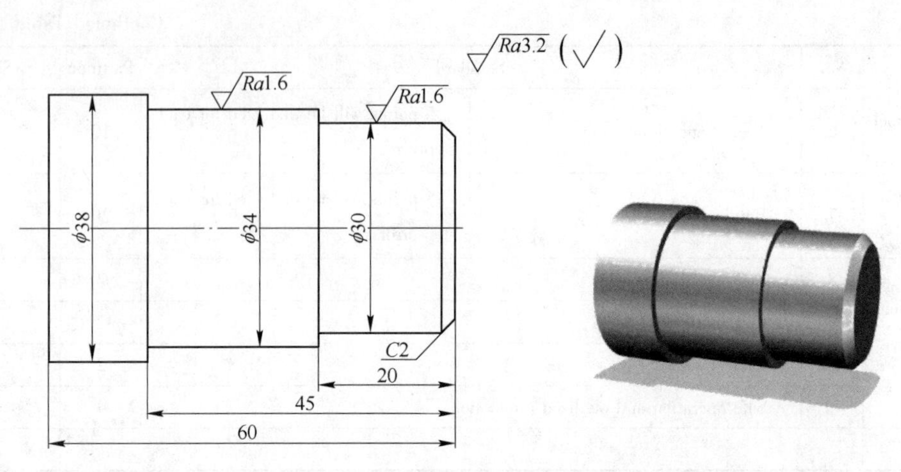

Figure 1−19　The part drawing

Table 1−9　The finishing program of the part

O131;	Program No.
G40 G97 G99 M03 S1000;	Cancel tool nose radius compensation, setting constant spindle speed control, spindle CW at 1000r/min
T0101;	Calling No. 1 tool and it's offset
M08;	Open cutting fluid
G00 Z2. 0;	Rapid positioning to point (*X*26. 0, *Z*2. 0)
X26. 0;	
G01 Z0 F0. 1;	Linear interpolation to the end face
G01 X30. 0 Z−2. 0;	Finishing chamfer
Z−20. 0;	Finishing *ϕ*30 outer circle
X34. 0;	Finishing the end face
Z−45. 0;	Finishing *ϕ*34 outer circle
X38. 0;	Finishing the end face
Z−60. 0;	Finishing *ϕ*38 outer circle
X40. 0;	Finishing the end face
G00 X100. 0;	Rapid positioning to point (*X*100. 0, *Z*100. 0)
Z100. 0;	
M30;	End of program and reset

2) Have the ability to use the CNC machine tool system panel to complete the manual input of CNC programs.

3) Have the ability to use numerical control machine operation panel to simulate numerical control program.

(3) Quality:

1) Consciously maintain safe operation and 8S work requirements.

2) Develop a serious and responsible work attitude.

Relevant Knowledge

1.3.1 Basic Knowledge of CNC Turning Programming

1.3.1.1 The Contents and Methods of CNC Programming

(1) The contents of CNC programming. The process of CNC programming includes the following aspects:

1) Analyze part drawings and make machining plans. According to the part drawing, the programmer analyzes the material, shape, size, accuracy, blank shape and heat treatment requirements of the part, defines the machining contents and requirements, selects the appropriate CNC machine tool, formulates the machining plan, determines the machining sequences, cutting routes, clamping methods, cutting tools and reasonable cutting amounts etc., in combination with the specifications, performance and functions of the CNC machine tools used, so to give full play to the efficiency of machine tools.

2) Numerical calculation. After the machining plan is determined, it is necessary to calculate the movement path of the tool center according to the geometric dimension and processing route of the part, so as to obtain the tool position data. Generally, CNC system has the functions of line interpolation and arc interpolation. For machining simpler parts consisting of arcs and lines, It only needs to calculate the coordinate values of intersection or cutting of adjacent geometric elements on the part contour, and get the coordinate values of the starting point, end point and arc center of each geometric element, which can meet the programming requirements. When the geometric shape of the part is not consistent with the interpolation function of the control system, it is necessary to carry out more complex numerical calculation, which is computer-aided calculation, otherwise it is difficult to complete.

3) Programming. After finishing the above process treatment and numerical calculation work, the programmer writes the part programming according to the function instruction code and program segment format specified by the CNC system. In addition, relevant process documents shall be filled in, such as CNC machining process card, CNC tool card, CNC program card, etc.

4) Preparation of control medium. The contents of the program is recorded on the control medium, as the input information of the numerical control device, which is input into the numerical control system by means of manual input or communication transmission of the program.

5) Program verification and first piece test cutting. Before formal machining, the programming should be verified and the first piece trial cut should be carried out. Generally, the dry running

function of the machine tool is used to check the correctness of the movement and track of the machine tool, so as to check the program. In the CNC machine tool with CRT graphic simulation display function, the cutting process of the workpiece can be simulated by displaying the tool path, and the program can be checked. However, these methods can only check whether the movement is correct, but cannot check the machining accuracy of the machined parts. Therefore, in order to the first trial cutting, when machining error is found, the causes of the error should be analyzed, and size compensation measures should be taken to correct it. The contents and steps of CNC programming are shown in Figure 1-20.

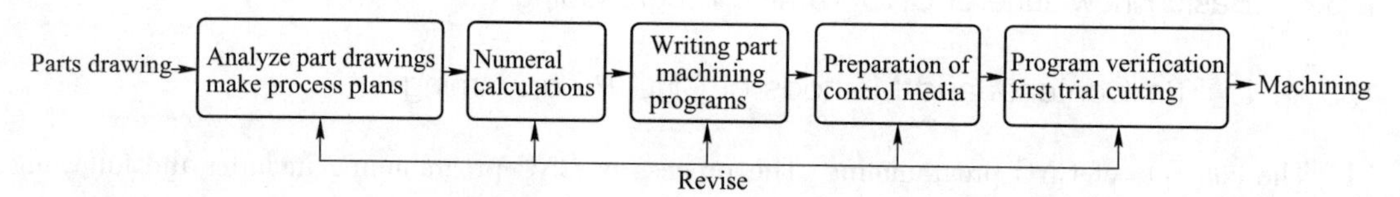

Figure 1-20 CNC programming flow chart

(2) The methods of CNC programming. Generally, there are two methods of CNC programming, they are manual programming and automatic programming.

1) Manual programming: Manual programming is a programming process which is mainly completed by human from analyzing part drawings, making process plans, mathematical processing of drawings, writing part processing program sheets, preparation of control medium to program verification. For the parts with simple shape, small amount of calculation and a few program sections, it can be realized by manual programming, and it is economical and timely. Therefore, manual programming is still widely used in point machining or contour machining composed of straight lines and arcs. The disadvantage of manual programming is that it takes a long time, prone to errors, and it is not competent for the programming of complex shape parts.

2) Automatic programming. Automatic programming means that in the programming process, except for the analysis of part drawings and the development of process plans, which are performed manually, the rest of the work are completed by computer assistance. When the computer is used for automatic programming, the mathematical processing, writing programs, and inspection programs are automatically completed by the computer. Since the computer can automatically draw the tool center motion trajectory, the programmer can check whether the program is correct in time and modify it if necessary. Obtaining the correct program, and because the computer automatic programming replaces the programmer to complete the cumbersome numerical calculations, the programming efficiency can be improved by dozens or even hundreds of times, so the programming problems of many complex parts that are difficult to solve by manual programming are solved. Therefore, the characteristic of automatic programming is that the programming efficiency is high and it can solve the programming problems of complex parts.

1.3.1.2 Structure and Format of CNC Program

In order to meet the needs of design, manufacturing, maintenance and popularization, there are two kinds of standardization developed by international organization for Standardization (ISO) and American Electronic Engineering Association (EIA) in terms of input code, coordinate system, processing instruction, auxiliary function and program format.

According to ISO standards, China formulated "Seven Unit Code Characters for Digital Control Machine Tools" (JB3050-82), "Naming of Digital Control Coordinates and Moving Directions" (JB3051-82), "Gode of Reparation Function G and Auxiliary Function M in the Format of Program Section of Digital Control Machine Tool Punch Belt" (JB3208-83). Because the standards used by CNC machine tool manufacturers have not been completely unified, and the codes, instructions and their meanings are not exactly the same, the CNC programming should be carried out according to the provisions of the CNC machine tool programming manual.

(1) Structure of CNC machining program. Every program consists of three parts: program number, program content and program end. The program content is composed of several program segments, which are composed of several words, and each word is composed of letters and numbers. Words form a program segment, and program segments form a program. The following is the finishing program of the part in Figure 1-21.

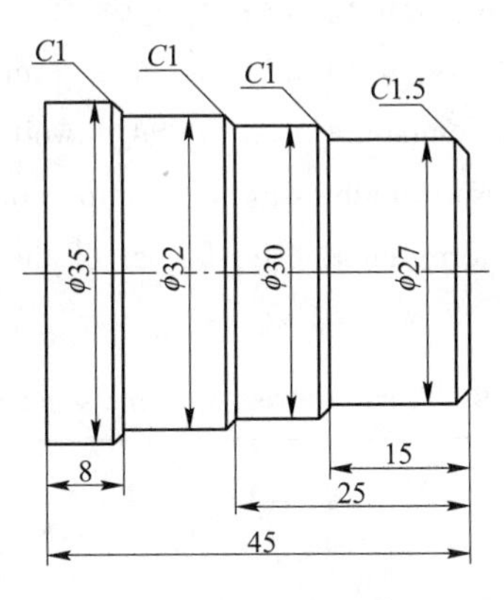

Figure 1-21 The part drawing

1) Program number. The program number is the beginning part of the program. In order to distinguish the programs in the memory, each program should have a program number. Before numbering, the program number address character is used. Different CNC systems have different program address characters. In FANUC system, the English letter "O" is used as the program number address, and "%" and "P" are used in other systems.

2) Program contents. The program content is the core of the whole program, which is composed of many program sections. Each program section is composed of one or more instructions, indicating all the actions to be completed by the machine tool.

3) Program end. Each program uses the program end instruction M02 or M30 as the symbol to end of the whole program.

```
O1121;                                                              Program number
  M03 S1000; (Spindle CW at 1000r/min)
  T0101; (Calling No. 1 tool and it's offset)
  M08; (Open cutting fluid)
  G00 Z5.0; (Rapid positioning to point Z5.0)
  X30.0; (Rapid positioning to point X30.0)
  G01 Z0 F0.1; (Linear interpolation to Z0, feed rate at 0.1mm/r)
  X32.0 Z-1.0; (Linear interpolation cutting chamfer)
  Z-20.0; (Linear interpolation cutting φ32 outer circle)           Program contents
  X34.0; (Cutting end face with linear interpolation)
  X38.0 Z-22.0; (Linear interpolation cutting chamfer)
  Z-50.0; (Linear interpolation cutting φ38 outer circle)
  G00 X100.0; (Rapid positioning to point X100.0)
  Z100.0; (Rapid positioning to point Z100.0)
  M30; (End of program and reset)                                   Program end
```

(2) Segment format. The machining program of parts is composed of program segments. Program segment format refers to the writing rules of words, characters and data in a program segment. There are usually three formats: word-address block format, block format using delimiters, and fixed block format. The most commonly used format is word address format.

The program segment format of word address form is composed of program segment number, program word and program segment terminator. The format of the word address block is shown in Table 1-10.

Table 1-10 Word address variable segment format

1	2	3	4	5	6	7	8	9	10
N___	G___	X___ U___ P___	Y___ V___ Q___	Z___ W___ R___	I_J_K_ R___	F___	S___	T___	M___
Program segment number	Prepare function	Dimension word				Feed function	Spindle function	Tool function	Auxiliary function

The instructions included in the program segment are not necessary in every program segment of the machining program. Instead, corresponding instructions are programmed according to the specific functions of each program segment. For example:

N10 G01 X30.0 Z-25.5 F0.2;

1) Program segment number. The number used to identify the program segment, which is located at the beginning of the program segment and consists of the address code N and the following digits. N10 indicates that the segment number of the program segment is 10.

2) Program word. Program word is usually composed of address character, number and symbol.

The function category of the word is determined by the address character. The arrangement order of the word is variable. The number of data bits can be more or less. Unnecessary words and the program words with the same program segment can be omitted.

3) End of segment. Each program segment is followed by an end character, indicating the end of the program segment. In the ISO standard code of international standardization organization, the ending symbol is "NL" or "LF"; in the EIA Standard Code of American Electronic Industry Association, the ending symbol is "CR"; in the FANUC numerical control system, the ending symbol is ";", in some numerical control systems, the ending symbol is " * " or not, press enter directly.

(3) Program function.

1) Preparatory function G. The preparation function G is used to control the command of the system action mode. The address G and two digits are used to represent 100 kinds from G00 to G99. G code is divided into modal code and non modal code. Modal code means that the G code function remains in one program segment until it is cancelled or replaced by another G code of the same group. Modeless code is only valid in segments with that code.

G codes are grouped according to their functions. The codes of the same function group can be replaced by each other and cannot be written in the same program segment. The common G code of CNC lathe is shown in Table 1-11.

Table 1-11 G codes commonly used in CNC lathe

Code	Group	Interpolation function
G00	01	Positioning
G01		Linear interpolation
G02		Circular interpolation clockwise direction (CW)
G03		Circular interpolation counterclockwise direction (CCW)
G04	00	Dwell
G20	06	Inch conversion
G21		Metric conversion
G40	07	Cancel tool nose radius compensation
G41		Tool nose radius left compensation
G42		Tool nose radius right compensation
G50	00	Coordinate system setting or max spindle speed setting
G65		Simple call
G70		Finishing cycle
G71		Stock removal in turning
G72		Stock removal in facing
G73		Pattern repeating
G74		End face peck
G75		Outer diameter drilling cycle
G76		Multiple thread cutting cycle

Continued Table 1-11

Code	Group	Interpolation function
G66	12	Macro modal call
G67		Cancel macro modal call
G90	01	Outer diameter cutting cycle
G92		Thread cutting cycle
G94		End Face turning cycle
G96	02	Constant surface speed control
G97		Cancel constant surface speed control
G98	05	Feed per minute
G99		Feed per revolution

2) Dimension word. The dimension word is used to determine the coordinate position of tool movement during machining.

The coordinate system of all coordinate points measured from the programmed origin is called absolute coordinate system. X, Y and Z are used to determine the absolute dimension of the linear coordinate of the end point.

The coordinate value in the coordinate system is calculated relative to the previous position (or starting point), which is called incremental (relative) coordinate. U, V and W represent the linear coordinate increment dimension used to determine the end point.

A, B and C are used to determine the angular coordinate dimension of the end point of the additional axis. I, J and K are used to determine the center coordinate of the arc, and R is used to determine the arc radius.

3) Feed function F. The feed function F is composed of the code F and the following numbers, which is used to specify the feed speed of cutting, that is the feed speed when the tool center moves, in mm/min or mm/r.

① Per minute feed mode G98:

Format: G98 F__;

The number after F indicates the feed rate of the spindle per minute, the unit is mm/min, and G98 is the modal command.

② Per revolution feed mode G99.

Format: G99 F__;

The number after F indicates the feed rate of the spindle per revolution, the unit is mm/r, and G99 is the modal command.

4) Spindle speed function S. The spindle speed function S is composed of address code S and the number behind it, which is used to specify the spindle speed, and the unit is r/min.

Format: S__;

【Example 1-1】S600 means the spindle speed is 600r/min.

On the machine tool with constant linear speed function, the S command also has the following functions:

① Max spindle speed G50.

Format: G50 S__;

The number after S indicates the maximum spindle speed in r/min.

【Example 1-2】 G50 S2500 indicates that the maximum speed of the main shaft is 2500r/min.

② Constant surface speed control G96.

Format: G96 S__;

The number after S indicates a constant linear speed in m/min.

【Example 1-3】 G96 S100 means that the linear speed of cutting point is controlled at 100m/min.

This command is used to turn the end face or the occasion with large diameter change. This function can ensure that when the diameter of the workpiece changes, the linear speed of the workpiece cutting surface remains unchanged, so as to ensure that the cutting speed remains unchanged and improve the processing quality.

③ Constant surface speed control cancels G97.

Format: G97 S__;

【Example 1-4】 G97 S1000 means to cancel the constant linear speed control, and the spindle speed is 1000r/min.

This command is used for turning thread or when the diameter of workpiece changes little. This function can set the spindle speed and cancel the constant line speed control.

5) Tool function T. The tool function T is used to specify the number of tools used in machining. The number of the tool function is the designated tool number. The number of the number is determined by the system used. The number for the CNC lathe also has the designated tool compensation function.

Format: T××××;

The first two digits represent the tool number and the last two digits represent the tool compensation number. T ×× 00 is to cancel tool compensation.

【Example 1-5】 T0303 indicates that No. 3 cutting tool and No. 3 cutting tool compensation value are selected. T0300 cancels the tool compensation of No. 3 tool position.

6) Auxiliary function M. The auxiliary function M is used to control the switch action of the miscellaneous device of the machine tool or system. It is composed of address code M and the following two digits. There are 100 kinds from M00 to M99. There are differences in the provisions of M code of various machine tools, so programming must be carried out according to the provisions of the manual. The commonly used M code is shown in Table 1-12.

Table 1-12 Common M functions of CNC lathe

Code	Function	Code	Function
M00	Program stop	M09	Cutting fluid off
M01	Optional stop	M30	End of program and return to start

Continued Table 1-12

Code	Function	Code	Function
M02	End of stop	M41	Low-range
M03	Spindle rotation	M42	Mid-range
M04	Spindle reversal	M43	High-range
M05	Spindle stop	M98	Calling of subprogram
M08	Cutting fluid on	M99	End of subprogram

1.3.2 CNC Turning Coordinate System

1.3.2.1 Standard Coordinate System and Directions of Motion

In order to describe the movement of the machine tool, simplify the programming method and ensure the interchangeability of the recorded data, the coordinate system and the movement direction of the machine tool are standardized. The international organization for standardization has unified the standard coordinate system, and the Ministry of Machinery Industry of the People's Republic of China has also promulgated the standard JB3051-82 "Naming of Coordinates and Moving Directions of Numerically Controlled Machine Tools", which regulates the coordinate system and moving directions of numerically controlled machine tools.

(1) Regulations on relative movement of machine tools. In order to make the programmer do not consider the specific movement of the workpiece and tools on the machine tool, only needs to determine the machining process of the machine tool according to the part drawing, which stipulates: always assume that the tool moves relative to the stationary workpiece.

(2) Coordinate system. In the process of CNC machining, the movement of machine tool is controlled by the instruction from CNC system. In order to determine the movement displacement and direction of machine tool, coordinate system is needed to realize. This coordinate system is called standard coordinate system, also known as machine coordinate system.

The CNC machine tool adopts the standard cartesian right-hand coordinate system, as shown in Figure 1-22. The thumb, index finger and middle finger of the right hand are perpendicular to each other. The direction of the thumb is the positive direction of the X-axis, the index finger is the positive direction of the Y-axis, and the middle finger is the positive direction of the Z-axis.

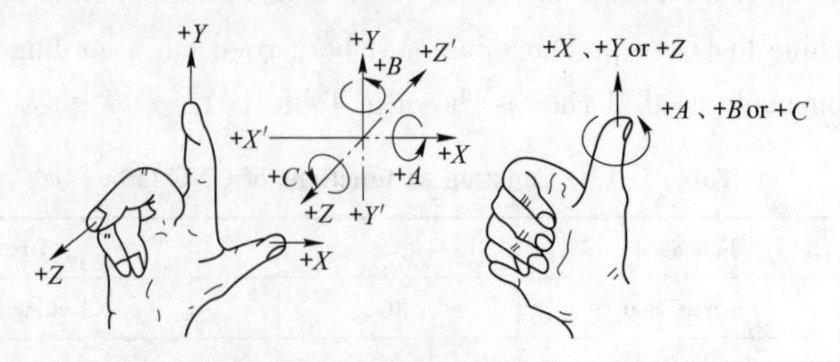

Figure 1-22 Right-handed cartesian coordinate system

The rotation coordinate system revolving around the X, Y and Z coordinates is respectively represented by A, B and C. According to the right-hand spiral rule, the thumb points to the positive direction of any axis in the X, Y and Z coordinates, and the rotation direction of the other four fingers is the positive direction of the rotation coordinates A, B and C.

(3) Regulation of movement directions. JB3051-82 stipulates that the positive direction of the movement of a certain part of the machine tool is to increase the distance between the tool and the workpiece.

1) Regulation of Z coordinate. The origin of Z coordinate is determined by the spindle which transmits the cutting force, i. e. the coordinate axis parallel to the spindle axis is Z coordinate. If there are multiple spindles, the spindle perpendicular to the workpiece clamping surface is selected as the main spindle, and the Z coordinate is parallel to the spindle axis. If the machine tool has no spindle, the direction perpendicular to the workpiece clamping plane is specified as Z coordinate. The positive direction of Z coordinate is the direction of the tool leaving the workpiece.

2) Regulation of X coordinate. The X coordinate is parallel to the clamping surface of the workpiece, generally horizontal. The X coordinate of the CNC lathe is in the radial direction of the workpiece and parallel to the transverse carriage. The direction of the tool leaving the workpiece rotation center is the positive direction of the X coordinate.

3) Regulation of Y coordinate. The Y coordinate axis is perpendicular to the X and Z coordinate axes. The positive direction of the Y coordinate axis is determined according to the positive directions of the X and Z coordinates according to the right-handed cartesian coordinate system.

1. 3. 2. 2 Machine Origin and Machine Reference

(1) Machine origin. The origin of machine tool is a fixed point on the machine tool, also known as the mechanical origin, which is the reference point for machining movement of CNC machine tools. It has been determined during machine assembly and debugging. The origin of CNC lathe is generally set at the intersection of chuck end face and spindle center line, as shown in Figure 1-23.

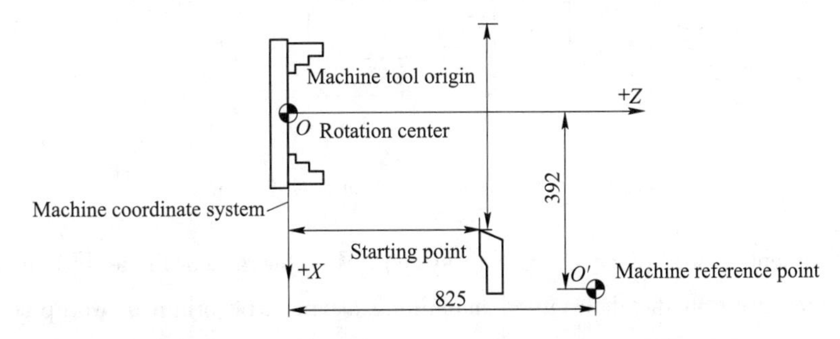

Figure 1-23 Coordinate system and reference point of CNC lathe

(2) Machine reference point. The reference point of machine tool is a fixed position point used

to detect and control the movement of machine tool. The position of the reference point of the machine tool is precisely adjusted by the manufacturer of the machine tool on each feed axis with a limit switch, and its coordinate value has been input into the CNC system. The reference point of machine tool has an accurate position relationship with the origin of machine tool, which is a known number. The reference point of the machine tool is determined by the travel limit switch and reference pulse of the machine tool. Generally, the reference point of the machine tool is located at the positive limit point of the travel, as shown in Figure 1-23. When the CNC machine is started, it is necessary to determine the origin of the machine first, and the movement of the origin of the machine is the operation of returning to the reference point.

Figure 1-23 shows the coordinate system of the flat bed and front tool rest CNC lathe. The Z axis is the main axis, and the direction to the tailstock is positive. The direction of the X-axis is the radial direction of the workpiece and parallel to the transverse sliding seat. The direction of the tool away from the center of the spindle is the positive direction of the X-axis.

1.3.2.3 Establishment of Workpiece Coordinate System

(1) Programming coordinate system. Programming coordinate system is a coordinate system for programming, which is established by the programmer according to the part drawing and processing technology. The origin of the programmed coordinate system shall be selected according to the part drawing and processing technology requirements, and shall be selected on the design basis or process basis of the part as far as possible. The direction of each axis of the programmed coordinate system should be consistent with that of the corresponding coordinate axis of the CNC machine tool used, as shown in Figure 1-24.

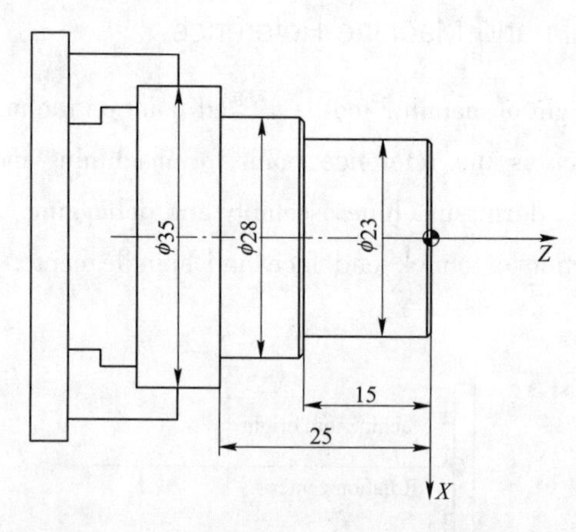

Figure 1-24 Workpiece origin of CNC lathe

(2) Establishment of workpiece coordinate system. Workpiece coordinate system refers to the coordinate system based on the determined machining origin. The origin of workpiece coordinate system is also known as the programming origin, which refers to the position of the corresponding programming origin in the machine coordinate system after the parts are clamped. The CNC lathe is generally set at the midpoint of the right end face of the workpiece, as shown in Figure 1-24.

The CNC machine tool processes the workpiece according to the position of the machining origin and the program requirements determined after the workpiece is clamped. When programming, the programmer can select the programming origin, establish the programming coordinate system and calculate the coordinate value according to the part drawing without considering the actual position of workpiece blank clamping. For the machining personnel, when the workpiece is clamped and the program is debugged, the programming origin shall be converted to the machining origin, and the position of the machining origin shall be determined, which shall be set in the CNC system. After the machining coordinate system is set, the coordinate value of the starting point of the tool can be determined according to the current position of the tool. During machining, the coordinate values of all dimensions of the workpiece are relative to the machining origin, so that the CNC machine can start machining according to the accurate machining coordinate system position.

1.3.3 CNC Programs input and Simulation

1.3.3.1 CNC Programs Input

There are two ways to input program on NC machine tool, one is manual input program, the other is to transmit import program through CF card of CNC machine tool.

(1) Manual input program

Press the key [编辑] → press the key [PROG] to enter the program interface → input the program name such as "O111" → press the key [INSERT] → press the key [EOB] → press the key [INSERT] →use the mouse or keyboard to input the contents of O111 program → press the key [RESET] to return to the starting point of the program after input.

Common functions of input and edit program are as follows:

1) Line feed: press the key [EOB]→ press the key [INSERT].

2) Input data: press the number/letter key, such as M03 S500, the data will be input to the input area; if the input is wrong, use the key [CAN] to delete the data in the input area.

3) Move cursor: press the key [⇧ PAGE] or [PAGE ⇩] to turn the page up and down; press [⇧] or [⇩] or [⇦] to [⇨] move the cursor up, down, left and right.

4) Delete, insert and replace: press the key [DELECTE] to delete the code of the cursor location, press the key [INSERT] to enter the contents of the area after the cursor code, press the key [ALTER], enter the code where the content of the area overrides the cursor.

(2) Import program using CF card. CF card import procedure steps are as follows: (FANUC Oi TD system).

1) Make sure the output device is ready.

2) Press the key [编辑]→ press the key [PROG]→ display the program.

3) Press [MENU]→ press [OPRT]→ press the right extension key.

4) Select [EQUP]→ select [M-Card]→ display the contents of the card.

5) Press the [F READ] key → input the M card program number → press the [F SET] key to confirm → input the program number in the machine tool → press the [O SET] key to confirm → press the [EXCU] key →press the key PROG to display the imported program on the CRT.

6) After the program transmission, press the [OPRT] key → press the right extension key → press the [EQUP] key → press the [CNC MEM] key to return to the original state.

(3) Program call.

Press the key 编辑 → press the key PROG to enter the program interface → press the [DIR] key to display all programs in the CNC system → input the program name such as "O111" → press the [O SERCH] key to call the O111 program.

1.3.3.2 CNC Program Simulation

The input program must be checked. It is usually used to check whether the program is correct by means of graphic simulation. The operation steps of graphic simulation are as follows:

Press the key 编辑 → press the key PROG → input the program number, press the key ⇓ to display the program → press the key 记忆 → press the GRPH key → press the key 机械空跑 → press the cycle start key 程序启动 CYCLE START, and observe the processing track of the program.

Note:

After the graphic simulation, the empty operation and locking function must be cancelled, and the full axis operation must be carried out at the same time.

The full axis operation steps are as follows:

Cancel 机械空跑 → press the key POS → press the [ABS] key → press the [OPRT] key → press the [W SET] key → press the [ALL AXES] key → the coordinates of CRT panel are consistent with the actual coordinates.

Call up the processing program → press the key 记忆 → press the cycle start key 程序启动 CYCLE START to automatically machine parts.

Implemention

(1) Step 1 Machine Tool Preparation:

1) Routine inspection before starting the machine tool.

2) System start up.

3) Manual return to machine reference point.

4) Manually operate the sliding plate feeding, spindle rotation and tool holder rotation, and

check whether the function is running operates normally.

(2) Step 2 Program Input:

Manually input the program in Table 1-9.

1) Select the working mode on the operation panel of CNC machine tool as [编辑], the Key [PROG] on the panel of CNC machine tool system, enter the program input status.

2) Enter the program number "O131", click the key [INSERT], enter the program number, click the key [EOB], and enter ";".

3) Input program contents in sequence: G40 G97 G99 M03 S1000; ... Until the end of the process.

4) Click the key [INSERT], return the cursor to the position of the program number to prepare for the program simulation.

(3) Step 3 Program simulation:

1) Program simulation.

Select the working mode on the operation panel of CNC machine tool as [连线], the Key [CUS GRPH] on CNC machine system panel, [GEOM key, press the key [机械空跑], press the key [程序启动 CYCLE START], and observe the processing track of the program.

2) Full axis operation.

Assessment

The assessment table is shown in Table 1-13.

Table 1-13 The assessment table

Item	No.	Standard		Partition	Score
Basic knowledge of programming (40%)	1	Program structure	Dividing the program structure correctly	3	
	2	Program number	Name the program correctly	3	
	3	Block number	Explain the function of block number	2	
	4	Preparatory function	Explain the function according to the specific program segment	6	
	5	Feed function	Explain the function according to the specific program segment	4	
	6	Spindle function S	Explain the function according to the specific program segment	4	
	7	Tool function T	Explain the function according to the specific program segment	4	
	8	Miscellaneous function M	Explain the function according to the specific program segment	6	

Continued Table 1-13

Item	No.	Standard		Partition	Score
Basic knowledge of programming (40%)	9	Machine coordinate system	Explain machine coordinate system, machine origin, reference point	6	
	10	Programming coordinate system	Determine the programming coordinate system based on the part drawing	2	
Program input and simulation (50%)	11	Program input	Input the program accurately within the specified time, 2 points deducted for each error	30	
	12	Program simulation	Program simulation operation is correct., 10 points in case of dangerous operation	15	
	13	Full axis operation	Wrong operation, no points	5	
Vocational ability (10%)	14	Learning ability		2	
	15	Communication skills		2	
	16	Team cooperation		2	
	17	Safe operation and civilized production		4	
Total					

Exercises

(1) Answer the following questions:

1) Explain the meaning of "N10 G01 X26.0 Z-30.0 F0.1;" in the program.

2) What is machine coordinate system? What is the workpiece coordinate system?

(2) Synchronous training:

The parts are shown in Figure 1-25 and Figure 1-26. The program names are O137 and O138 respectively, as shown in Table 1-14. Input the programs and simulate them on the CNC lathe.

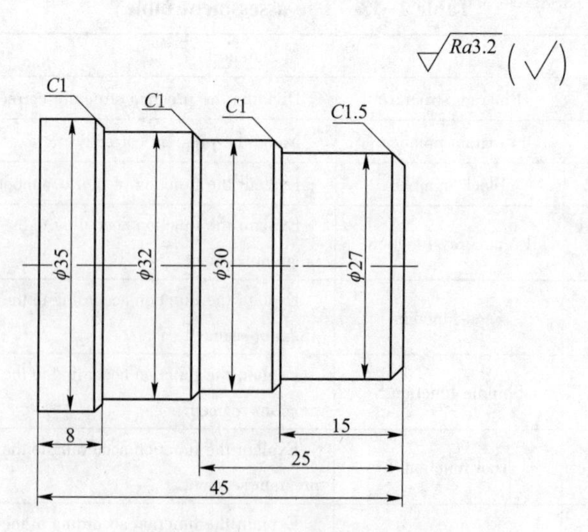

Figure 1-25 Synchronous training 1

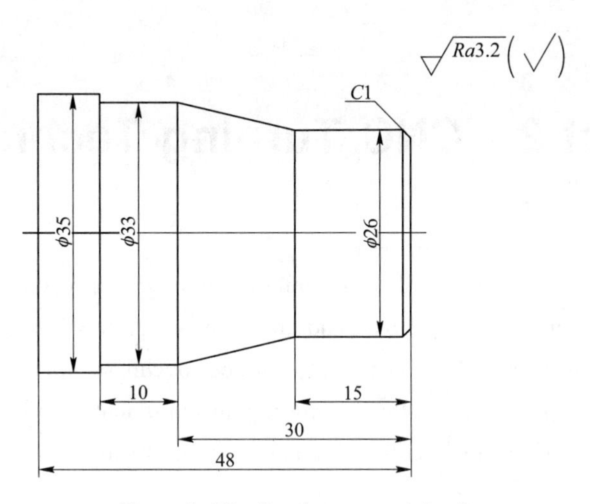

Figure 1-26 Synchronous training 2

Table 1-14 Synchronous training processing procedure

O137;	O138
G40 G97 G99 M03 S1200;	G40 G97 G99 M03 S1200;
T0101;	T0101;
M08;	M08;
G00 Z5.0;	G00 Z5.0;
X24.0;	X24.0;
G01 Z0 F0.1;	G01 Z0 F0.1;
G01 X27.0 Z-1.5;	G01 X26.0 Z-1.0;
Z-15.0;	Z-15.0;
X30.0 C1.0;	X33.0 Z-30.0;
Z-25.0;	W-10.0;
X32.0 C1.0;	X35.0;
Z-37.0;	Z-48.0;
X35.0 C1.0;	G00 X100.0;
Z-45.0;	Z100.0;
G00 X100.0;	M30;
Z100.0;	
M30;	

Project 2　CNC Turning Technology

The CNC lathe can automatically complete the internal and external cylindrical surfaces, conical surfaces, arc surfaces, end surfaces, thread and other contour cutting processing, the dimensional accuracy of processing can reach IT5～IT6 level, surface roughness can reach $Ra1.6\mu m$ or less, the production efficiency is 3～5 times that of ordinary machine tools, and it is especially suitable for machining complex shafts or disk parts. It is one of the most widely used CNC machining methods.

Task 2.1　CNC Turning of Stepped Shafts

Introduction

As shown in Figure 2-1, the part is made of hard aluminum alloy LY15, and the blank is ϕ40mm. It is produced in one piece. This task needs to prepare the program, machine the part using the CNC lathe, and then measure the part.

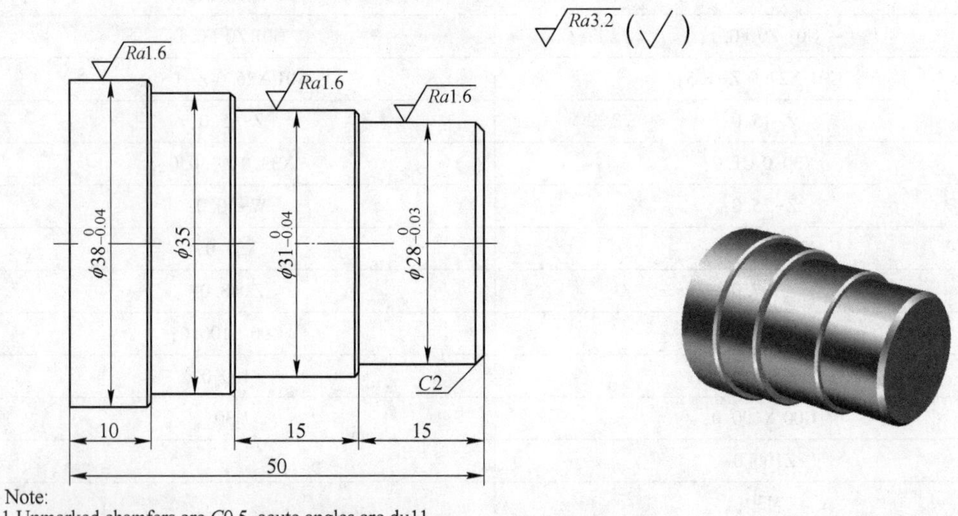

Figure 2-1　The part drawing

Competences

(1) Knowledge:

1) Know the cutting amount, turning tools, CNC machining process files, turning methods of stepped shafts.

2) Interpret G20/G21, G00/G01 instructions.

3) Describe the process of external turning tool setting.

4) Know the structure of vernier calipers.

(2) Skill:

1) Have the ability to write stepped shaft machining programs.

2) Learn how to set the coordinate system of the external turning tool.

3) Understand the inspection method of vernier caliper.

(3) Quality:

1) Correctly implement the safety technical operation procedures and cultivate safety awareness.

2) Develop a serious and responsible work attitude and quality awareness.

Get ready

2.1.1 Process Preparation

CNC turning technology is based on the characteristics of CNC lathes, comprehensively using process knowledge, and performing process analysis on the process parameters, tool parameters, and cutting parameters of the processed parts before programming, to solve the problems encountered in the process of CNC turning, and achieve high quality, high yield and low consumption of CNC machining.

2.1.1.1 Division of CNC Turning Operations

(1) Selection of processing methods. Although the structure and shape of the rotating body parts is various, most of them are composed of planes, cylindrical surfaces, curved surfaces, and threads. Each surface has a variety of processing methods, and the actual selection should be based on the machining accuracy, surface roughness, material, structural shape, size, production type and other factors of the part to determine the CNC surface turning method and processing plan.

1) Except for hardened steel, common metals with a machining accuracy of IT8~IT9 and a surface roughness of *Ra*1.6~3.2μm can be processed by ordinary CNC lathes according to the plan of roughing, semi-finishing and finishing.

2) Except for hardened steel, common metals with a machining accuracy of IT6~IT7 and a surface roughness of *Ra*0.2~0.63μm can be processed by precision CNC lathes according to the rough, semi-finish, fine, and fine plan.

3) Except for hardened steel, common metals with a machining accuracy of IT5 and a surface roughness $Ra<0.2$μm can be processed by high-grade precision CNC lathes according to the rough, semi-finished, precision, and precision solutions.

(2) Division of processes. The machining process of a part usually includes a cutting process, a heat treatment process, and an auxiliary process. For CNC machine tools, the principle of process

concentration is adopted, so that each process includes as much processing content as possible, which can reduce the number of processes, shorten the process route, and provide production efficiency. The method of dividing processes according to the principle of process concentration is as follows:

1) Divided according to the positioning method of part clamping. Taking the process completed in one installation as a procedure, it is suitable for parts with less processing contents and reaches the state of inspection for inspection after processing.

2) Divided by the tools used. The process of machining with one tool is a process.

3) Divided by rough and finishing. The process completed in the rough machining is a process, and the part completed in the finishing process is a process.

4) Divided by processing area. Taking the process of completing the same profile as a process, such as the shape, cavity, curved surface or plane, each part is processed as a process.

(3) Arrangement of Non-NC Turning Process for Rotary Parts. Parts have surfaces that are not suitable for CNC turning, such as keyways, splines, involute teeth, etc. , and have special requirements such as shot peening, rolling, and polishing. It is necessary to arrange corresponding non-NC turning processes. When the surface hardness and accuracy of parts are high, heat treatment is usually arranged after CNC turning and then grinding is arranged.

(4) Arrangement of processing sequence. For machining parts on CNC lathes, the principle of roughing and finishing is used to arrange the machining sequence of parts, which can gradually improve the surface machining accuracy and reduce the surface roughness.

1) Roughing stage. The main task is to remove most of the margin on each surface in a short time, in order to improve the processing efficiency.

2) Semi-finishing stage. Finish the processing of the secondary surface, and prepare for the finishing work of the primary surface, as far as possible to meet the uniformity requirements of the finishing allowance.

3) Finishing stage. To ensure that the main surface meets the requirements specified in the drawings, the key is to ensure the processing quality.

2.1.1.2 Common Clamping Methods of CNC Lathe

CNC lathes often use three-jaw chucks, four-jaw chucks, one-clamp one-top, and center frame to clamp workpieces. The three-jaw chuck has a small clamping force and a fast clamping speed, and it is suitable for clamping small and medium cylindrical, regular three-sided or regular hexagonal workpieces. The four-jaw chuck has a large clamping force and high clamping accuracy, and it is not affected by the wear of the jaw. But the three-jaw chuck and the top of the machine's tailstock are used to clamp the workpiece. The positioning accuracy is high, the clamping is reliable, and it is suitable for clamping longer shaft parts. The center frame cooperates with a three-jaw chuck or a four-jaw chuck to clamp the workpiece, which can prevent bending and deformation, and is suitable for clamping slender shaft parts.

2.1.1.3 Selection of Cutting Parameters

During the CNC programming, the cutting parameters of each process needs to be determined and in the form of instructions written into the program. The purpose of selecting the cutting parameters is to ensure the shortest cutting time, the highest productivity and the lowest cost while ensuring the machining quality and tool durability. It includes the amount of back engagement a_p, the feed F and the spindle speed n (cutting speed v).

(1) The determination of the back engagement a_p. The back engagement is mainly determined by the stiffness of the machine tool, fixture, tool, and workpiece. During roughmaching, if the conditions allow, choose a larger back feed as much as possible to reduce the number of passes and improve productivity. When finishing, usually choose a smaller back engagement to ensure the processing accuracy and surface degree roughness.

(2) The determination of the feed F. The feed refers to the distance that the tool moves in the forward direction for each revolution of the spindle. It is selected according to the processing accuracy of the part, the surface roughness requirements, the material properties of the tool and the workpiece. In the rough machining, under the premise of ensuring the rigidity of the tool, machine tool, and workpiece, etc., in order to shorten the cutting time, the maximum feed is selected as much as possible. During finishing, the feed is mainly limited by the surface roughness. When the surface roughness is higher, a smaller feed should be selected. When the material of the workpiece is soft, a larger feed can be selected. Otherwise, a smaller feed should be selected.

(3) The determination of the spindle speed n. The spindle speed should be selected according to the allowed cutting speed. The cutting speed is selected under the condition that the durability of the tool and the cutting load do not exceed the rated power of the machine. When rough turning, the back engagement and the feed are both large, so choose a lower cutting speed, choose higher cutting speed when finishing turning. The relationship between cutting speed and spindle speed is as follows:

$$n = 1000v/\pi d$$

Among them n——Spindle speed (r/min);

d——Diameter of surface to be processed (mm);

v——Cutting speed (m/min).

(4) General principles of cutting parameters selection. Whether the selection of cutting parameters is reasonable has a very important role in achieving high quality, high yield, low cost and safe operation. The general principles are as follows:

1) In rough turning, productivity improvement is generally the main consideration, while economy and processing costs are also considered. First, choose a large back engagement, and then choose a larger feed. Increasing the feed is beneficial for chip breaking. Finally, according to the two selected parameters, the process system rigidity, tool life, choose a reasonable cutting speed under the condition of the machine's power permit, to reduce the tool consumption and the processing cost.

2) When semi-finishing or finishing, the machining accuracy and surface roughness are high, and the machining allowance is not large and uniform. The machining efficiency, economy and machining cost should be taken into consideration while ensuring the machining quality. In order to increase the cutting speed as much as possible, to ensure the machining accuracy and surface roughness of parts, usually select a smaller back engagement and feed, and select high cutting tool materials and reasonable geometric parameters.

The specific value of the cutting parameters can be determined by referring to the cutting amount manual and the allowable cutting amount range given in the machine manual. The reference of CNC cutting parameters is shown in Table 2-1.

Table 2-1 Reference table for selection of cutting parameters

Workpiece material	Machining method	Back engagement /mm	Cutting speed /$m \cdot min^{-1}$	Feed rate /$mm \cdot r^{-1}$	Tool material
Carbon steel σ_b>600MPa	Roughing	5~7	60~80	0.2~0.4	YT
	Roughing	2~3	80~120	0.2~0.4	
	Finishing	0.2~0.3	120~150	0.1~0.2	
	Threading		70~100	thread lead	
	Drill the center hole		500~800r/min		W18Cr4V
	Drilling		~30	0.1~0.2	
	Cutting (width<5mm)		70~110	0.1~0.2	YT
Alloy steel σ_b=1470MPa	Roughing	2~3	50~80	0.2~0.4	YT
	Finishing	0.1~0.15	60~100	0.1~0.2	
	Cutting (width<5mm)		40~70	0.1~0.2	
Cast iron below 200HBS	Roughing	2~3	50~70	0.2~0.4	YG
	Finishing	0.3~0.15	70~110	0.1~0.2	
	Cutting (width<5mm)		50~70	0.1~0.2	
Aluminum	Roughing	2~3	600~1000	0.2~0.4	YG
	Finishing	0.2~0.3	800~1200	0.1~0.2	
	Cutting (width<5mm)		600~1000	0.1~0.2	
Brass	Roughing	2~4	400~500	0.2~0.4	YG
	Finishing	0.1~0.15	450~600	0.1~0.2	
	Cutting (width<5mm)		400~500	0.1~0.2	

2.1.1.4 Turning Tools

(1) Turning tool materials. High-speed steel, hard alloy, ceramic, cubic boron nitride, and diamond are commonly used as cutting tool materials in metal cutting. At present, the most commonly used cutting tools in numerical control processing are high-speed steel cutting tools and cemented carbide cutting tools.

1) High speed steel. High-speed steel is a high-alloy tool steel that contains more alloy elements such as tungsten, chromium, vanadium, and molybdenum, and has good comprehensive properties. Its strength is the highest among the existing tool materials, and its toughness is also the best. The manufacturing process of high-speed steel is simple, and it is easy to sharpen into sharp cutting edges. Forging and heat treatment have small deformation. At present, it still occupies a major position in the manufacture of complex tools such as twist drills, taps and forming tools.

① Ordinary high speed steel. Ordinary high speed steel has certain hardness and wear resistance, high strength and toughness. For example, W18Cr4V is widely used in manufacturing various complex tools. The cutting speed is generally not too high, and it is 40 ~ 60m/min when cutting ordinary steel. Not suitable for high speed cutting and cutting of hard materials.

② High performance high speed steel. High performance high speed steel is smelted by adding some carbon content, vanadium content, and adding cobalt, aluminum and other elements to ordinary high-speed steel. For example, $W12Cr4V4M_0$, its durability is 1.5 to 3 times that of ordinary high-speed steel. But the overall performance of this type of steel is not as good as ordinary high-speed steel.

2) Cemented carbide. Cemented carbide is made of refractory metal carbides such as TiC, WC, NbC, etc., and metal binders such as Co, Ni, etc. by powder metallurgy. According to different chemical compositions, it can be divided into 4 categories.

① Tungsten cobalt (WC+Co).

The alloy code is YG, corresponding to the international standard K, suitable for cutting short-cut ferrous metals, non-ferrous metals and non-metal materials. The higher the cobalt content, the better the toughness is suitable for rough processing, the lower cobalt content is suitable for finishing.

② Tungsten titanium and cobalt (WC+TiC+Co).

The alloy code is YT, corresponding to the international P class. This type of alloy has high hardness and heat resistance, and is mainly used for processing plastic materials such as steel with long chips. The TiC content in the alloy is high, the wear resistance and heat resistance are improved, but the strength is reduced. Therefore, rough processing generally chooses a grade with less TiC content, and finish processing chooses a grade with more contents.

③ Tungsten, titanium, tantalum (niobium), cobalt (WC+TiC+TaC(NbC)+Co).

The alloy code is YW, corresponding to the international M class. This kind of cemented carbide is not only suitable for semi-finishing of cold-hardened cast iron, non-ferrous metals and alloys, but also for semi-finishing and finishing of high manganese steel, quenched steel, alloy steel and heat-resistant alloy steel.

④ Titanium carbide based (WC+TiC+Ni+Mo).

The alloy code is YN, which corresponds to the international P01 class. Generally used for finishing and semi-finishing. It is especially suitable for large, long and high precision parts, but it is not suitable for roughing and low-speed machining under impact load.

(2) Turning tool types. Commonly used tools for CNC turning include external turning tool, grooving tool, internal turning tool, external thread turning tool, internal thread turning tool, center drill, drill bit, etc. , as shown in Figure 2-2.

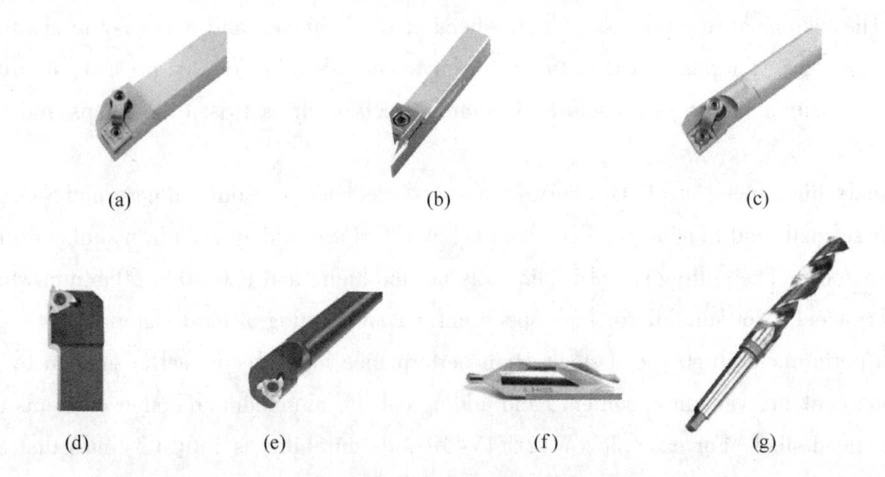

Figure 2-2 Common turning tools

(a) Cylindrical turning tool; (b) Grooving tool; (c) Internal turing tool; (d) External thread turning tool; (e) Internal thread turning tool; (f) Center drill; (g) Twist drill

Cylindrical turning tool (shown Figure 2-2(a)) is used for cutting cylindrical, conical and end faces. Grooving tool (shown Figure 2-2(b)) is used for grooving. Internal turning tool (shown Figure 2-2(c)) is used for machining inner surfaces. External thread turning tool (shown Figure 2-2(d)) for cutting external thread. Internal thread turning tool (shown Figure 2-2(e)) for cutting internal thread. Center drill (shown Figure 2-2(f)) for drill center hole. Twist drill (shown Figure 2-2(g)) is used for rough machining of holes.

(3) Turning tools with indexable clamps. In CNC turning, turning tools with indexable clamps are often used. The machine clamp indexable turning tool is to clamp the grinded indexable blade on the tool bar with a clamping component. In the process of use, when one cutting edge is blunt, loosen the clamping mechanism and turn the blade to all other cutting edges for cutting. When all cutting edges are worn, remove them and replace them with new blades of the same type.

1) Blade clamping methods. The clamping methods of the indexable turning tool of the machine clamp are generally the upper pressing type (code C), the upper pressing and pin hole clamping type (code M), pin hole clamping type (code P) and screw clamping (Code S).

2) Blade shapes. Generally, 80° convex triangle (W type), square (S type) and 80° prism (C type) inserts are commonly used for turning, and 55° (D type), 35° (V type), and round (R type) are commonly used for contouring, as shown in Figure 2-3. Under the condition that the rigidity and power of the machine tool allow, the insert with a larger tip angle is used for roughing or large allowance machining, and the insert with a smaller tip angle is used for finishing or small allowance machining.

3) Tool forms. There are 15 to 18 types of shank heads according to the main declination,

Figure 2-3 Blade shape

straight and elbow. For machining right-angle stepped shafts, a toolholder with a main deflection angle greater than or equal to 90° is selected. For rough turning, a shank with a main deflection angle of 45°~90° can be used, and for fine turning, a shank with 45°~75° can be used. For cutting in the middle of the workpiece and contour turning, a shank of 45°~107.5° is used.

4) Cutting directions

Cutting direction is divided into three types: R (right-handed), L (left-handed), and N (left-right-handed). The selection is based on the front or back of the lathe tool post, whether the rake face is up or down, the direction of rotation of the spindle and the required feed direction.

(4) Tool setting point and tool changing point.

1) Tool setting point. Tool setting point refers to the reference point for determining the relative position of the tool and the workpiece by tool setting. For CNC machine tools, at the beginning of processing, it is very important to determine the relative position of the tool and the workpiece. This relative position is achieved by confirming the tool setting point. The tool setting point can be set on the part being processed, or it can be set at a certain position on the fixture that has a certain size relationship with the part positioning reference. The tool setting point is often selected at the machining origin of the part.

2) The tool position point. The tool position point refers to the positioning reference point of the tool. When programming, the entire tool is usually regarded as a point, that is, the tool position point, which is the reference point representing the tool position. Figure 2-4 shows the location of common tools. The radius and length of each tool are different. The tool setting operation is the operation that makes the "tool position point" coincide with the "tool setting point".

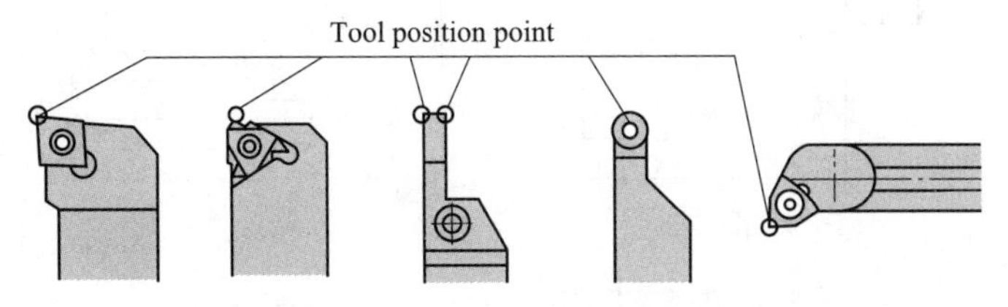

Figure 2-4 Tool position of common tools

3) The tool change point. The tool change point refers to the position of the tool holder when the tool is changed. To prevent damage to parts and other parts during tool change, the tool change point is usually set outside the contour of the part or fixture being processed, and a certain amount of safety is left.

2.1.1.5 CNC Machining Process Files

The CNC machining process files are specific descriptions of CNC machining. The purpose is to allow the operator to be more clear about the contents of the machining program, the clamping method, the tools used in each machining location, and other technical issues. It is not only the basis for CNC machining and product acceptance, but also the rules that operators follow and execute. CNC machining technology files mainly include CNC machining tool cards, CNC machining procedure cards, and part machining program sheets.

(1) The CNC machining tool card. In CNC machining, the requirements for tools are very strict. The tool card mainly reflects the tool number, tool name and specification, blade model and material, quantity, machining surface, etc. It is the basis for assembling and adjusting tools.

(2) The CNC machining process card. The CNC machining process card briefly describes the machine model and program number. The main contents include the step number, the step content, the tools used in each step, and the cutting amount.

(3) The CNC machining program sheet. The CNC machining program sheet is the main basis of CNC machining. It is compiled by programmers according to the machine-specific instruction codes and records the CNC machining process, process parameters, and displacement data.

2.1.1.6 Turning Methods of the Stepped Shaft

When the diameter difference between two cylinders adjacent to the stepped shaft is small, it can be completed by one turning with a turning tool. The machining route is $A \rightarrow B \rightarrow C \rightarrow D \rightarrow E$, as shown in Figure 2-5 (a).

When the diameter difference between two cylinders adjacent to the stepped axis is large, layer cutting is used. The roughing route is $A1 \rightarrow B1$, $A2 \rightarrow B2$, $A3 \rightarrow B3$, and the finishing route is $A \rightarrow B \rightarrow C \rightarrow D \rightarrow E$, as shown in Figure 2-5 (b).

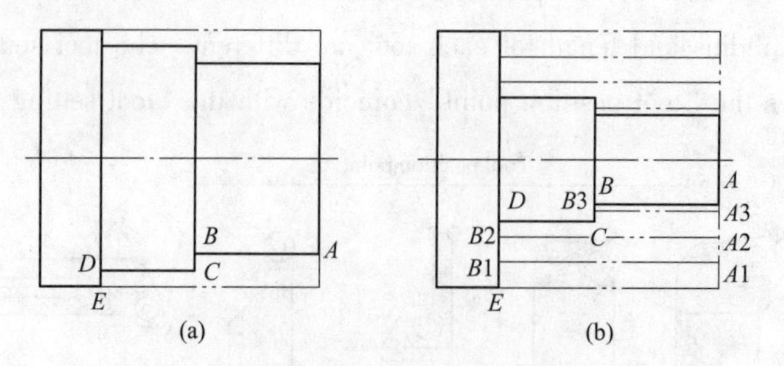

Figure 2-5 Turning method of the step shaft

(a) One turning; (b) Layer cutting

2.1.2 Programming

2.1.2.1 Inch Conversion G20 and Metric Conversion G21

The G20 command indicates that the coordinate size is input in inch. The G21 command indicates that the coordinate size is input in metric.

G20 or G21 is usually specified at the beginning of the program, and G20 and G21 cannot be converted in the middle of the program. After the system is powered on, the CNC retains the G20 or G21 from the last shutdown. Usually the factory setting is G21 mode.

2.1.2.2 Positioning G00

(1) Function: The tool moves quickly from the position of the tool to the target point by point control.

(2) Format:

G00 X(U)__ Z(W)__;

Among them, X(U) and Z(W) are the coordinate values of the target point.

Note:

1) When this command is executed, the tool moves from the point to the target point at the speed set in advance by the manufacturer, and the movement speed cannot be set by the F command.

2) G00 is a modal command. It will be replaced only when it encounters the same group of commands such as G01, G02, G03.

3) X and Z are followed by absolute coordinate values, and U and W are followed by incremental coordinate values.

(3) Diameter and radius programming

The programming method in which the *X* coordinate value is expressed by the diameter value of the rotating part is called diameter programming. As the diameter of the part is usually expressed by the diameter on the drawing, it is more convenient to use the diameter programming. The programming method in which the *X* coordinate value is expressed by the radius value of the rotating part is called radius programming, which is in line with the rectangular coordinate system representation method and is seldom used.

【Example 2-1】The tool moves quickly from the starting point *E* to the target point *A*1, and the starting point *D* moves quickly to the target point *E* (Figure 2-6)。

E→*A*1

Absolute programming: G00 X35.0 Z5.0; *E*→*A*1

Incremental programming: G00 U-65.0 W-25.0; *E*→*A*1

D→*E*

Absolute programming: G00 X100.0 Z30.0; *D*→*E*

Incremental programming: G00 U35.0 W100.0; *D*→*E*

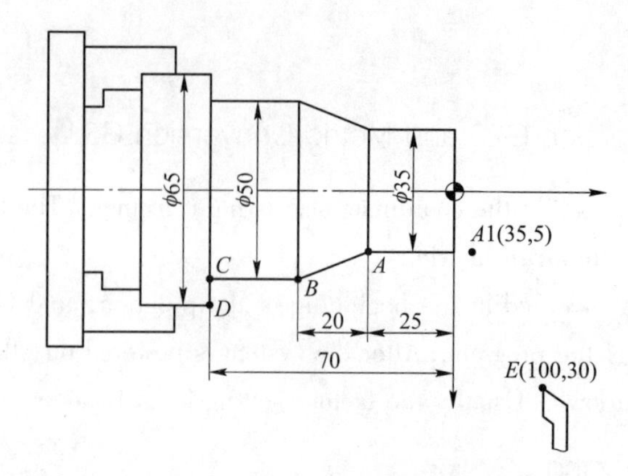

Figure 2-6　G00/G01 application

2.1.2.3　Linear interpolation G01

(1) Function: This command makes the tool move straight from the point to the target point at a given speed.

(2) Format:

G00 X(U)＿ Z(W)＿ F＿;

Among them, X(U) and Z(W) are the coordinate values of the target point, F is the feed rate.

Note:

The feed rate of the G01 is determined by F. If there is no F before the G01 block, and there is no F in the current G01 block, the machine will not move.

【Example 2-2】Finish machining the parts shown in Figure 2-6. Select the right end O point as the programming origin. The programming are shown in Table 2-2.

Table 2-2　The programming

Absolute programming		Increment programming	
O0001;		O0002;	
N10	G40 G97 G99;	N10	G40 G97 G99;
N20	T0101;	N20	T0101;
N30	M03 S1000;	N30	M03 S1000;
N40	G00 X35.0 Z5.0;	N40	G00 U-65.0 W-25.0;
N50	G01 Z-25.0 F0.1;	N50	G01 W-30.0 F0.1;
N60	X50.0 Z-45.0;	N60	U15.0W-20.0;
N70	Z-70.0;	N70	W-25.0;
N80	X65.0;	N80	U15.0;
N90	G00 X70.0;	N90	U35.0;
N100	Z30.0;	N100	W100.0;
N110	M30;	N110	M30;

【Example 2-3】The high-step shaft is shown in Figure 2-7. The blank diameter is 40mm, the back engagement is 2. 5mm, the finishing allowance is 0. 5mm, and the rough and finish turning program is as shown in Table 2-3.

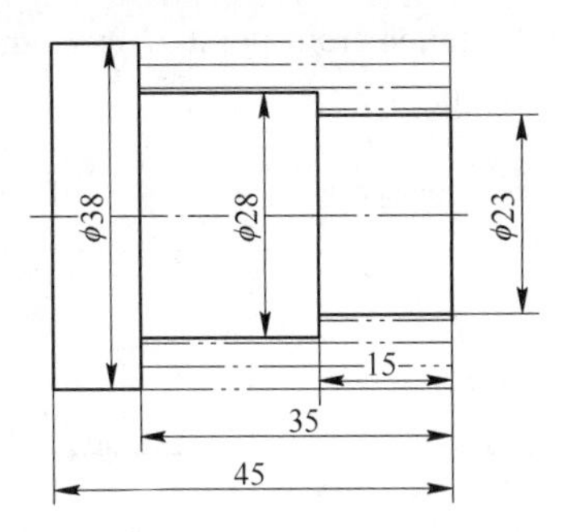

Figure 2-7 High-step shaft

Table 2-3 Rough and finishing programmings for the high-step shaft

O121;	Program No.
G40 G97 G99 M03 S600 F0. 2;	G00 Z5. 0;
T0101;	X24. 0;
M08;	G01 Z-15. 0;
G00 Z5. 0;	X25. 0;
X39. 0;	G00 Z5. 0;
G01 Z-45. 0 F0. 2;	X23. 0;
X40. 0;	G01 Z-15. 0 F0. 1 S1000;
G00 Z5. 0;	X28. 0;
X34. 0;	Z-35. 0;
G01 Z-35. 0;	X38. 0;
X35. 0;	Z-45. 0;
G00 Z5. 0;	X40. 0;
X29. 0;	G00 X100. 0;
G01 Z-35. 0;	Z100. 0;
X30. 0;	M30;

(3) G01 extended functions

1) Function: It is to automatically interpolate the right angle or fillet between adjacent trajectories, as shown in Figure 2-8.

2) Format:

G01 X(U)_ Z(W)_R_F_;

G01 X(U)_ Z(W)_C_F_;

Among them X(U), Z(W)——intersection coordinates of adjacent lines (as point *D* in Figure 2-8);

R——Arc radius of fillet;

C——Distance from point *D* to chamfer start point *B*.

3) Note: *R*, *C* is the non modal code.

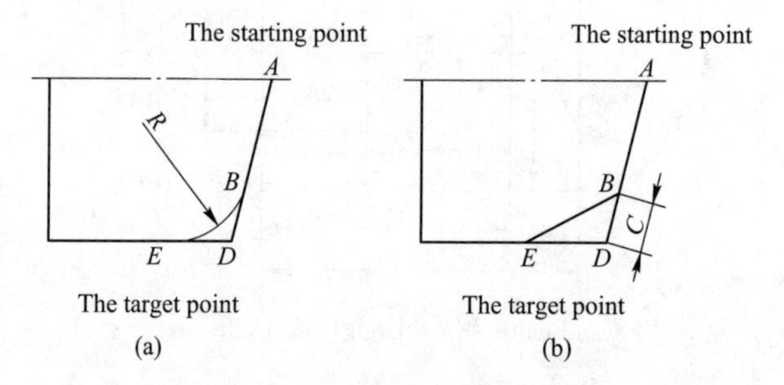

Figure 2-8 G01 extended functions

(a) Fillet with G01; (b) The right angle with G01

【Example 2-4】Use the functions of chamfer and fillet to write a finishing program for the part shown in Figure 2-9.

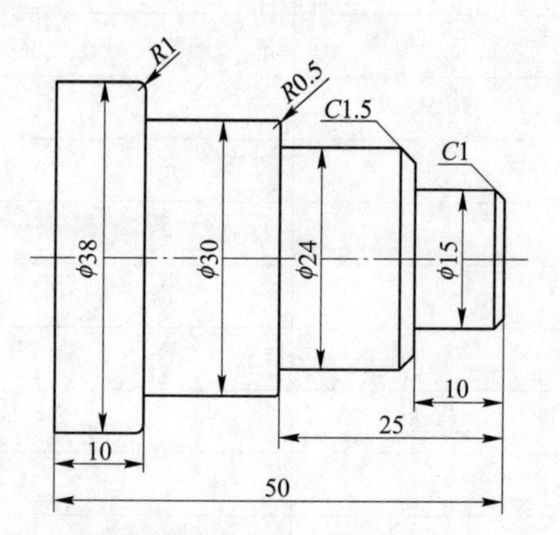

Figure 2-9 The part drawing

The finishing program is shown in Table 2-4.

Table 2-4 Finishing program

O0006;	Program No.
G40 G97 G99 M03 S1000;	Z-25.0;
T0101;	X30.0 R0.5;
M08;	Z-40.0;
G00 Z5.0;	X38.0 R1.0;
X0;	Z-50.0;
G01 Z0 F0.1;	G00 X100.0;

Continued Table 2-4

X15.0 C1.0;	Z100.0;
Z-10.0;	M30;
X24.0 C1.5;	

2.1.3 Operation

2.1.3.1 Workpiece Clamping and Alignment

A three-jaw self-centering chuck is used to clamp the outer circle of the bar, and after the outer circle is aligned, the workpiece is clamped. Note that the work piece must be installed firmly.

The method of workpiece alignment usually uses a dial indicator. The commonly used clock face dial indicator is shown in Figure 2-10. The dial indicator is an indicator gauge, which can be used to measure the size, shape, and position error of the workpiece in addition to the calibration.

Figure 2-10 Structure of the dial indicator

1—Portable measuring rod; 2—Pointer; 3—Speedindicator; 4—Dial; 5—Bezel; 6—Sleeve; 7—Measuring rod; 8—Measuring head

The use of dial indicator is as follows:

(1) Before use, check the flexibility of the measuring rod. That is, when the measuring rod is gently pushed, the measuring rod must move flexibly in the sleeve, and the pointer can return to the original scale position after each relaxation.

(2) When using the dial indicator, it must be fixed on a reliable holder such as on a multimeter or magnetic stand.

(3) Do not let the measuring head hit the part suddenly.

(4) Do not subject the dial indicator to severe vibration and shock.

Use a dial indicator to align it as shown in Figure 2-11. The specific steps are as follows:

(1) Preparation stage: insert the clock dial indicator into the hole of the magnetic force gauge seat and lock it, check the probe's elasticity, and check whether the probe and pointer cooperate normally.

(2) Measurement stage: The dial gauge probe is perpendicular to the rotation axis of the work-piece. Turn the three-jaw chuck by hand, and tap the workpiece to adjust it according to the

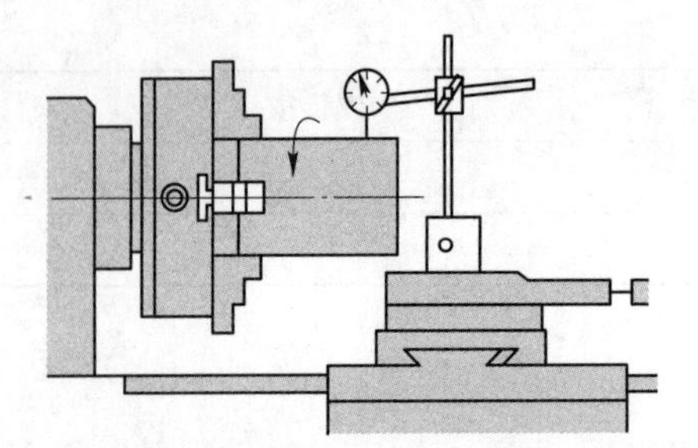

Figure 2-11 Parts alignment

swing direction of the pointer of the dial indicator, so that the rotation axis of the workpiece is the *Z axis of the workpiece coordinate system.* It coincides with the central axis of the spindle of the CNC lathe.

Note:

(1) When clamping the workpiece, the reference should be unified as much as possible to reduce positioning errors and improve machining accuracy.

(2) When clamping the processed surface, a copper layer should be wrapped on the processed surface to avoid pinching the surface of the workpiece.

(3) The clamping position should be selected on the workpiece with a strong and rigid surface.

2.1.3.2 Tool Installation

The installation steps of the external turning tool of the machine clamp are as follows:

(1) Insert the blade into the tool body, screw in the screws, and tighten.

(2) Before attaching the tool holder to the tool holder, clean the tooling surface and the turning tool holder.

(3) The length of the turning tool on the tool holder is about 1.5 times the height of the tool bar. Too long a protrusion will affect the rigidity of the tool bar.

(4) The turning tool tip should be at the same height as the center of the workpiece.

(5) The center of the tool holder should be perpendicular to the feeding direction.

(6) Press the turning tool with at least two screws to secure the tool holder.

2.1.3.3 External Turning Tool Setting Operation

The tool setting operation can be completed by manual or hand wheel. The steps for external turning tool setting are as follows:

(1) *Z* compensation.

1) Moving the external turning tool to cut the end face.

① Press the manual key [手动] →press key [X-] or [X+] →the machine tool moves in the *X* direction. In the same way, the machine tool moves along the *Z* direction to approach the blank.

② Press the key [MDI] →press the key [PROG] →enter MDI interface→input "M03 S600" →press the key →press the key →press the cycle start key →the spindle rotates forward.

③ The tool approaches the workpiece →press the key , cut the end face of the workpiece.

④ Press the key , the Z coordinate remains unchanged, and the tool is retracted in the positive direction along the X-axis.

2) Setting Z direction correction.

① Press the key →press the key [OFFSET]→press the key [GEOM]→move the cursor to the selected tool position, such as G01. The interface is shown in Figure 2-12 (a).

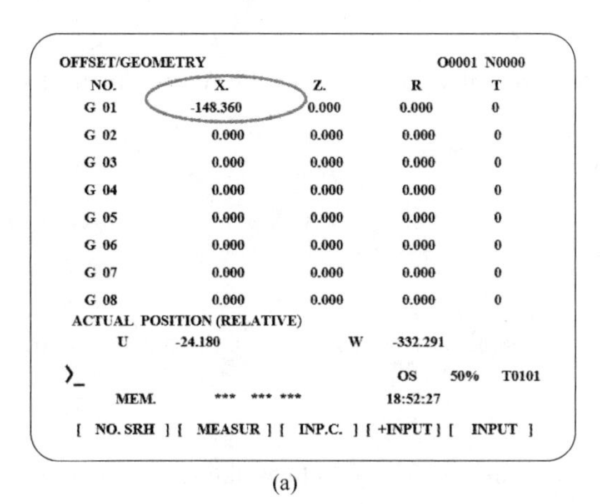

OFFSET/GEOMETRY O0001 N0000

NO.	X.	Z.	R	T
G 01	-148.360	0.000	0.000	0
G 02	0.000	0.000	0.000	0
G 03	0.000	0.000	0.000	0
G 04	0.000	0.000	0.000	0
G 05	0.000	0.000	0.000	0
G 06	0.000	0.000	0.000	0
G 07	0.000	0.000	0.000	0
G 08	0.000	0.000	0.000	0

ACTUAL POSITION (RELATIVE)
U -24.180 W -332.291
>_ OS 50% T0101
MEM. *** *** *** 18:52:27
[NO. SRH] [MEASUR] [INP.C.] [+INPUT] [INPUT]

(a)

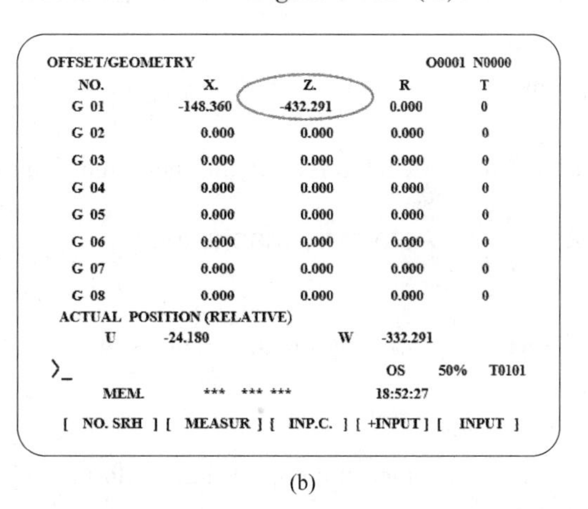

OFFSET/GEOMETRY O0001 N0000

NO.	X.	Z.	R	T
G 01	-148.360	-432.291	0.000	0
G 02	0.000	0.000	0.000	0
G 03	0.000	0.000	0.000	0
G 04	0.000	0.000	0.000	0
G 05	0.000	0.000	0.000	0
G 06	0.000	0.000	0.000	0
G 07	0.000	0.000	0.000	0
G 08	0.000	0.000	0.000	0

ACTUAL POSITION (RELATIVE)
U -24.180 W -332.291
>_ OS 50% T0101
MEM. *** *** *** 18:52:27
[NO. SRH] [MEASUR] [INP.C.] [+INPUT] [INPUT]

(b)

Figure 2-12 Parameter input interface

(a) Z compensation; (b) X compensation

② Input Z0→press the key [MEASUR].

3) Inspection of tool setting operation.

After the Z direction correction is completed, the tool moves forward in the X-axis direction away from the workpiece (the Z value does not change), press the key →enter tool number→ press the cycle start key , at this time, the absolute value of the Z coordinate displayed on the CRT screen is zero.

(2) X compensation.

1) Cutting outer diameter.

① The spindle rotates forward.

② The tool approaches the workpiece→press the key →The machine tool moves in the negative direction of the Z axis, and the tool cuts the outer circle of the workpiece.

③ Pess the key , the X axis coordinate remains unchanged, and the tool is retracted in the positive direction along the Z axis.

2) Measuring cutting diameter.

Spindle stop → measure the trial cutting outer circle and note the diameter value.

3) Setting X direction correction.

① Press the key [OFS/SET] →press the key [OFFS]→press the key [GEOM]→move the cursor to the selected tool position, such as G01. The interface is shown in Figure 2-12 (b).

② Enter the X diameter value (such as X33. 539) →press the key [MEASUR] .

4) Inspection of tool setting operation.

After the X-direction correction is completed, the tool moves in the positive direction of the Z axis away from the workpiece (the X value does not change). Press the key [MDI] →enter the tool number→press the cycle start key [程式启动 CYCLE START], at this time, the absolute value of the X coordinate displayed on the CRT screen is the measured diameter.

2. 1. 3. 4 Automatic Machining

Calling the processing program→press the key [记忆] →press the cycle start key [程式启动 CYCLE START], machine parts automatically .

Before the automatic processing, perform the full-axis operation, and check the empty operation and lock button status.

2. 1. 4 Measurement

2. 1. 4. 1 Vernier Calipers

(1) Application. Vernier calipers can measure dimensions such as length, width, thickness, inner diameter and outer diameter, hole distance, height and depth, etc. , and are widely used universal measuring tools.

(2) Structure. As shown in Figure 2-13.

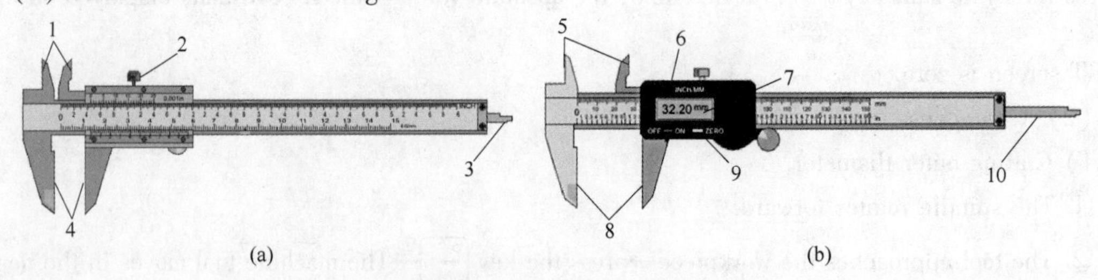

Figure 2-13 Vernier caliper

(a) Ordinary Vernier Caliper; (b) Digital Vernier Caliper

1, 5—Inner measuring jaw; 2—Fastening screw; 3—Depth gauges; 4, 8—Outer measuring jaw; 6—Metric/Imperial Button; 7—LCD; 9—Switch zero setting; 10—Depth gauges

(3) Measurement methods. When measuring, hold the workpiece in the left hand, hold the main ruler in the right hand, move the vernier ruler with your thumb, so that the workpiece to be measured is located between the measuring claws. When the measuring claws are in close contact with the measuring claws, tighten the fastening screws to read, as shown in Figure 2-14.

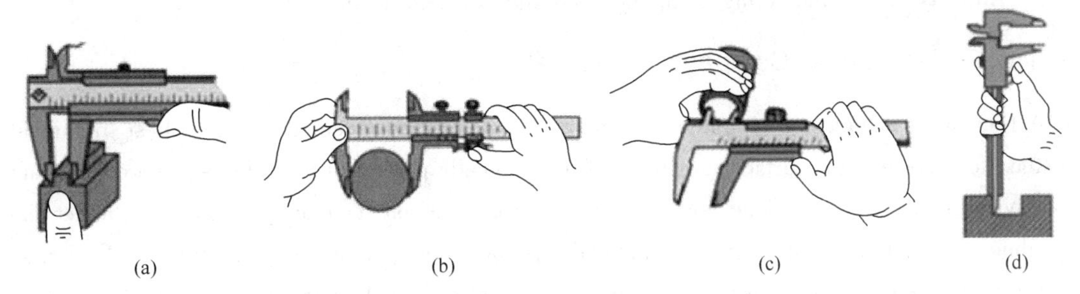

(a) (b) (c) (d)

Figure 2-14 Correct measurement method

(a) Measuring width; (b) Measuring outer diameter; (c) Measuring inner diameter; (d) Measuring depth

(4) Reading. Digital vernier calipers can be read directly on the LCD screen. Ordinary vernier calipers can be divided into three types: 0. 10mm, 0. 05mm and 0. 02mm according to their measurement accuracy. At present, vernier calipers with an accuracy of 0. 02mm are commonly used in machining. The vernier caliper uses the zero line of the vernier as the baseline for reading. Taking Figure 2-15 as an example, the reading method is divided into three steps:

1) Read the integer, that is, read the millimeter on the left ruler of the zero line of the cursor as an integer value (19mm).

2) Read the decimal, that is to find the vernier scribe line on the vernier ruler aligned with the main ruler. Multiply the number of grids between the aligned vernier scribe line and the vernier zero line by the accuracy of the caliper to a decimal value (0. 52mm) .

3) Add the integer value and the decimal value to obtain the actual size (19. 52mm).

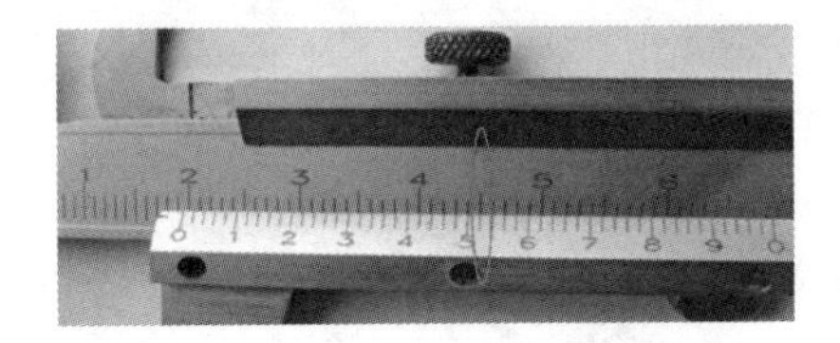

Figure 2-15 0. 02mm vernier caliper reading

(5) Note.

1) Before measurement, wipe the measuring claw and the surface of the workpiece to be cleaned, then close the two measuring claws to make them fit, and check whether the zero lines of the main ruler and vernier rule are aligned. If it is not aligned, correct the reading based on the original error after the measurement or adjust the vernier caliper to zero before using.

2) When the measuring jaw is in contact with the workpiece to be measured, the force should

not be too large, so as not to deform or wear the jaw and reduce the accuracy of the measurement.

3) When measuring the size of the part, the line connecting the two measuring surfaces of the caliper should be perpendicular to the surface to be measured, and should not be skewed.

4) The surface of the blank cannot be measured with a vernier caliper.

5) After use, wipe the vernier calipers clean and put them in the box.

2.1.4.2 Roughness Comparison Template

(1) Roughness comparison template. The roughness comparison method is the earliest traditional method for detecting the surface roughness of mechanically processed workpieces. The comparison method is to use the workpiece and the roughness comparison template to evaluate whether the roughness is qualified. This detection method has low efficiency and poor accuracy. Roughness comparison samples are shown in Figure 2-16, also called roughness comparison plates, roughness comparison blocks or roughness comparison samples.

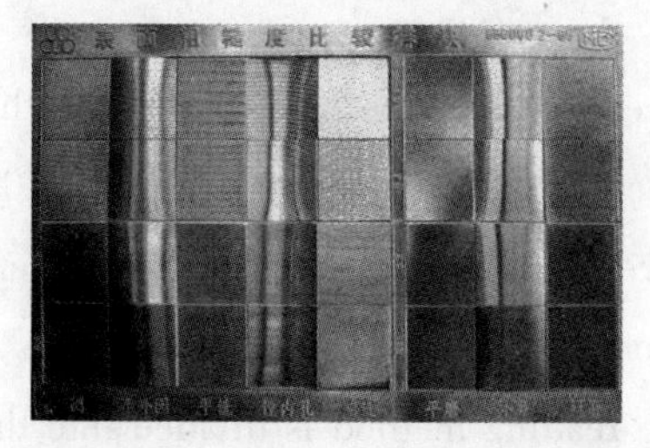

Figure 2-16 Roughness comparison template

(2) Roughness gauge. For parts with relatively high surface quality, a rough gauge can be used for inspection. It has the characteristics of high measurement accuracy, wide measurement range, easy operation, portability, and stable work. It can be widely used in the detection of various metal and non-metal processed surfaces, as shown in Figure 2-17.

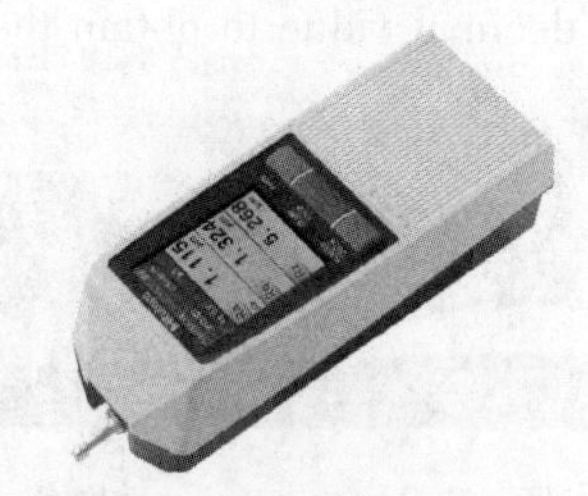

Figure 2-17 Portable Roughness Tester

Implemention

(1) Step 1 Drawing analysis:

The part material shown in Figure 2-1 is LY15, with a total length of 50mm. The machining surfaces include $\phi28^{0}_{-0.03}$、$\phi31^{0}_{-0.04}$、$\phi35$、$\phi38^{0}_{-0.04}$ cylindrical surfaces, *C*2 chamfer, etc., and the surface roughness is *Ra*1.6 and *Ra*3.2 respectively.

(2) Step 2 Making the machining plan:

1) Machining plan

① The three-jaw self-centering chuck is used to mount the card blank, and the length of the extended chuck is about 60mm.

② Machining the outer contour of the part to the required size.

2) Tool selection

One tool is used for roughing and finishing of the part. Therefore, an external turning tool with a tip angle of 55° and a main deflection angle of 90° is selected. The CNC machining tool card is shown in Table 2-5.

Table 2-5 CNC machining tool card

Part name		Stepped shaft		Drawing No.	2-1		
S No.	Tool No.	Tool name	Quantity	Machined surface	Tool nose radius R/mm	Tool tip orientation T	Note
1	T01	90° external turning tool	1	Rough and finish turning the outer diameter	0.4	3	
Edit		Check	Approve	Date		Total 1 Page	Page 1

3) Machining process

The machining route of the part is shown in Figure 2-18.

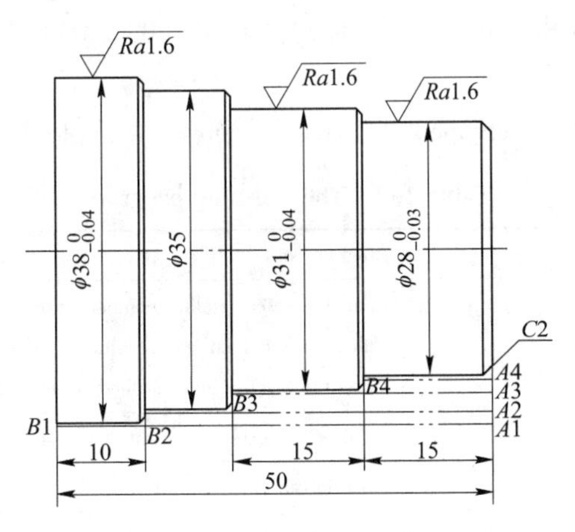

Figure 2-18 Stepped shaft machining route

Coordinate points:

$A1$(39, 0), $A2$(36, 0), $A3$(32, 0), $A4$(29, 0);

$B1$(39, -50), $B2$(36, -40), $B3$(32, -30), $B4$(29, -15).

The CNC machining process card is shown in Table 2-6.

Table 2-6 CNC machining process card

Company name	Tianjin polytechnic college			Part name		Drawing No.	
				Stepped shaft		2-1	
Program No.	Fixture name		Machine tool	CNC system		Workshop	
O211	Three-jaw chuck		CKA6140	FANUC SERIES Oi		CNC Training Center	
Steps	Contents	Tool	Spindle speed n/r · min^{-1}	Feed rate F/mm · r^{-1}	Back engagement a_p/mm	Note	
1	Clamp and align the workpiece					Manual	
2	Setting external turning tool	T01					
3	Rough turning outer diameter with allowance of 1mm	T01	600	0. 2	1. 5	O211	
4	Finishing outer diameter	T01	1000	0. 1	0. 5		
Edit		Check		Approve	Date	Total 1 Page	Page 1

(3) Step 3 Programming:

1) Numerical calculation

For single-piece and small-lot production, the finishing size of the part is generally the average of the limit sizes.

Programming dimension = basic dimension + (upper deviation + lower deviation)/2

The programming size of the outer circle $\phi28^{0}_{-0.03} = 28+(0-0.03)/2 = 27.985$

The programming size of the outer circle $\phi31^{0}_{-0.04} = 31+(0-0.04)/2 = 30.98$

The programming size of the outer circle $\phi38^{0}_{-0.04} = 38+(0-0.04)/2 = 37.98$

2) Programming

The finishing program of the stepped shaft part is shown in Table 2-7.

Table 2-7 The finishing program

O211;	Program No.
G40 G97 G99 M03 S600 F0. 2;	Cancel the tool radius compensation, cancel the constant speed of the spindle, set the feed per revolution, the spindle rotates forward, and the speed is 600r/min, and the feed is set to 0. 2mm/r
T0101;	Change T01 external turning tool, substitute tool compensation No. 01
M08;	Turn on the cutting fluid
G00 Z5. 0;	Rapid feed to the starting point
X39. 0;	
G01 Z-50. 0	Cut to point *B*1
G01 X40. 0	Resignation
G00 Z5. 0;	
X36. 0;	Rapid feed to the starting point

Continued Table 2-7

G01 Z-40.0;	Cut to point *B*2
G00 X40.0;	Resignation
Z5.0;	
G01 X32.0	Rapid feed to the starting point
Z-30.0;	Cut to point *B*3
X33.0;	Resignation
G00 Z5.0;	
X29.0;	Rapid feed to the starting point
G01 Z-15.0;	Cut to point *B*4
X30.0;	Resignation
G00 Z5.0;	
X0;	Rapid feed to the starting point
G01 Z0;	Feed to the starting point of finishing
X27.985 C2;	Sequence of finishing shape
Z-15.0;	
X30.98 C1;	
Z-30.0;	
X35.0 C1;	
Z-40.0;	
X37.98 C1;	
Z-50.0;	
G00 X100;	Quickly retract to the tool change point (100, 100)
Z100.0;	
M30;	The program ends and returns to the starting point

(4) Step 4 Part machining:

Machine system startup → Return to reference point → Clamp and align the workpiece → Install the tool → Input the program in Table 2-7 → Simulation the program → Tool setting→ Automatic machining.

(5) Step 5 Dimension measurement:

1) Use vernier calipers to measure outer diameter $\phi 28^{0}_{-0.03}$, $\phi 31^{0}_{-0.04}$, $\phi 35$, $\phi 38^{0}_{-0.04}$.

2) Use vernier calipers to measure the length 50mm, 10mm, 15mm.

3) Inspect the roughness with roughness comparison board.

Assessment

The assessment table is shown in Table 2-8.

Table 2-8 The assessment table

Item	No.	Standard		Partition	Score
Machining plan (10%)	1	Machining steps	Meet the requirements of NC turning	3	
	2	Dimension	Calculate dimensions	3	
	3	Tool selection	The cutter selection is reasonable and meets the machining requirements	4	
Programming (30%)	4	Program No.	No program no point	2	
	5	Program segment number	No program segment number no point	2	
	6	Cutting parameters	Unreasonable choice of cutting parameters no point	4	
	7	Origin and coordinates	Mark the origin and coordinates of the program, otherwise there is no point	2	
	8	Program contents	Deduct 2 points for each segment that does not meet the requirements of program logic and format	20	
			Deduct 5 points if the procedure content is not in accordance with the process		
			Deduct 5 points in case of dangerous instruction		
Operation (30%)	9	Program input	Deduct 2 points for each wrong section	10	
	10	Machining operation 30min	Workpiece coordinate system origin setting error no point	20	
			Misoperation no point		
			Overtime no point		
Dimension measurement (20%)	11	External profile	$\phi38^{0}_{-0.04}$ deduct 3 points for 0.01 out of tolerance	3	
	12		$\phi35$ no grade reduction	2	
	13		$\phi31^{0}_{-0.04}$ deduct 3 points for 0.01 out of tolerance	3	
	14		$\phi28^{0}_{-0.03}$ deduct 3 points for 0.01 out of tolerance	3	
	15	Length	50、15、15、10	4	
	16	Chamfer	*C*2	1	
	17	Surface roughness	*Ra*1.6 over tolerance no points	3	
	18		*Ra*3.2 over tolerance no points	1	
Vocational ability (10%)	19	Learning ability		2	
	20	Communication skills		2	
	21	Cooperation ability		2	
	22	Safe operation and civilized production		4	
Total					

Exercises

(1) Answer the Following Questions:

1) Describe the general principles of cutting parameters selection.

2) Which tools are commonly used in CNC turning?

3) Describe how to clamp and align the workpiece briefly.

(2) Synchronous Training:

The parts are shown in Figure 2-19 and Figure 2-20, please programming and machining.

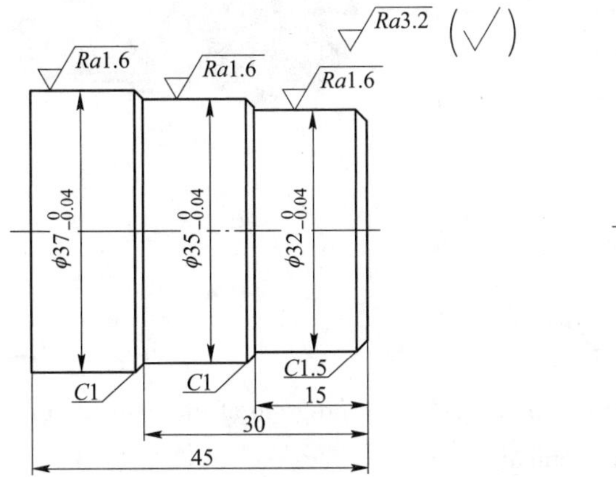

Figure 2-19 Synchronous training 1

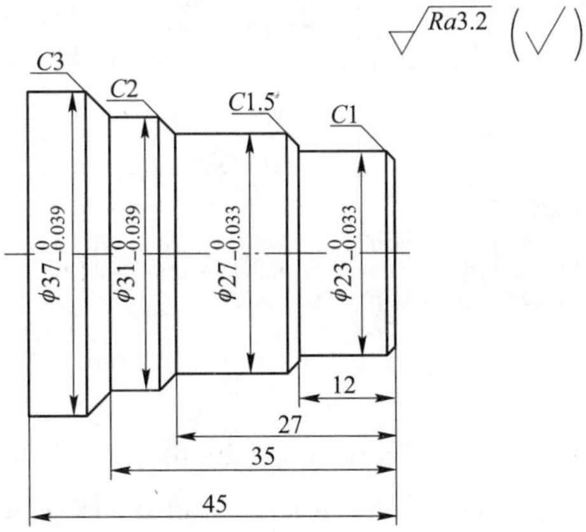

Figure 2-20 Synchronous training 2

Task 2.2 CNC Turning of Conical Surface Parts

Introduction

As shown in Figure 2-21, the part is made of hard aluminum alloy LY15, and the blank is ϕ40mm. It is produced in one piece. This task needs to prepare the program, machine the part using the CNC lathe, and then measure the part.

Competences

(1) Knowledge:

1) Know the turning technology and parameters of conical surface.

2) Interpret G71/G70, G40/G41/G42 instructions.

3) Know the measuring tools of conical surface.

(2) Skill:

1) Learn how to use the formula to calculate taper size.

2) Have the ability to use G71/G70, G40/G41/G42 instructions to write the conical surface machining program.

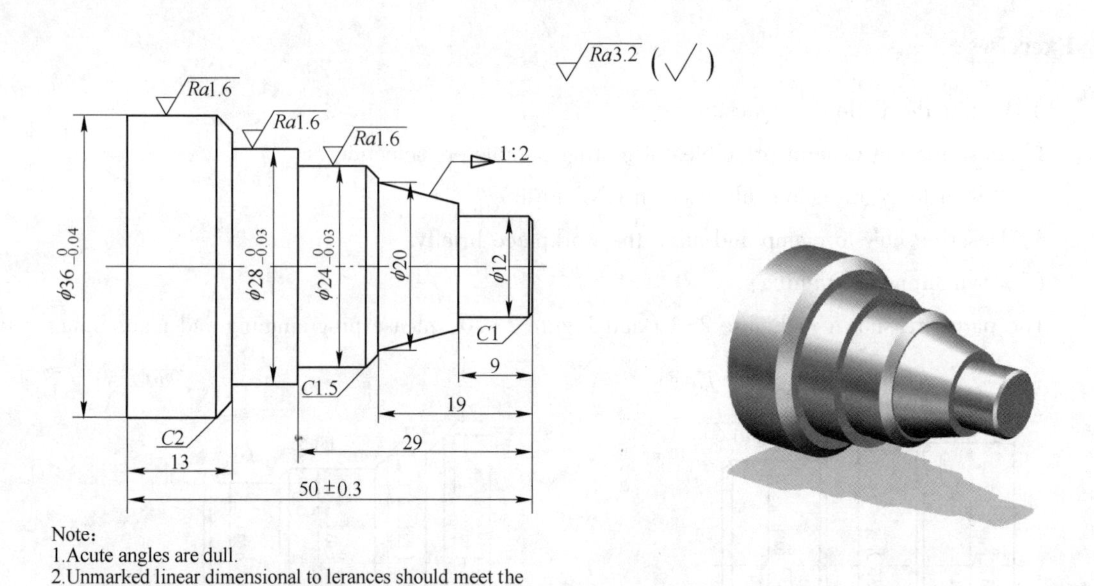

Figure 2-21 The part drawing

3) Have the ability to operate the CNC lathe to verify machining program of the conical surface.

4) Understand the inspection methods of the conical surface.

(3) Quality:

1) Carry out safety technical operation procedures correctly, develop safety awareness.

2) Cultivate serious and responsible work attitude and quality awareness.

Relevant Knowledge

2.2.1 Process Preparation

The conical surface has the advantages of high coaxiality and convenient disassembly. When the conical angle is smaller ($\alpha<3°$), it can transfer large torque, so it is widely used in mechanical manufacturing.

2.2.1.1 CNC Turning Methods of The Conical Surface

Figure 2-22 (a) shows the turning mode of the cone, the cutting path is parallel to the conical surface, and the cutting amount is even. As shown in Figure 2-22 (b), the cutting amount is uneven and the programming is relatively simple.

2.2.1.2 Conical Surface Parameters and Size Calculation

The commonly used conical surface parameters are shown in Figure 2-23.

(1) Maximum diameter of cone D, referred to as the big end diameter.

(2) Minimum diameter of cone d, referred to as the small end diameter.

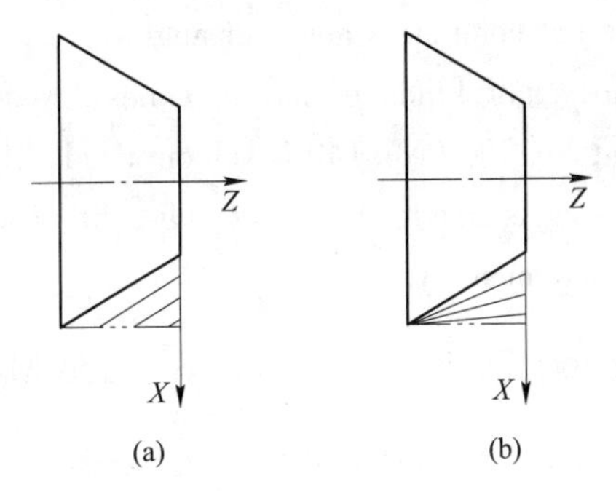

Figure 2-22 Turning method of conical surface

(a) Even cutting; (b) Uneven cutting

(3) Cone length L, axial distance between the maximum cone diameter and the minimum cone diameter.

(4) Taper C, the taper is the ratio of the difference between the maximum diameter and the minimum diameter of the cone and the length L of the cone, i. e. $C=\frac{D-d}{L}$.

【Example 2-5】As shown in Figure 2-24, the taper of the conical surface is 1 : 2.5, please calculate maximum diameter D of the conical surface.

In Figure 2-24, the minimum diameter $d=22$, length of the cone $L=15$, taper $C=1:2.5=0.4$, according to the formula $C=\frac{D-d}{L}$, the maximum diameter of the cone is obtained.

$$D=d+CL=22+0.4\times15=28$$

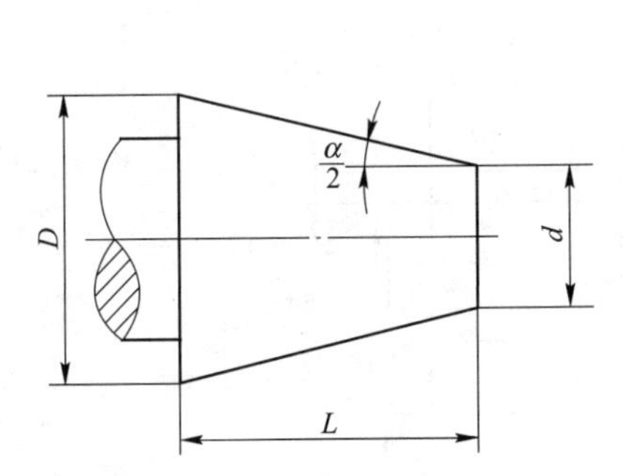

Figure 2-23 Taper parameters

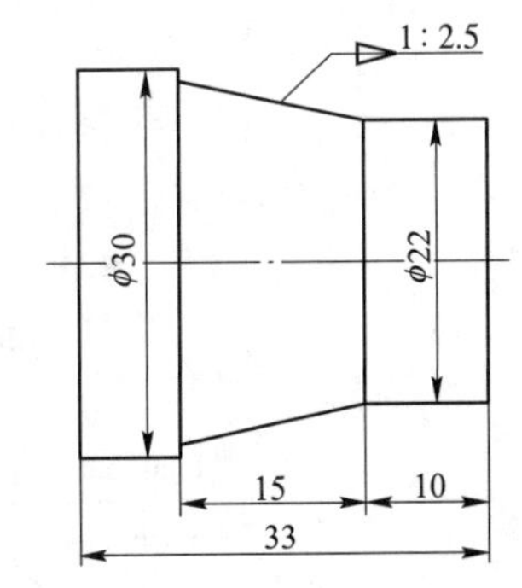

Figure 2-24 Calculation of conical surface

2.2.1.3 Standard Tool Cone

In order to manufacture and use conveniently and reduce the production cost, the cone on the common tools and cutters is standardized. There are two kinds of cones for standard tools.

(1) Morse cone. Morse cone is one of the most widely used in the mechanical manufacturing industry, such as lathe spindle taper hole, center, bit handle, reamer handle, etc. Morse taper is divided into seven types: 0, 1, 2, 3, 4, 5 and 6. The smallest is 0 and the largest is 6. Different

Morse cone numbers have different cone sizes and half angles.

(2) Metric cone. There are eight kinds of metric cones: No. 4, No. 6, No. 80, No. 100, No. 120, No. 140, No. 160 and No. 200. No. 140 is seldom used. Their numbers represent the diameter of large end, and the taper is fixed, that is, $C = 1 : 20$. The advantage of metric cone is that the taper is constant and easy memory.

2.2.1.4 Selection of Cutting Tools for Conical Surface Machining

When machining the conical surface, the external turning tool is also used, which is the same as the tool when turning the stepped shaft. However, when turning the inverted cone, it is necessary to select a tool with a larger deflection angle, so that the cutting edge of the tool cannot collide with the cone surface.

2.2.2 Programming

2.2.2.1 Stock Removal in Turning G71

(1) Function.

This instruction only needs to specify the depth of cut for rough machining, machining allowance for finishing and finishing route, the CNC system can automatically give rough machining route and machining times, and complete the rough machining of the inner and outer contour. Figure 2-25 shows the G71 instruction cycle route in which A is the starting point of tool cycle. When the rough machining compound cycle is executed, the tool moves from point A to point C. After the rough turning cycle, the tool returns to point A.

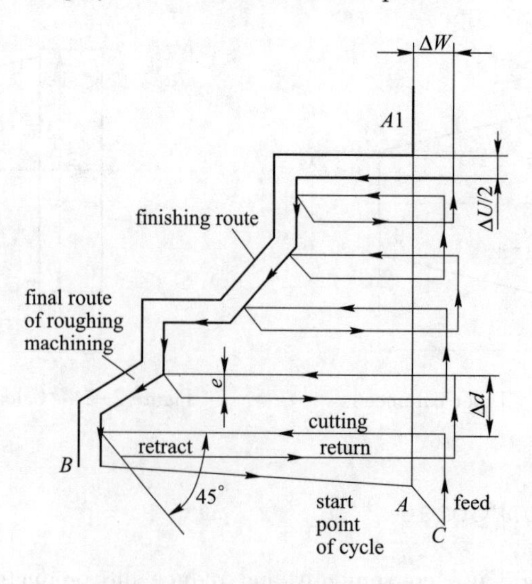

Figure 2-25 Tool path of G71

(2) Format:

$$G71\ U(\Delta d)R(e);$$

$$G71\ P(ns)Q(nf)\quad U(\Delta u)\quad W(\Delta w);$$

Among them　Δd——Depth of cut (radius designation);

e——Escaping amount (radius designation);

ns——Sequence number of the first block for the program of finishing shape;

nf——Sequence number of the last block for the program of finishing shape;

Δu——Distance and direction of finishing allowance in X direction (diameter designation);

Δw——Distance and direction of finishing allowance in Z direction.

(3) Note:

1) When rough machining is performed by G71, the F and S instructions contained in "ns" and "nf" block are invalid for rough turning cycle.

2) The subprogram cannot be called from the block between sequence number "ns" and "nf".

3) The outline of the part must conform to the directions of the X-axis and Z-axis and increase or decrease monotonously at the same time.

4) The first sentence of the finishing line must be feed in X direction with G00 or G01.

(4) Application: rough machine of raw materials.

2.2.2.2　Finishing Cycle G70

(1) Function: remove finishing allowance.

(2) Format:

G70 P(ns)Q(nf);

(3) Note:

1) When using G71 roughing, the F and S commands contained in the ns~nf block are invalid for the roughing cycle.

2) After cutting finished by G70, the tool returns to the cycle starting point of G71.

(4) Application. G70 is used for finishing machining, cutting the machining allowance left after G71 rough machining.

【Example 2-6】As shown in Figure 2-26, the blank is ϕ40mm long bar stock. Use G71/G70 instruction to write the part program of rough turning and finish turning. The part program is shown in Table 2-9.

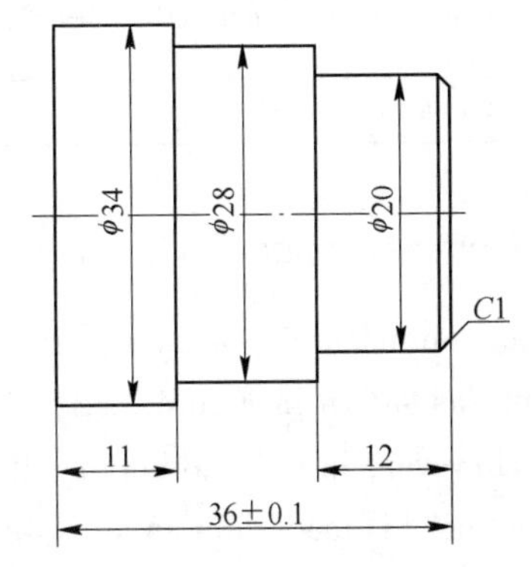

Figure 2-26　G71/G70 instruction example

Table 2-9 Part program and explanation

O2001;	Program No.
G40 G97 G99 M03 S600 F0. 2;	Cancel the tool nose radius compensation, setting constant spindle speed control, feed rate 0. 2mm/r, spindle CW at 600r/min
T0101;	Replace T01 external turning tool, substitute tool compensation No. 01
M08;	Open cutting fluid
G00 Z5. 0; X40. 0;	Rapid positioning to point (*X*40. 0, *Z*5. 0)
G71 U1. 5 R0. 5; G71 P10 Q20 U0. 5 W0. 05;	Calling stock removal in turning
N10 G00 X0; G01 Z0; X18. 0; X20. 0 Z-1. 0; Z-12. 0; X28. 0 Z-25. 0; X34. 0 N20 Z-40. 0;	Sequence of finishing shape
G00 X100. 0; Z100. 0;	Rapid positioning to point (*X*100. 0, *Z*100. 0)
M05;	Spindle stop
M00;	Program stop
M03 S1000 F0. 1;	Spindle CW at 1000r/min, calling No. 1 tool and it's offset, feed rate 0. 1mm/r
T0101;	Substitute tool compensation No. 01
G00 Z5. 0 X40. 0;	Rapid positioning to the starting point of the composite cycle of rough machining (*X*40. 0, *Z*5. 0)
G70 P10 Q20;	Calling finishing cycle in turning
G00 X100. 0; Z100. 0;	Rapid positioning to the change point (*X*100. 0, *Z*100. 0)
M30;	End of program and reset

2. 2. 2. 3 Tool Nose Radius Compensation G41/G42/G40

(1) Influence of tool nose arc radius on machining accuracy. In the ideal state, the tool point of a sharp turning tool is its sharp point, as shown in Figure 2-27. Pint *A* is the imaginary tip. This point is also used for tool setting. However, in order to improve the tool life and reduce the surface roughness of the workpiece, the tool nose is made into an arc with a small radius in the actual processing. *BC* arc is shown in Figure 2-27.

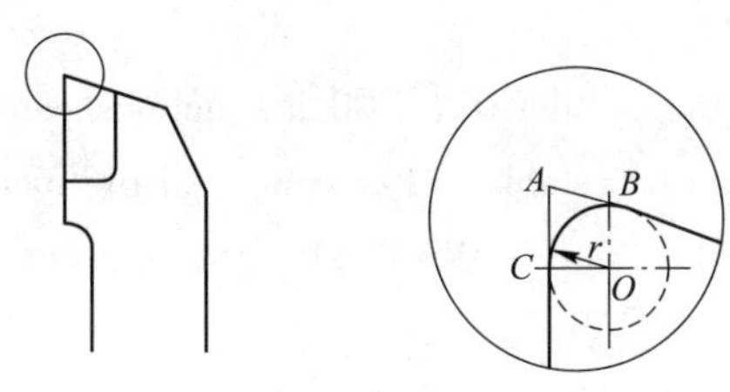

Figure 2-27 Schematic diagram of imaginary tool tip

As shown in Figure 2-28, when turning the outer cylinder surface and end face, the actual tool path is consistent with the contour of the workpiece without any error. When turning the cone, the contour of the workpiece is a solid line, and the actual turning shape is a dotted line in Figure 2-28, resulting in undercut error δ. If the accuracy requirement of workpiece is not high or there is finishing allowance left, this error can be ignored. Otherwise, the influence of the tool tip on the shape of workpiece should be considered.

When cutting arc contour, there will be over cutting or under cutting due to the existence of tool nose radius, as shown in Figure 2-29.

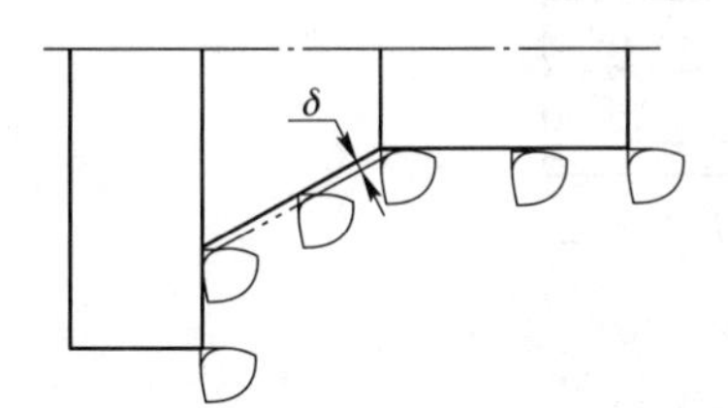

Figure 2-28 The error of turning cone

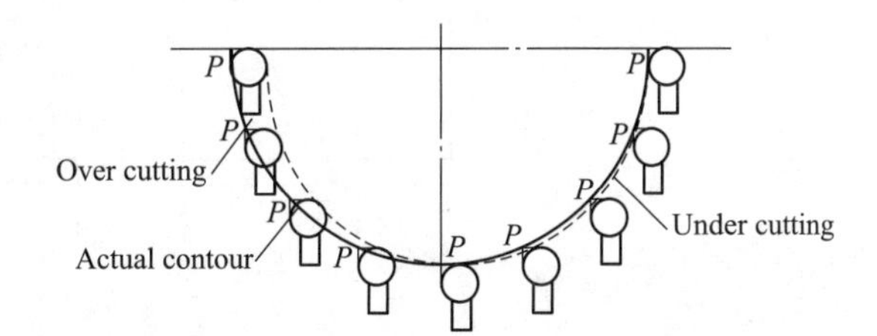

Figure 2-29 Error caused by turning arc

(2) Compensation method. The CNC system with the tool nose radius compensation function can prevent the above phenomenon. When programming, only need to program according to the contour of the part with the imaginary position of the tool nose. In the process of automatically machining, the CNC system will compensate the tool nose radius according to the parameter values or compensation instructions set in the compensation register.

(3) Tool nose radius compensation instructions (G41、G42、G40). G41 is the tool nose radius compensation left, G42 is the Tool nose radius compensation right, and G40 is the tool nose radius compensation cancel. The format is as follows:

$$\left.\begin{matrix}G41\\G42\\G40\end{matrix}\right\}\left.\begin{matrix}G01\\G00\end{matrix}\right\}\ X(U)\_\ Z(W)\_\ F\_;$$

Among them X(U)Z(W)——Terminal coordinates of the tool nose radius compensation establish or cancel;

F——Specifies the feed rate of G01.

Note:

1) G41, G42, G40 instructions and G01, G00 instructions can appear in the same program block and tool compensation can be established or cancelled by linear motion.

2) In the block where G41, G42 and G40 is located, at least one value of X or Z changes, otherwise an alarm will occur.

3) G41 and G42 can't be used at the same time, that is, in the program if G41 is available in the previous program block, G42 can't be used again. G42 can only be used when G40 command is used to cancel tool nose radius compensation.

4) Before calling a new tool, G40 instruction must be used to cancel tool nose radius compensation, otherwise an alarm will occur.

(4) Selection of G41 and G42. The tool and machining directions are shown in Figure 2-30. Looking along the direction of tool movement, the workpiece is on the left side of the tool, which is called left compensation, using G41 instruction. The workpiece is on the right side of the tool, which is called right compensation, using G42 instruction.

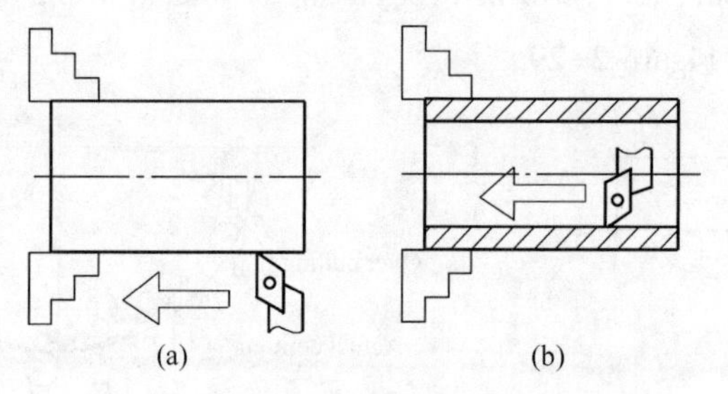

Figure 2-30 Selection of G41 and G42

(a) G42; (b) G41

(5) Tool nose radius compensation parameters and setting.

1) Tool nose radius. After the tool nose radius compensation is performed, the tool automatically deviates from the workpiece contour by a tool nose radius. Therefore, the value of the radius of the tool nose must be entered into the memory of the system. For the specific operation method, see the machining operation section.

2) Determination of tool tip orientation. The direction of the tool automatic deviation from the workpiece contour is also different when the tool radius compensation is performed at different tool nose arc positions. Therefore, it is necessary to input the parameters representing the shape and position of the turning tool into the memory. For the specific operation method, refer to the machining operation section. The shape and position parameters of the turning tool are called the tool tip orientation T. As shown in Figure 2-31, there are 9 types, which are represented by parameters 0~9, and P is the theoretical tool point. The common tool tip orientation T of a CNC lathe with a front tool post is: $T=3$ for external turning tool, $T=2$ for boring tool.

【Example 2-7】 Use the tool nose radius instructions to write the program. Part drawing is shown in Figure 2-32.

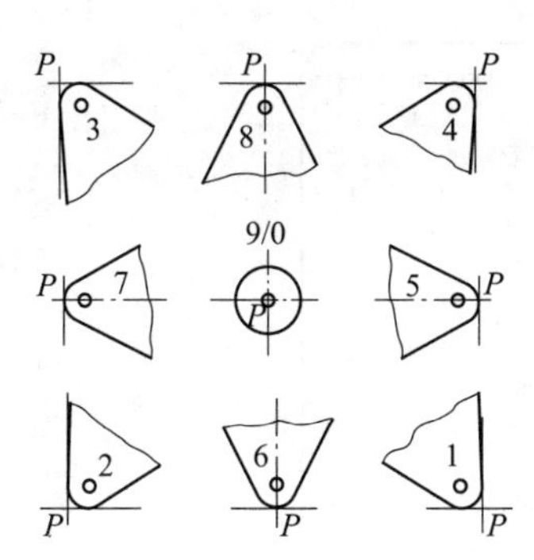

Figure 2-31 Tool tip orientation

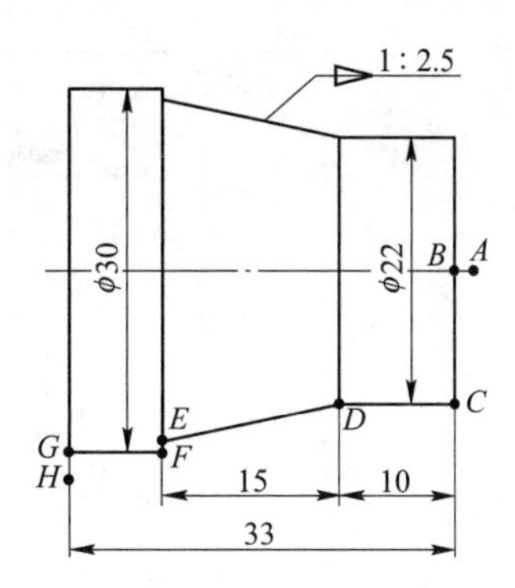

Figure 2-32 Example of tool nose radius compensation

The tool compensation process is divided into three steps: the first step is to establish the tool compensation before machining, as shown in section $A \rightarrow B$; the second step is to perform tool compensation during machining, as shown in section $B \rightarrow G$; the last step is to cancel tool compensation after machining completed, as shown in section $G \rightarrow H$. The machining program of using the tool nose arc radius compensation is shown in Table 2-10.

Table 2-10 Example of radius compensation for tool nose arc

G00 Z5. 0;	Rapid positioning to point *A*
X0;	
G42 G01 X0 Z0;	Line interpolation to point *B*, call tool nose radius compensation right
X22. 0;	Line interpolation to point *C*
…;	…
G01 Z-33. 0;	Line interpolation to point *G*
G40 X41. 0;	Line interpolation to point *H*, and cancel tool nose radius compensation ($G \rightarrow H$)

2. 2. 3 Operation

2. 2. 3. 1 Tool Nose Arc Compensation Parameter Input

Press the key [OFS/SET] → click [OFFSET] → click [GEOM] → enter the interface of tool compensation parameters.

Move the cursor to the position of the *R* → input the radius of the tool tip (e. g. T0101 = 0. 4) → click [INPUT], move the cursor to the position of the *T* → enter the tool position number (e. g. *T* = 3) → click [INPUT], as shown in Figure 2-33, to complete the input of tool compensation parameters.

2. 2. 3. 2 Dimensional Accuracy Control Method

For the first trial cutting workpiece, use the program pause instruction (M00) to stop the machine after rough cutting, and measure whether the workpiece size meets the requirements. If there is

OFFSET/GEOMETRY O0001 N0000

NO.	X.	Z.	R	T
G 01	-148.360	-432.291	0.400	3
G 02	0.000	0.000	0.000	0
G 03	0.000	0.000	0.000	0
G 04	0.000	0.000	0.000	0
G 05	0.000	0.000	0.000	0
G 06	0.000	0.000	0.000	0
G 07	0.000	0.000	0.000	0
G 08	0.000	0.000	0.000	0

ACTUAL POSITION (RELATIVE)

U -24.180 W -332.291

>_ OS 50% T0101

MEM. *** *** *** 18:52:27

[NO.SRH] [MEASUR] [INP.C.] [+INPUT] [INPUT]

Figure 2-33 Tool compensation parameters

any deviation, it should be corrected in time before finishing machining. The correction method is as follows.

(1) Press the key OFS SET to display the screen on CTR as shown in Figure 2-34.

(2) Use the key PAGE ⇧ or PAGE ⇩ to move the cursor to the position where you want to set the compensation number.

(3) Input X and Z values.

As shown in Figure 2-21, after rough turning with 90° external turning tool (tool number: T01), the diameter of the workpiece is 0.02mm larger than the target dimension, and the end face is 0.03mm longer. Press the key OFS SET and move the cursor to the "W01" corresponding to "T01". Enter -0.02 at the corresponding X-axis position, click [INPUT], enter -0.03 at Z-axis position, click [Input] to complete the tool wear compensation, as shown in Figure 2-34. If the dimensions in each direction are correspondingly small, enter "+" data. Through the above modification, after finishing, the workpiece can meet the target dimension requirements.

OFFSET/WEAR O0001 N0000

NO.	X.	Z.	R	T
W 01	-0.020	-0.030	0.400	3
W 02	0.000	0.000	0.000	0
W 03	0.000	0.000	0.000	0
W 04	0.000	0.000	0.000	0
W 05	0.000	0.000	0.000	0
W 06	0.000	0.000	0.000	0
W 07	0.000	0.000	0.000	0
W 08	0.000	0.000	0.000	0

ACTUAL POSITION (RELATIVE)

U -50.736 W -432.291

>_ OS 50% T0101

MEM. *** *** *** 18:52:27

[NO.SRH] [MEASUR] [INP.C.] [+INPUT] [INPUT]

Figure 2-34 Inputting of tool wear compensation parameters

2.2.4 Measurement

The commonly used measurement of taper inspection includes taper gauges, vernier universal angle rulers and a sine gauges.

2.2.4.1 Taper Gauge

(1) Structure. The taper gauge is a measuring tool for checking the internal and external taper of a workpiece. It is divided into taper plug gauge and taper ring gauge. As shown in Figure 2-35. Taper plug gauge is mainly used to inspect the large diameter, taper and contact rate of products, which belongs to special comprehensive inspection tool. Taper plug gauge can be divided into size plug gauge and color plug gauge. Taper plug gauge specification: 3~300mm.

Figure 2-35 Taper gauge

(2) Usage of taper gauge. The steps to check the taper fitting rate with taper ring gauge are as follows.

1) Wipe the taper ring gauge and the taper surface of the part, and observe that there is no fiber on the taper surface.

2) Apply 3 thin red lead or blue oil on the axial direction of the taper surface (thickness within 0.01mm).

3) Tighten the ring gauge and rotate it 90 degrees.

4) Take out the ring gauge and observe the contact trace to judge the quality of the inner cone.

The more the contact area, the better quality the taper is, and vice versa, it is not good. Generally, the contact rate is required to be more than 75% when checking the taper with standard gauge.

2.2.4.2 Vernier Universal Angle Ruler

(1) Structure. Vernier universal angle ruler is a measuring tool used to measure the internal and external angles of part. Its structure is shown in Figure 2-36. It consists of ruler body, 90° angle ruler, vernier, brake head, base ruler, ruler, clamp block, etc. There are two scale values of 5′ and 2′.

(2) Usage of vernier universal angle ruler. Before the measurement, the zero position should be calibrated first. When the bottom edge of the angle ruler and the base ruler have no gap contact

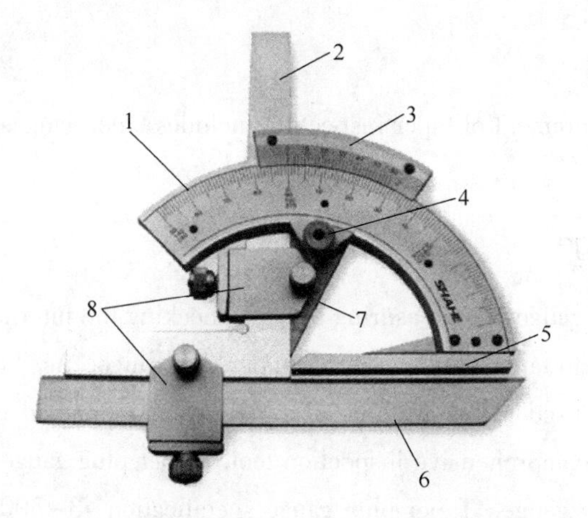

Figure 2-36 Vernier universal angle ruler

1—Main ruler; 2—Angle ruler; 3—Vernier; 4—Brake head; 5—Base ruler; 6—Ruler; 7—Sector plate; 8—Clamp block

with the ruler, the main ruler and the "0" line of the vernier are aligned, which is the zero of the universal angle ruler. After adjusting the zero position, any angle within the range of 0° ~320° can be tested by changing the mutual positions of the base ruler, angle ruler and ruler. The reading method is basically the same as the vernier caliper. First read out the whole degree before the zero line of the vernier, and then read out the value of the angle "minute" on the vernier. The sum of the two values is the angular value of the part.

When measuring the angle of the part with the universal angle ruler, the direction of the base ruler and the bus bar of the angle of the part should be the same, and the part should be in good contact with the full length of the two measuring surfaces of the measuring angle ruler, so as to avoid the measurement errors.

2.2.4.3 Sine Bar

(1) Structure of sine bar. Sine bar is a kind of precise measuring tool which uses trigonometry to measure angle. The sine bar is composed of the main body and two precision cylinders with the same diameter. The center distance of the two precision cylinders is required to be very accurate. There are two sizes of center distance: 100mm and 200mm. In order to facilitate the positioning and orientation of the inspected workpiece on the main body plane, the main body is equipped with a back baffle and a side baffle. As shown in Figure 2-37.

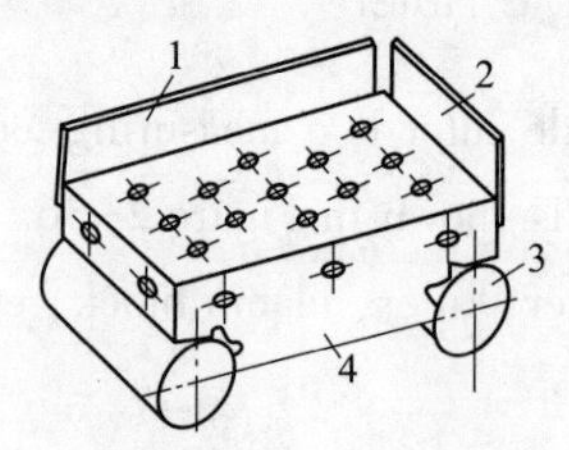

Figure 2-37 Sine bar

1—Slide baffle; 2—Back baffle; 3—Cilinder; 4—Main body

(2) Measuring principle. Place the cone workpiece with a cone angle α on the working surface of the sine bar. One cylinder of the sine bar is in contact with the flat plate, and the other cylinder is padded with gauge blocks. As shown in Figure 2-38, L is the center distance of sine bar, H is the height of gauge block. The height of the measuring block can be calculated according to the conical angle of the part using the formula: $\sin\alpha = H/L$. Then use a dial indicator (or micrometer) to check the height of both ends of the generatrix on the conical surface of the workpiece. If the height of both ends is equal, the conical angle of the workpiece is correct. Sine bar is generally used to measure the angle less than 45°. When measuring the angle less than 30°, the accuracy can reach 3″~5″.

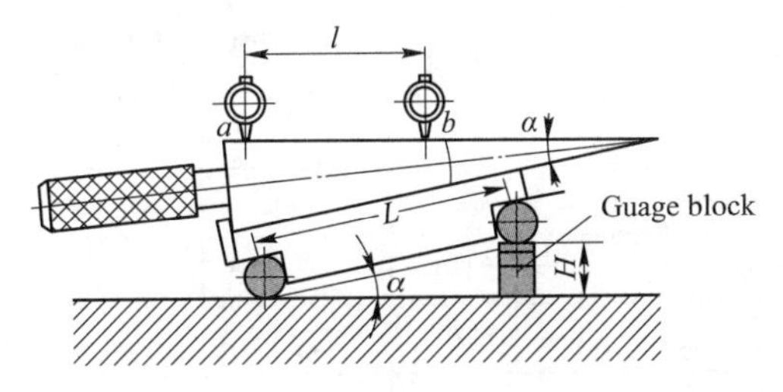

Figure 2-38 Measuring principle

Implemention

(1) Step 1 Drawing analysis:

The part material shown in Figure 2-21 is LY15, with a total length of 50mm. The machining surfaces include $\phi36^{0}_{-0.04}$、$\phi28^{0}_{-0.03}$、$\phi24^{0}_{-0.03}$、$\phi12$ cylindrical surfaces, taper surface, chamfer, etc., and the surface roughness is $Ra1.6$ and $Ra3.2$ respectively.

(2) Step 2 Making the machining plan:

1) Machining plan:

① Three jaw chuck to install the workpiece, and the workpiece extend out of the chuck about 60mm.

② Rough and finishing machining the workpiece until the size meet the dimension requirements.

2) Tool selection:

The tool card of conical surface is shown in Table 2-11.

Table 2-11 CNC machining tool card

Part name		Conical surface part		Drawing No.		2-21	
S No.	Tool No.	Tool name	Quantity	Machined surface	Tool nose radius R/mm	Tool tip orientation T	Note
1	T01	90° external turning tool	1	Rough and finish turning outer diameter	0.4	3	
Edit		Check		Approve	Date	Total 1 Page	Page 1

3) Machining process:

The CNC machining process card is shown in Table 2-12.

Table 2-12 CNC machining process card

Company name	Tianjin polytechnic college		Part name		Drawing No.	
			Conical surface Part		2-21	
Program No.	Fixture name	Machine tool	CNC system		Workshop	
0221	Three-jaw chuck	CKA6140	FANUC SERIES Oi		CNC training center	
Steps	Contents	Tool	Spindle speed n/r · min^{-1}	Feed rate F/mm · r^{-1}	Back engagement a_p/mm	Note
1	Clamp and align the workpiece					
2	Setting external turning tool	T01				Manual
3	Setting grooving tool	T02				
4	Rough turning outer diameter with allowance of 1mm	T01	600	0. 2	1. 5	0221
5	Finish turning outer diameter	T01	1000	0. 1	0. 5	
Edit	Check	Approve		Date	Total 1 Page	Page 1

(3) Step 3 Programming:

1) Taper dimension calculation:

Maximum diameter of cone $D=20$, length of cone $L=18-8=10$, taper $C=1/2=0.5$.

According to the formula $C=\dfrac{D-d}{L}$, it can be calculated that the minimum diameter of cone $d=D-CL=20-10\times0.5=15$.

2) Machining program:

Conical surface part program and explanation is shown in Table 2-13.

Table 2-13 The program of the conical surface parts

0221;	Program No.
G40 G97 G99 M03 S600 F0. 2;	Cancel tool nose radius compensation, setting constant spindle speed control, feed rate 0. 2mm/r, spindle CW at 600r/min
T0101;	Replace T01 external turning tool, substitute tool compensation No. 01
M08;	Open cutting fluid
G42 G00 Z5. 0;	Rapid positioning to point (X40. 0, Z5. 0), call tool nose radius compensation
X40. 0;	
G71 U1. 5 R0. 5;	Calling stock removal in turning
G71 P10 Q20 U0. 5 W0. 05;	

Continued Table 2-13

N10 G00 X0;	Sequence of finishing shape
G01 Z0;	
X12.0 C1.0;	
Z-9.0;	
X15.0;	
X20.0 Z-19.0;	
X23.985 C1.5;	
Z-29.0;	
X27.985;	
Z-37.0;	
X35.98 C2;	
N20 Z-50.0;	
G40 G00 X100.0;	Rapid positioning to point (*X*100.0, *Z*100.0), cancel tool nose radius compensation
Z100.0;	
M05;	Spindle stop
M00;	Program stop
M03 S1000 F0.1;	Spindle CW at 1000r/min, calling No. 1 tool and it's offset, feed rate 0.1mm/r
T0101;	Substitute tool compensation No. 01
G42 G00 Z5.0;	Rapid positioning to point (*X*40.0, *Z*5.0)
X40.0;	
G70 P10 Q20;	Calling finishing cycle in turning
G40 G00 X100.0;	Rapid positioning to point (*X*100.0, *Z*100.0)
Z100.0;	
M30;	End of program and reset

(4) Step 4 Part machining:

Machine system startup → Return to reference point → Clamp and align the workpiece → Install the tool → Input the program in Table 2-13 → Simulation the program → Tool setting→ Automatic machining.

(5) Step 5 Dimension measurement:

1) Use vernier calipers to measure the outer diameters $\phi36^{0}_{-0.04}$、$\phi28^{0}_{-0.03}$、$\phi24^{0}_{-0.03}$、$\phi12$.

2) Use vernier calipers to measure length dimension 8, 18, 27, 10, 50±0.3.

3) Measure taper with universal angle ruler.

4) Inspect the roughness with roughness comparison board.

Assessment

The assessment table is shown in Table 2-14.

Table 2-14　The assessment table of conical surface

Item	No.	Standard		Partition	Score
Machining plan (10%)	1	Machining steps	Meet the requirements of NC turning process	3	
	2	Dimension	Calculate dimensions	3	
	3	Tool selection	The cutter selection is reasonable and meets the machining requirements	4	
Programming (30%)	4	Program number	No program no point	2	
	5	Program segment number	No program segment number no point	2	
	6	Cutting parameters	Unreasonable choice of cutting parameters no point	4	
	7	Origin and coordinates	Mark the origin and coordinates of the program, otherwise there is no point	2	
	8	Program contents	Deduct 2 points for each segment that does not meet the requirements of program logic and format	20	
			Deduct 5 points if the procedure content is not in accordance with the process		
			Deduct 5 points in case of dangerous instruction		
Machining (20%)	9	Program input	Deduct 2 points for each wrong section	10	
	10	Machining operation 30min	Workpiece coordinate system origin setting error no point	10	
			Misoperation no point		
			Overtime no point		
Dimension measurement (30%)	11	External profile	$\phi 12^{0}_{-0.027}$ deduct 3 points for 0.01 out of tolerance	3	
	12		$\phi 24^{0}_{-0.033}$ deduct 3 points for 0.01 out of tolerance	3	
	13		$\phi 28^{0}_{-0.033}$ deduct 3 points for 0.01 out of tolerance	3	
	14		$\phi 36^{0}_{-0.039}$ deduct 3 points for 0.01 out of tolerance	3	

Continued Table 2-14

Item	No.	Standard		Partition	Score
Dimension measurement (30%)	15	Taper	Out of tolerance no point	6	
	16	Length	50±0. 3	3	
	17		8 18 27 10	4	
	18	Chamfer	*C*1 *C*1. 5	3	
	19	Surface roughness	*Ra*1. 6 no grade reduction	1	
	20		*Ra*3. 2 no grade reduction	1	
Vocational ability (10%)	21	Learning ability		2	
	22	Communication skills		2	
	23	Team cooperation		2	
	24	Safe operation and civilized production		4	
Total					

Exercises

(1) Answer the following questions:

1) Briefly describe the functions of G71 and G70 instruction briefly.

2) Briefly describe G71, G70 instruction formats and the meaning of each parameter briefly.

3) Briefly describe the influence of tool nose arc radius on machining accuracy briefly.

4) Briefly describe the usage method of tool nose arc radius compensation instruction briefly.

5) Briefly describe the measuring tools commonly used for conical surface briefly.

(2) Synchronous training:

The parts drawing are shown in Figure 2-39 and Figure 2-40. The material is hard aluminum alloy. The blank is ϕ40mm bar material. Please machining and measuring.

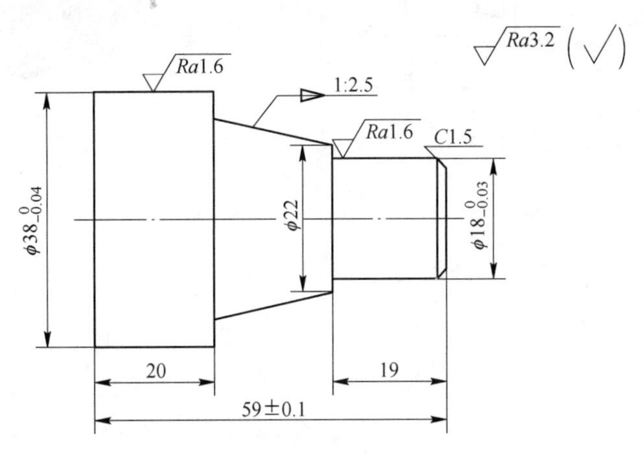

Figure 2-39 Synchronous training 1

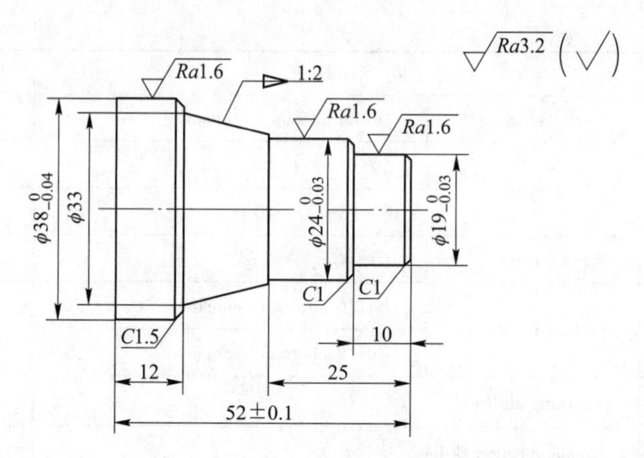

Figure 2-40　Synchronous training 2

Task 2.3　Grooving and Cutting of Parts

Introduction

As shown in Figure 2-41, the part is made of hard aluminum alloy LY15, and the blank is ϕ40mm. It is produced in one piece. This task needs to prepare the program, machine the part using the CNC lathe, and then measure the part.

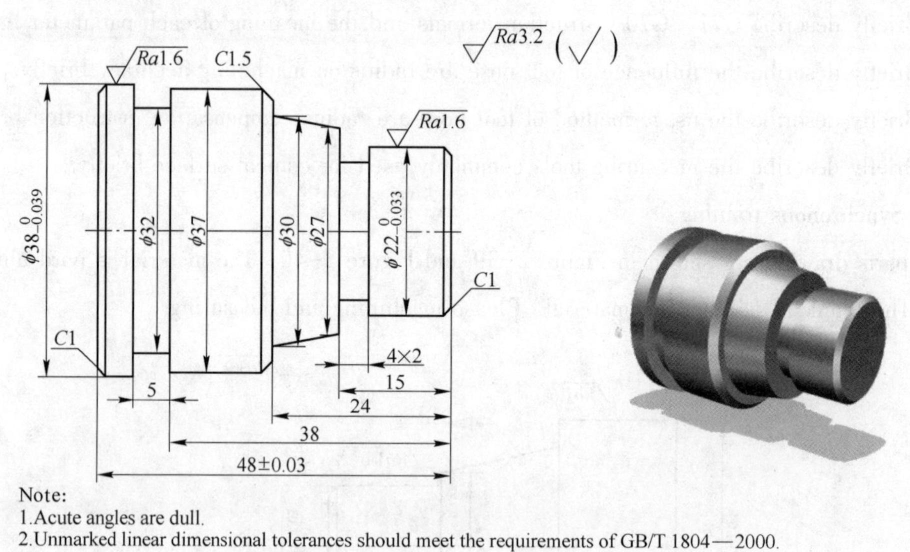

Figure 2-41　The part drawing

Competences

(1) Knowledge:

1) Know the types and markings of grooves in the parts.

2) Describe the machining path of grooving.

3) Understand G04 command.

4) Understand the structure of outside micrometer.

(2) Skill:

1) Have the ability to write machining program of grooving and cutting.

2) Learn how to set the coordinate system of grooving tools.

3) Understand the inspection method of outside micrometer.

(3) Quality:

1) Carry out safety technical operation procedures correctly, and cultivate safety awareness.

2) Cultivate serious and responsible work attitude and quality awareness.

Relevant Knowledge

2.3.1 Process Preparation

Grooving and cutting are one of the important contents of CNC machining. Generally, the external threads of shaft parts are equipped with undercut grooves, overtravel grooves of grinding wheels, etc. The internal threads of sleeve parts also often have internal grooves.

2.3.1.1 Features of Grooving

(1) High cutting force. Due to the friction between the chips and the tool and the workpiece during the grooving process, the plastic deformation of the metal being cut is large. Therefore, under the same cutting conditions, the cutting force is 20%~25% greater than the cutting force during general cylindrical turning.

(2) Large cutting deformation. The cutting edges that participate in the cutting at the same time include the main cutting edge of the grooving tool and the left and right cutting edges. When the chips are discharged, friction and squeezing on both sides of the groove cause cutting deformation.

(3) More concentrated cutting heat. When grooving, the plastic deformation is large, the friction is violent, and a large amount of cutting heat is generated, which aggravates the wear of the tool.

(4) Tool rigidity is poor. The main cutting edge of the grooving tool is narrow, generally 2~6mm. As shown in Figure 2-42, the tool head is narrow, so the rigidity of the tool is poor, and vibration is easy to occur during the cutting process.

Figure 2-42 The grooving and cutting tool

2.3.1.2 Machining of Groove

(1) Machining of narrow grooves. When the blade width is equal to the groove width, the groove is called a narrow groove. Generally, G01 is used to cut straight to the bottom of the groove, stay at the bottom of the groove for a few seconds, smooth the bottom of the groove, and then use G01 to retract the tool, as shown in Figure 2-43.

(2) Machining of wide grooves. When the blade width is less than the groove width, the groove is called a wide groove. Rough cutting is often carried out with a row of blades when processing wide grooves, as shown in Figure 2-44 ①→②→③, and then use a grooving tool to cut along the side of the groove to the bottom of the groove, as shown in Figure 2-44④. Process the bottom of the groove to the other side of the groove, as shown in Figure 2-44⑤, then exit along the side, as shown in the Figure 2-44.

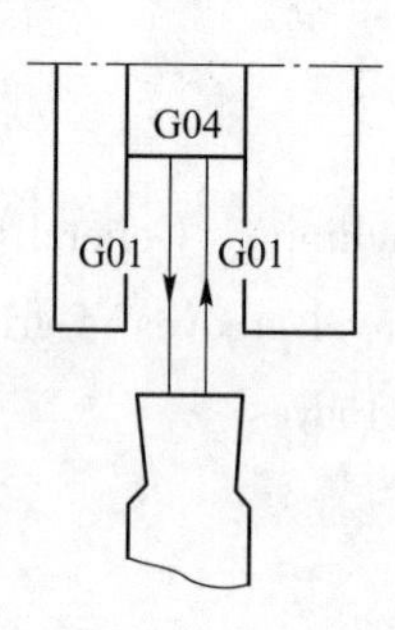

Figure 2-43 Machining routes of narrow grooves

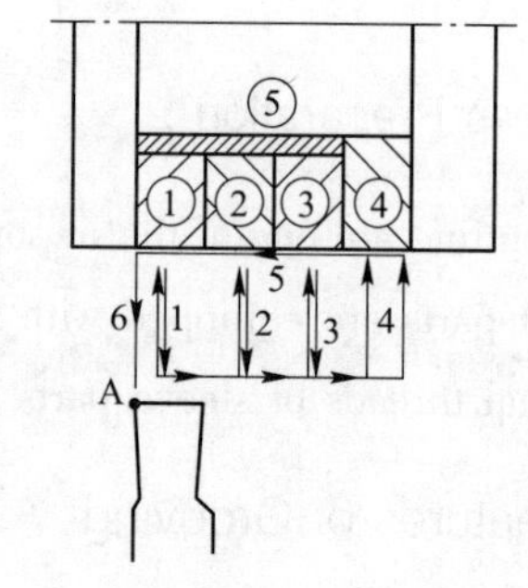

Figure 2-44 Machining routes of wide grooves

2.3.1.3 Selection of Cutting Parameters for Grooving

(1) Back engagement. When cutting transversely, the back engagement of the grooving is equal to the width of the main cutting edge of the cutting tool.

(2) Feed rate. If the tool rigidity, strength and heat dissipation conditions are poor, the feed should be appropriately reduced. When the feed is too large, the tool is easy to break. When the feed is too small, the tool and the workpiece produce strong friction and cause vibration. When cutting steel with a high speed steel slotter, f=0.05~0.1mm/r; when turning cast iron, f=0.1~0.2mm/r. when machining steel with a carbide cutting tool, f=0.1~0.2mm/r, when machining cast iron, f=0.15~0.25mm/r.

(3) Cutting speed. v_c=30~40m/min when cutting steel with a high-speed rigid turning tool; v_c=15~25m/min when machining cast iron. v_c=80~120m/min when cutting steel with cemented carbide. v_c=60~100m/min when machining cast iron.

2.3.2 Programming

2.3.2.1 Dwell G04

(1) Function.

The tool performs short-term non-feed finishing of the part relative to the part, which is mainly used for the machining of grooves to reduce the surface roughness and ensure the cylindricalness of the workpiece.

(2) Format.

G04 P____;　　　unit: ms

G04 X____;　　　unit: s

G04 U____;　　　unit: s

Among them　P, X, U——Pause time.

Note:

1) Decimal points can be used after X and U. For example, G04 X5.0 indicates that the previous program should be suspended for 5 seconds before executing the following program.

2) Decimal point is not allowed after P. For example, G04 P1000 means pause for 1s.

2.3.2.2 Narrow and Wide Slot Programmings

【Example 2-8】As shown in Figure 2-45, the blank is made of aluminum alloy, and the outer circle and chamfer of the workpiece have been processed to the size of the drawing. A program for the narrow groove part is written. The reference program is shown in Table 2-15.

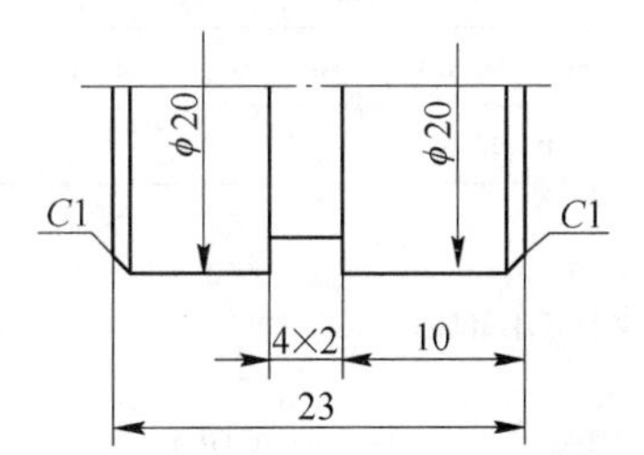

Figure 2-45　An example of narrow groove programming

Table 2-15　Partial program for narrow groove machining (cutting tool width 4mm)

Block	Explanation
T0202;	Replace T02 boring tool, substitute tool compensation No. 02
G00 X21.0;	Move quickly to the starting position
Z-14.0;	
G01 X16.0;	Feed
G04 X2.0;	Suspend for 2 seconds at the bottom of the tank
G01 X21.0;	Retract

【Example 2-9】As shown in Figure 2-46, the blank material is aluminum alloy, and the workpiece's outer circle and chamfer have been machined to the drawing size, and a program for the wide groove part is written. The reference program is shown in Table 2-16.

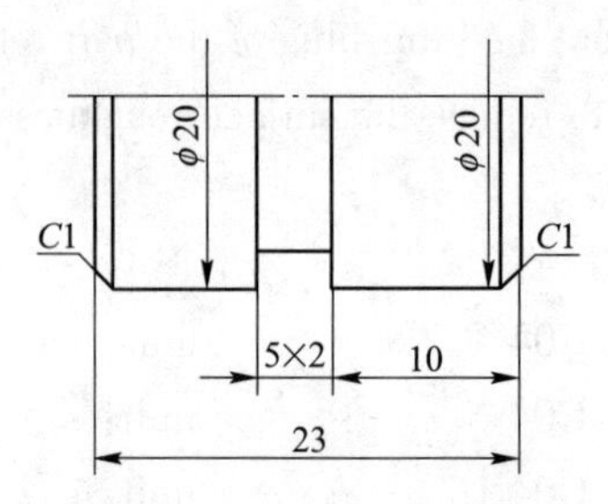

Figure 2-46　An example of wide groove programming

Table 2-16　Partial program for narrow groove machining (grooving tool width 4mm)

Block	Explanation
T0202;	Replace the T02 grooving tool and substitute the No. 02 tool compensation
G00 X21. 0;	Move quickly to the starting position
Z-15. 0;	
G01 X17. 0;	Feed, 0. 5mm difference from bottom of groove
X21. 0;	Retract
W1. 0;	1mm to the right
X16. 0;	Cut to the bottom of the groove
W-1. 0;	Feed 1mm to the left, bottom of finishing groove
X21. 0;	Retract

2. 3. 2. 3　The Left Chamfer Programming

【Example 2-10】As shown in Figure 2-46, write a program for left chamfer. The reference program is shown in Table 2-17.

Table 2-17　Reference program for left chamfer

Block	Explanation
G00 Z-27. 0;	Grooving tool quickly positioned at the end of the workpiece
X22. 0;	Positioned closer to the outer circle
G01 X18. 0 F0. 08;	Cut the outer circle to *X*18. 0
X22. 0;	Withdrawal, 2mm in one direction in *X* direction
W2. 0;	Grooving knife 2mm to the right
X18. 0 W-2. 0;	Cut the left chamfer with the right part of the grooving knife
X-1. 0;	Cut off the workpiece

2. 3. 3　Operation

2. 3. 3. 1　Precautions for Installing Grooving Tool

(1) The cutting edge must be the same height as the center line of the part.

(2) The cutting edge is parallel to the axis of the part and can't be skewed. Otherwise, the side wall of the workpiece is not straight. Skew caused serious tool broken off.

2. 3. 3. 2 The Method of Grooving Tool Setting

(1) *X* compensation.

1) The spindle rotates forward.

2) Select the handwheel , press the key , make the tool close to the outer surface of the workpiece, as shown in Figure 2-47 (a).

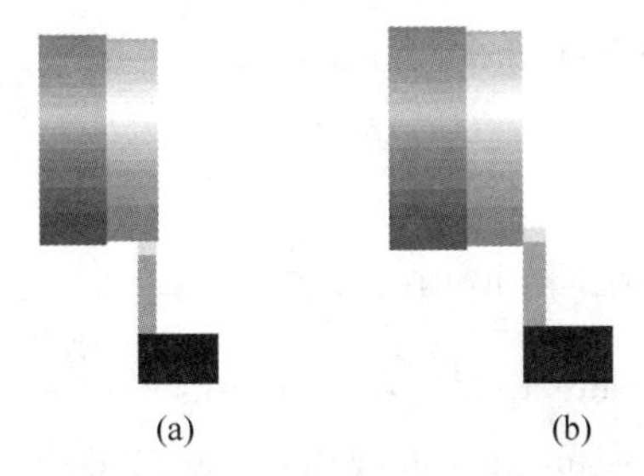

Figure 2-47 Grooving setting process

(a) *X* compensation; (b) *Z* compensation

3) Press the key →press the key [OFFSET]→press the key [GEOM]→move the cursor to the selected tool position, such as G02. The interface is shown in Figure 2-48 (a).

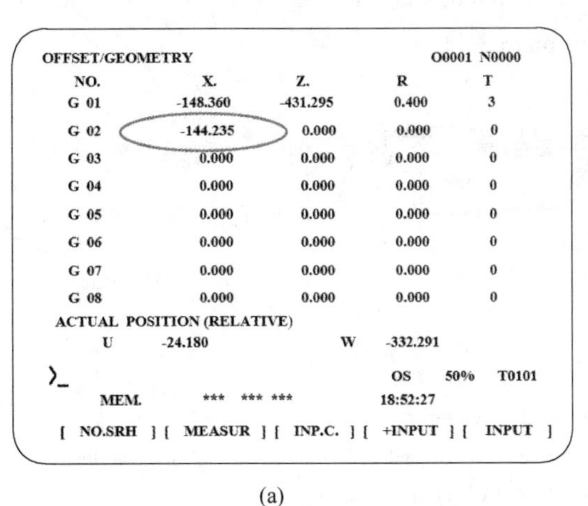

OFFSET/GEOMETRY O0001 N0000

NO.	X.	Z.	R	T
G 01	-148.360	-431.295	0.400	3
G 02	-144.235	0.000	0.000	0
G 03	0.000	0.000	0.000	0
G 04	0.000	0.000	0.000	0
G 05	0.000	0.000	0.000	0
G 06	0.000	0.000	0.000	0
G 07	0.000	0.000	0.000	0
G 08	0.000	0.000	0.000	0

ACTUAL POSITION (RELATIVE)

U -24.180 W -332.291

>_ OS 50% T0101

MEM. *** *** *** 18:52:27

[NO.SRH] [MEASUR] [INP.C.] [+INPUT] [INPUT]

(a)

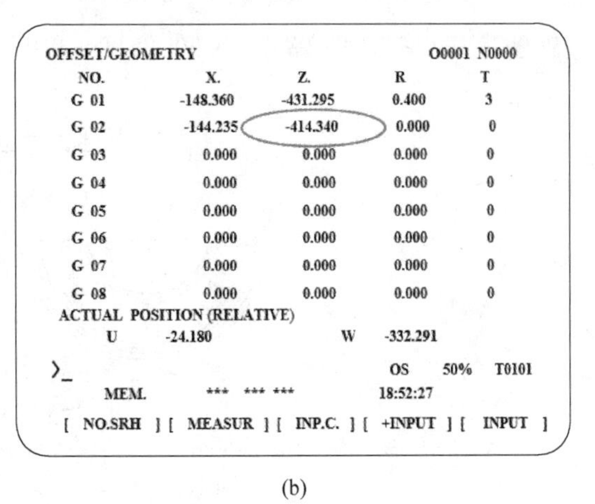

OFFSET/GEOMETRY O0001 N0000

NO.	X.	Z.	R	T
G 01	-148.360	-431.295	0.400	3
G 02	-144.235	-414.340	0.000	0
G 03	0.000	0.000	0.000	0
G 04	0.000	0.000	0.000	0
G 05	0.000	0.000	0.000	0
G 06	0.000	0.000	0.000	0
G 07	0.000	0.000	0.000	0
G 08	0.000	0.000	0.000	0

ACTUAL POSITION (RELATIVE)

U -24.180 W -332.291

>_ OS 50% T0101

MEM. *** *** *** 18:52:27

[NO.SRH] [MEASUR] [INP.C.] [+INPUT] [INPUT]

(b)

Figure 2-48 Parameters input interface

(a) *X* direction compensation input; (b) *Z* direction compensation input

4) Input the *X* measurement value, which is the measured outer circle value after the outer circle tool turning, such as *X*39. 875→press the key [MEASUR], complete the *X* direction compensation.

(2) *Z* compensation.

1) The spindle rotates forward.

2) Select the handwheel, press the key, make the tool close to the end face of the workpiece, as shown in Figure 2-47 (b).

3) Press the key →press the key [OFFS]→press the key [GEOM]→move the cursor to the selected tool position, such as G02. The interface is shown in Figure 2-48 (b).

4) Input *Z*0→ press the key [MEASUR], complete the *Z* direction compensation.

Note:

During operation, when the grooving tool is close to the outer circle or the end surface, the feed rate of the handwheel is gradually reduced to the minimum, and the handwheel is slowly shaken to enter the tool until debris appears on the contact surface.

2.3.4 Measurement

2.3.4.1 Application of Outside Micrometer

Outside micrometer is an even more rigid measuring instrument than vernier caliper. Commonly outside micrometer is used to measure the outer diameter, the shoulder thickness, the plate thickness and wall thickness.

2.3.4.2 Components of Outside Micrometer

The structure of the outside micrometer is shown in Figure 2-49. It is mainly composed of a ruler frame, a fixed measuring anvil, a micrometer screw, a fixed scale sleeve, a differential tube, a ratchet knob, a locking screw and a heat insulating plate.

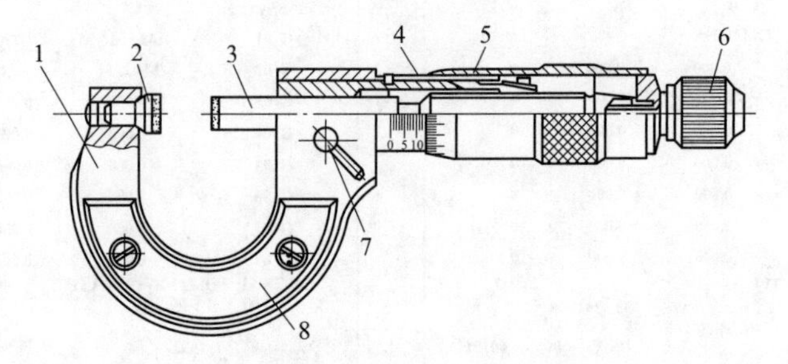

Figure 2-49 Outside micrometer

1—A ruler frame; 2—A fixed measuring anvil; 3—A micrometer screw; 4—A fixed scale sleeve; 5—A differential tube; 6—A ratchet knob; 7—A locking screw; 8—A heat insulation plate

2.3.4.3 Usage of Outside Micrometer

(1) When using the micrometer, handle it gently, and the measured object should be wiped clean.

(2) Loosen the micrometer locking device, calibrate the zero position, and turn the knob to make the distance between the anvil and the micrometer screw slightly larger than the object to be measured.

(3) Holds the micrometer ruler frame with one hand, place the object to be measured between the anvil and the end face of the micrometer screw, and turns the knob with the other hand. When the screw is close to the object, change the force measuring device until you hear clicking on the sound, and then turn it gently for 0.5～1 turn.

(4) Tighten the locking device to prevent the screw from rotating when the micrometer is moved, and the reading can be obtained.

2.3.4.4 Reading of Outside Micrometer

As shown in Figure 2-50, micrometer reading method is divided into following three steps:

(1) Using the end face of the thimble as reference line. We can read the value on the sleeve. (8.5mm)

(2) Take the center line on the sleeve as the directrix, read the markings on the thimble. (0.380mm)

(3) The values of the twoparts are added to the measured actual size. (8.880mm)

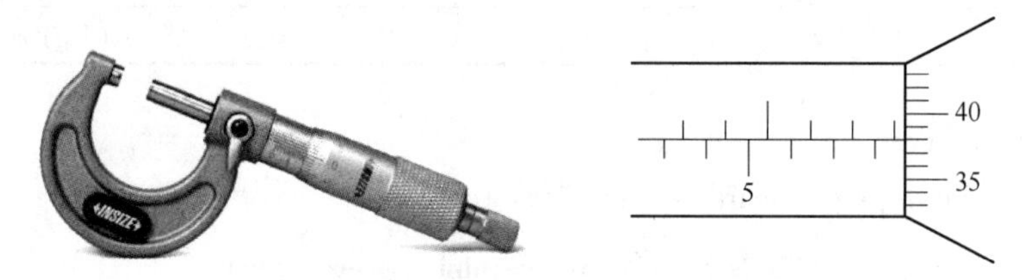

Figure 2-50 Reading of the outside micrometer

Note:

(1) Don't measure moving part because the micrometer may be severely damaged.

(2) The surface of the part to be measured should be wiped clean before measurement.

(3) Apply only moderate force to the knurled thimble when you take a measurement.

(4) Before you store a micrometer, back the spindle away from the anvil, wipe all exterior surfaces with a clean, soft cloth and coat the surfaces with butter.

Implemention

(1) Step 1 Drawing analysis:

The part material shown in Figure 2-40 is LY15, with a total length of 48mm. The machining surfaces include $\phi22^{0}_{-0.033}$、$\phi37$、$\phi38^{0}_{-0.039}$ cylindrical surfaces, two grooves with a width of 4×2 and a width of 5mm, chamfers, etc., and the surface roughness is *Ra*1.6 and *Ra*3.2 respectively.

(2) Step 2 Making the machining plan:

1) Machining plan.

① Three-jaw self-centering chuck is used to install the card, and the part protrudes about 65mm from the chuck.

② Cut the outer contour to size requirements for rough and finished turning.

③ Process 4×2 narrow grooves and 5mm wide grooves.

④ Cut the left chamfer, then cut off the workpiece.

2) Tool selection.

The CNC machining tool card is showed as Table 2-18.

Table 2-18 CNC machining tool card

Part name		Grooving part		Drawing No.		2-41	
S No.	Tool No.	Tool name	Quantity	Machined surface	Tool nose radius R/mm	Tool tip orientation T	Note
1	T01	90° external turning tool	1	Rough and finish turning the outer diameter	0.4	3	
2	T02	Grooving tool	1	Grooving and cutting			4mm
Edit		Check	Approve	Date		Total 1 Page	Page 1

3) Machining process.

The CNC machining process card is showed as Table 2-19.

Table 2-19 CNC machining process card

Company name	Tianjin polytechnic college		Part name		Drawing No.	
			Grooving part		2-41	
Program No.	Fixture name	Machine tool	CNC system		Workshop	
O231	Three-jaw chuck	CK6140	FANUCSERIES Oi		CNC Training Center	
Steps	Contents	Tool	Spindle speed n/r · min^{-1}	Feed rate F/mm · r^{-1}	Back engagement a_p/mm	Note
1	Clamp and align the workpiece					Manual
2	Setting external turning tool	T01				
3	Setting grooving tool	T02				
4	Rough turning outer diameter with allowance of 1mm	T01	600	0.2	1.5	
5	Finish turning outer diameter	T01	1000	0.1	0.5	O231
6	Grooving	T02	400	0.05	4	
7	Cutting off the workpiece	T02	400	0.05	4	
Edit	Check	Approve	Date		Total 1 Page	Page 1

(3) Step 3 Programming:

The part programming is showed as Table 2-20.

Table 2-20　The part programming

O231;	Program No.
G40 G97 G99 M03 S600 F0.2;	Cancel the tool nose radius compensation, set constant spindle speed control, feed rate 0.2mm/r, spindle CW at 600r/min
T0101;	Replace T01 external turning tool, substitute tool compensation No. 01
M08;	Turn on the cutting fluid
G00 Z5.0;	Rapid positioning to point (*X*42.0, *Z*5.0)
X42.0;	
G71 U1.5 R0.5;	Calling stock removal in turning
G71 P10 Q20 U0.5 W0.05;	
N10 G00 X0;	Sequence of finishing shape
G01 Z0;	
X19.984;	
X21.984 Z-1.0;	
Z-15.0;	
X27.0;	
X30.0 Z-24.0;	
X34.0;	
X37.0 W-1.5;	
Z-43.0;	
X37.981;	
Z-53.0;	
N20 X40.0;	
G00 X100.0;	Rapid positioning to point (*X*100.0, *Z*100.0)
Z100.0;	
M05;	Spindle stop
M00;	Program stop
M03 S1000 F0.1;	Spindle CW at 1000r/min, feed rate 0.1mm/r
T0101;	Substitute tool compensation No. 01
G42 G00 Z5.0;	Rapid positioning to point (*X*42.0, *Z*5.0), call tool nose radius compensation
X42.0;	
G70 P10 Q20;	Calling finishing cycle in turning
G40 G00 X100.0;	Rapid positioning to point (*X*100.0, *Z*100.0), cancel tool nose radius compensation
Z100.0;	
M05;	Spindle stop
M00;	Program stop
M03 S400 F0.05;	Spindle CW at 400r/min, feed rate 0.05mm/r
T0202;	Change T02 grooving tool, substitute tool compensation No. 02

Continued Table 2-20

O231;	Program No.
G00 Z5. 0;	Rapid positioning to point (*X*28. 0, *Z*5. 0)
X28. 0;	
Z-15. 0;	Rapid positioning to point *Z*-15. 0
G01 X17. 984;	Cutting the narrow groove 4×2
G04 X2. 0;	
G01 X27. 0;	
G00 X39. 0;	Rapid positioning to point *X*39. 0
Z-43. 0;	Rapid positioning to point *Z*-43. 0
G01 X32. 5;	Cutting the wide groove
X38. 0;	
W1. 0;	
X32. 0;	
W-1. 0;	
X38. 0;	
G00 X100. 0;	Rapid positioning to point (*X*100. 0, *Z*100. 0)
Z100. 0;	
M05;	Spindle stop
M00;	Program stop
M03 S400 F0. 05;	Spindle CW at 400r/min, feed rate 0. 05mm/r
T0202;	Substitute tool compensation No. 02
G00 Z5. 0;	Rapid positioning to point (*X*39. 0, *Z*5. 0)
X39. 0;	
Z-52. 0;	Rapid positioning to point *Z*-52. 0
G01 X35. 981;	Cut the left chamfer
X37. 981;	
W1. 0;	
X35. 981 W-1. 0;	
X-1. 0;	Cut the whole part
G00 X100. 0;	Rapid positioning to point (*X*100. 0, *Z*100. 0)
Z100. 0;	
M30;	End of program and reset

(4) Step 4 Part machining:

1) The three-jaw chuck is used to clamp the workpiece, with an extension of 60mm, to align the workpiece.

2) Install the external turning tools and grooving tools in the No. 1 and No. 2 positions of the tool

holder respectively.

3) Perform the tool setting operation of the external turning tool and grooving tool respectively.

4) Input the program in Table 2-20.

5) Simulation the programming.

6) Automatic machining.

(5) Step 5 Dimension measurement:

1) Measure the dimensions of the part with an outside micrometer.

2) Measure the dimensions of grooves with a vernier calipers.

3) Measure the total length of the part with a vernier calipers.

4) Measure the roughness of the part with roughness comparison board.

Assessment

The assessment table is shown in Table 2-21.

Table 2-21 The assessment table

Item	No.	Standard		Partition	Score
Machining plan (10%)	1	Machining steps	Meet the requirements of NC turning process	3	
	2	Dimension	Calculate dimensions	3	
	3	Tool selection	The cutter selection is reasonable and meets the machining requirements	4	
Programming (30%)	4	Program number	No program no point	2	
	5	Program segment number	No program segment number no point	2	
	6	Cutting parameters	Unreasonable choice of cutting parameters no point	4	
	7	Origin and coordinates	Mark the origin and coordinates of the program, otherwise there is no point	2	
	8	Program contents	Deduct 2 points for each segment that does not meet the requirements of program logic and format	20	
			Deduct 5 points if the procedure content is not in accordance with the process		
			Deduct 5 points in case of dangerous instruction		
Machining (30%)	9	Program input	Deduct 2 points for each wrong section	10	
	10	Machining operation 30min	Workpiece coordinate system origin setting error no point	20	
			Misoperation no point		
			Overtime no point		

Continued Table 2-21

Item	No.		Standard	Partition	Score
Dimension measurement (20%)	11	External profile	$\phi22^{0}_{-0.033}$ deduct 3 points for 0. 01 out of tolerance	3	
	12		$\phi38^{0}_{-0.039}$ deduct 3 points for 0. 01 out of tolerance	3	
	13	Taper	No grade reduction	2	
	14	Length	48±0. 3	3	
	15	Chamfer	*C*1. 5	1	
	16	Grooves	Two grooves	6	
	17	Surface roughness	*Ra*1. 6 no grade reduction	1	
	18		*Ra*3. 2 no grade reduction	1	
Vocational ability (10%)	19	Learning ability		2	
	20	Communication skills		2	
	21	Team cooperation		2	
	22	Safe operation and civilized production		4	
Total					

Exercises

(1) Answer the following questions:

1) What are the characteristics of grooving?

2) Briefly describe the path for cutting narrow and wide grooves.

3) How to choose the cutting amount for grooving?

4) What should be paid attention to when installing the grooving tool holder?

5) What are the components of an outside micrometer?

(2) Synchronous training:

The parts are shown in Figure 2-51 and Figure 2-52, please comply the program and machining.

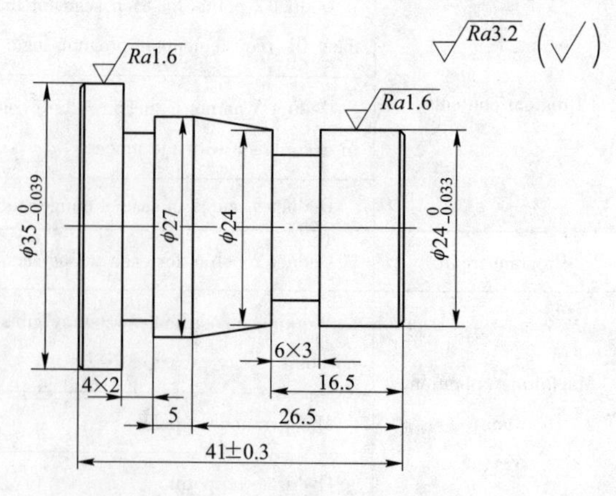

Figure 2-51 Synchronous training 1

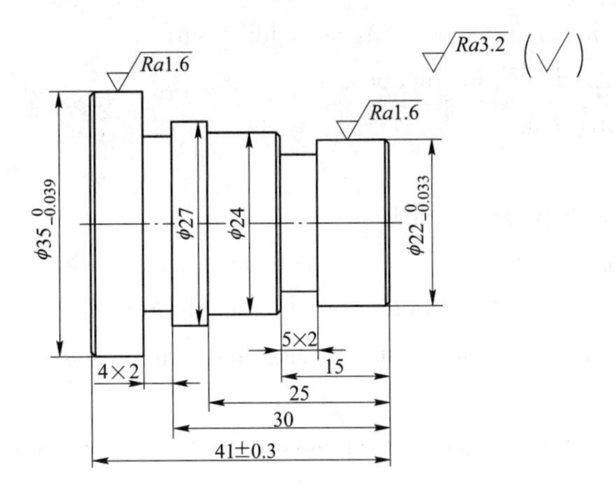

Figure 2-52 Synchronous training 2

Task 2.4 CNC Turning of Arc Surface Parts

Introduction

As shown in Figure 2-53, the part is made of hard aluminum alloy LY15, and the blank is φ40mm. It is produced in one piece. This task needs to prepare the program, machine the part using the CNC lathe, and then measure the part.

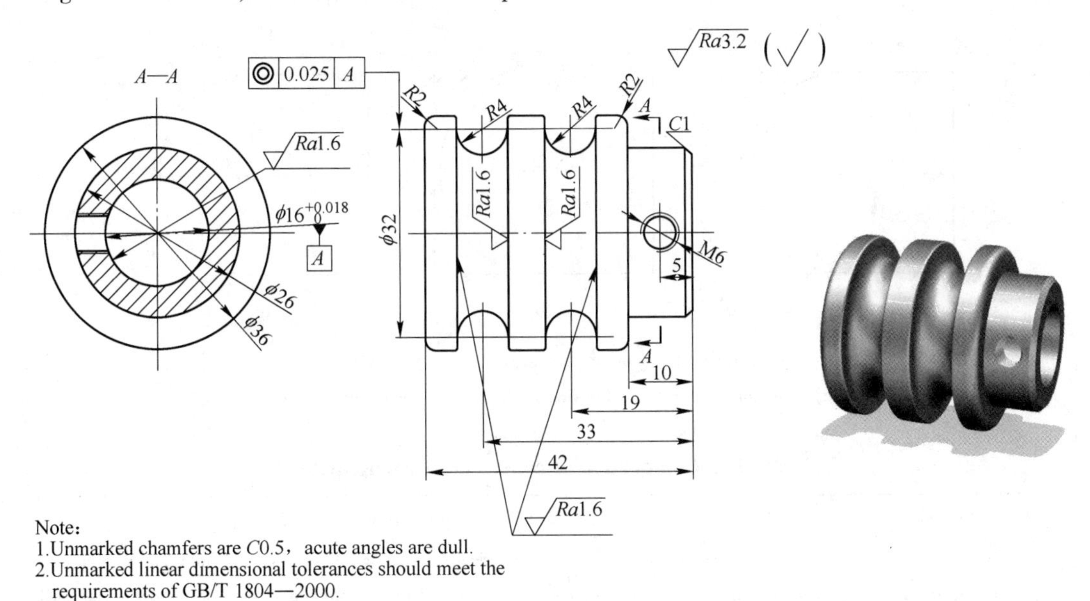

Figure 2-53 The part drawing

Competences

(1) Knowledge:

1) Familiar with the machining technology of molding surface.
2) Explain G02/G03 and G73 instructions.
3) Know arc machining tool.
(2) Skill:
1) Can read typical shaft parts drawings.
2) Have the ability to use G02/G03 commands to write simple arc surface part programs.
3) Have the ability to use G73 instruction to write simple forming surface part programs.
4) Learn how to measure parts quality using gages and templates.
(3) Quality:
1) Establish safety awareness, quality awareness and efficiency awareness.
2) Cultivate students' pioneering and innovative spirit.

Relevant Knowledge

2. 4. 1 Process Preparation

2. 4. 1. 1 Turning Method of Arc

(1) Convex arc turning method.

Convex arc turning methods include line cutting method, profiling method and combination method, as shown in Figure 2-54, and their respective turning characteristics are shown in Table 2-22.

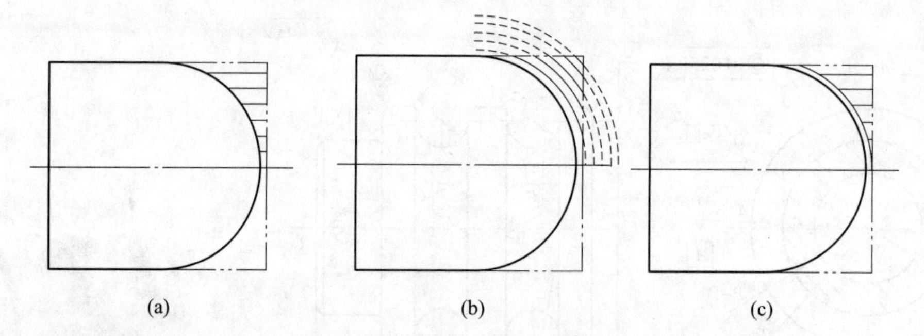

Figure 2-54 Convex arc surface turning methods
(a) Line cut method; (b) Copying method; (c) Combination method

Table 2-22 Comparison of convex arc turning methods

Method	Characteristics
Line cut method	The cutting path is short, the numerical calculation is complex, and the finishing allowance is not uniform. The cutting path cannot exceed the arc surface, otherwise the arc surface will be injured.
Copying method	The numerical calculation is simple, the finishing allowance is uniform, the empty cutting path is much.
Combination method	The cutting path is in between the two, and the remaining amount of the finished turning is uniform.

(2) Concave arc turning method.

As shown in Figure 2-55, concave arc turning methods are concentric circle turning method and stepped turning method, and their characteristics are shown in Table 2-23.

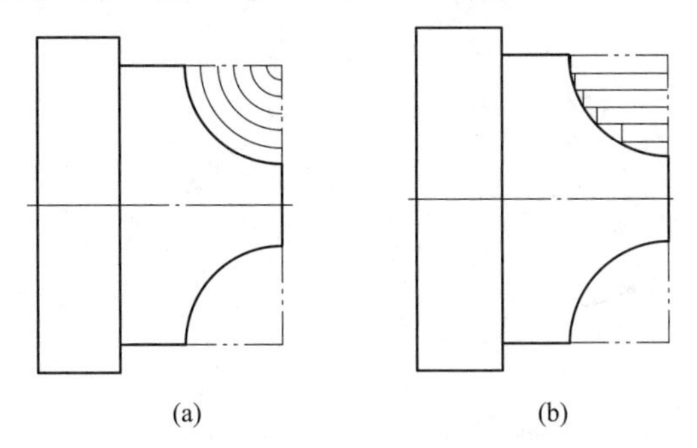

Figure 2-55 Concave arc turning methods

(a) Concentric arc form; (b) Stepped form

Table 2-23 Comparison of concave arc turning methods

Method	Characteristics
Concentric arc form	The numerical calculation is simple, the programming is convenient, the machining allowances is uniform.
Stepped form	The cutting path is short, the numerical calculation is complex, and the programming workload is increased.

As shown in Figure 2-56, there are three types of concave arc turning methods: concentric arc form, equal chord length form and stepped form. Their characteristics are shown in Table 2-24.

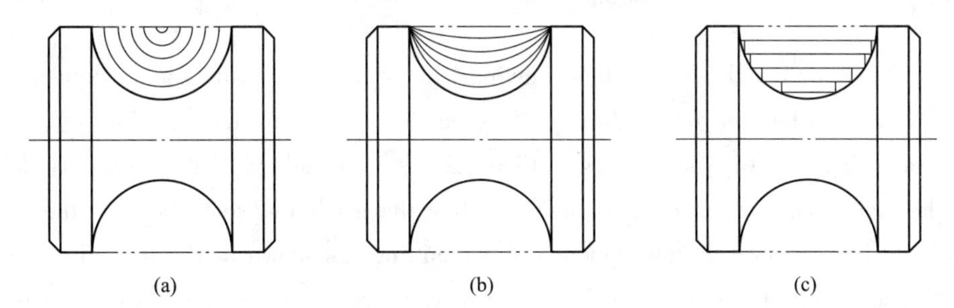

Figure 2-56 Concave arc 2 turning methods

(a) Concentric arc form; (b) Constant chord form; (c) Step form

Table 2-24 Comparison of concave arc 2 turning methods

Method	Characteristics
Concentric arc form	The cutting path is short, the machining allowances is uniform.
Constant chord form	Calculation and programming is the simplest, the cutting path is the longest.
Step form	The cutting force distribution is reasonable and the cutting rate is the highest.

2.4.1.2 Arc Turning Tool

There are four types of common outer arc surface as shown in Figure 2-57, and three types of commonly used turning tools for machining arc surface as shown in Figure 2-58.

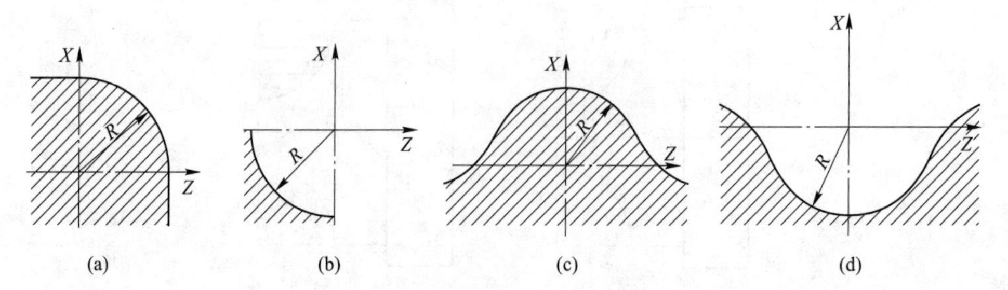

Figure 2-57 Type of arc surfaces

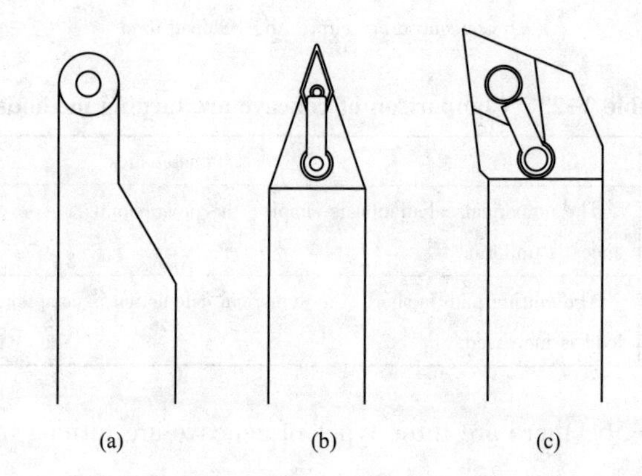

Figure 2-58 Arc turning tools

In Figure 2-57 (a) and (b), ordinary cylindrical turning tools with a main deflection angle greater than 90° can be used for machining. The tools are shown in Figure 2-58 (c).

When processing the two types of arcs in Figure 2-57 (c) and (d), the secondary deflection angle of the tool should be considered when selecting the machining tool, because the excessive declination angle will cause the phenomenon of overcutting, as shown in Figure 2-59.

When the secondary deflection angle allows, you can use the two tools shown in Figure 2-58 (a) arc turning tool or Figure 2-58 (b) sharp tool for machining.

When selecting a tool for processing, the minimum auxiliary deflection angle can be calculated according to the arc radius and arc depth, as shown in Figure 2-60. The auxiliary deflection angle of the selected tool must be greater than the calculated angle to ensure that overcutting will occur.

2.4.2 Programming

2.4.2.1 Circular Interpolation G02/G03

(1) Function. G02 is circular interpolation clockwise direction, and G03 is circular interpolation

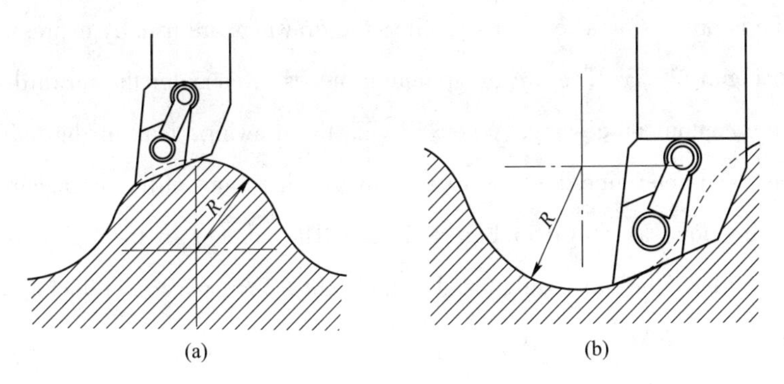

Figure 2-59 Overcutting

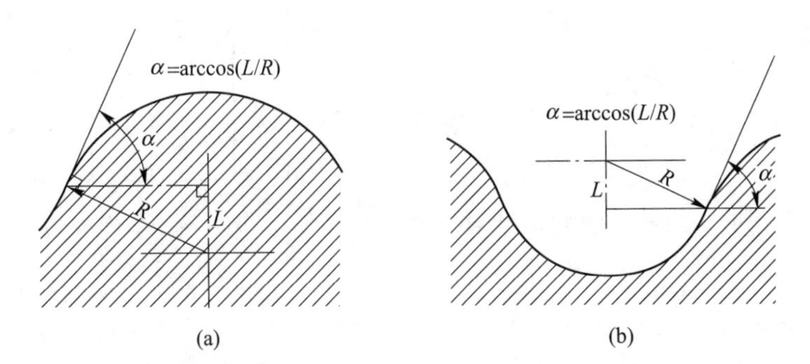

Figure 2-60 Calculation of tool deflection angle

counterclockwise direction.

(2) Format.

1) Format 1: Program with arc radius R

G02/G03 X(U)_Z(W)_R_F_;

Among them X, Z——The absolute coordinates of the end point of the arc;

U, W——The incremental coordinates of the end of the arc with respect to the beginning of the arc;

R——Arc radius, center angle $0°\sim180°$ is positive value, greater than $180°$ is negative value.

2) Format 2: Program the center position with I, K

G02/G03 X(U)_Z(W)_I_K_F_;

Among them I, K——The increment of the center of the arc relative to the start of the arc.

(3) Note.

1) Method for judging arc forward and reverse directions. The judgment of the arc direction is determined by the right-handed coordinate system: looking along the negative direction ($-Y$) of the vertical coordinate axis of the plane where the arc is located (XOZ), the clockwise direction is G02 and the counterclockwise direction is G03, As shown in Figure 2-61.

Since the lathe is processing a revolving profile, the drawings are usually expressed symmetrically, as shown in Figure 2-53. The arc programming needs to consider the forward and reverse directions of the arc contour processing. When viewing the drawing, look at the half of the contour on the centerline, and determine the forward and reverse directions of the arc according to the contour processing direction, as shown in Figure 2-62. The contour in the figure is generally processed from right to left, the processing trend of the right *R*2 arc is counterclockwise, and the processing trend of the two *R*4 arcs is clockwise.

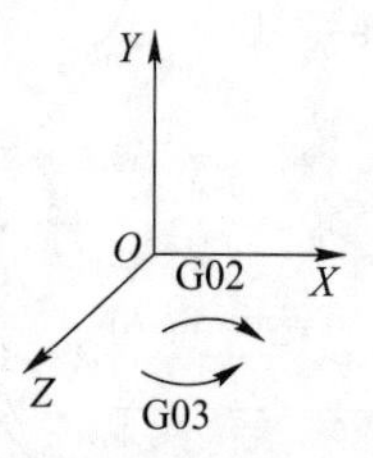

Figure 2-61 The method of judging the arc direction

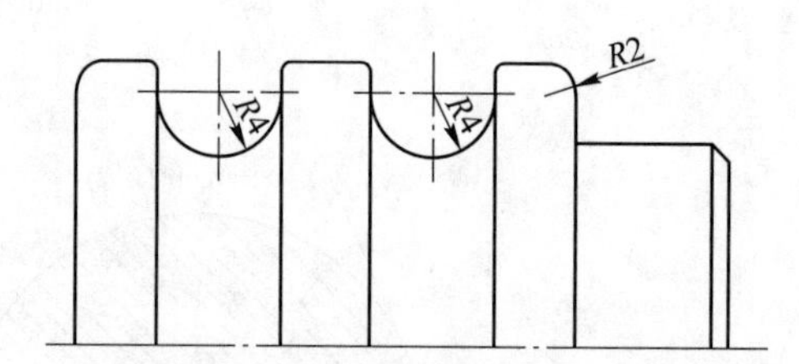

Figure 2-62 Arc machining direction

2) Values of I and K.

No matter it is programmed in absolute or incremental dimensions, I and K are the incremental values of the center of the circle relative to the starting point of the arc, I is the incremental value in the *X* direction, I = coordinate of the center of the circle X-starting point *X*. K is the incremental value in the *Z* direction, K = center coordinate Z-starting point coordinate *Z*. When I and K are zero, they can be omitted. If I, K and R are specified at the same time, R takes precedence, and I and K are invalid.

【Example 2-11】 As shown in Figure 2-63, use arc interpolation instruction to write arc part finishing program. The machining program is shown in Table 2-25.

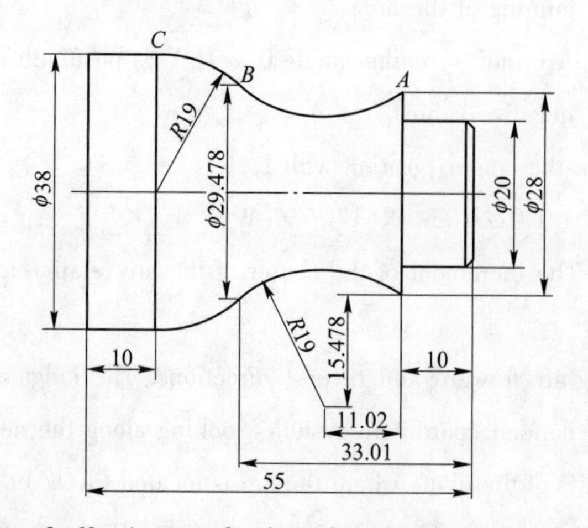

Figure 2-63 An example of circular interpolation instruction

Table 2-25 Arc machining program

R	I, K
G01 X0 Z0 F0.1;	G01 X0 Z0 F0.1;
G01 X20.0 C1.0;	G01 X20.0 C1.0;
G01 Z-10.0;	G01 Z-10.0;
G01 X28.0;	G01 X28.0;
G02 X29.478 Z-33.01 R19.0;	G02 X29.478 Z-33.01 I15.478 K-11.02;
G03 X38.0 Z-45.0 R19.0;	G03 X38.0 Z-45.0 I-14.739 K-11.99;
G01 Z-55.0;	G01 Z-55.0;

2.4.2.2 Pattern Repeating G73

(1) Function.

This instruction can effectively turn fixed graphics, such as casting, forging, or rough turning workpieces. This instruction only needs to specify the number of roughing cycles, finishing allowances and finishing routes. The system automatically calculates the cutting depth of roughing, gives the roughing route, and completes the roughing of each surface. The rough cycle of G73 command is shown in Figure 2-64.

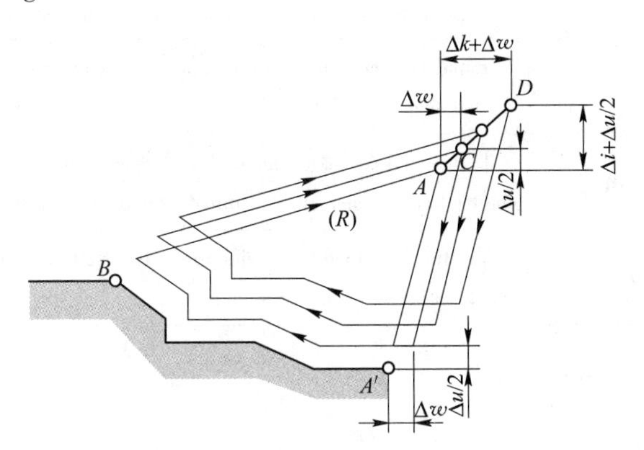

Figure 2-64 G73 instruction cycle course

(2) Format.

G73 U(Δi) W(Δk) R(d);

G73 P(ns) Q(nf) U(Δu) W(Δw);

Among them Δi——Distance and direction of relief in the *X* axis direction (Radius designation);

Δk——Distance and direction of relief in the *Z* axis direction;

d——The number of division;

ns——Sequence number of the first block for the program of finishing shape;

nf——Sequence number of the last block for the program of finishing shape;

Δu——Distance and direction of finishing allowance in *X* direction, diameter designation, generally set 0.5mm;

Δw——Distance and direction of finishing allowance in Z direction, diameter designation, generally set 0.05~0.1mm.

(3) Note.

1) The G73 and G71 commands are rough machining commands, the difference is that parts with arbitrary shapes and contours can be processed.

2) G73 can also machine unmoved bar stock, but there are many idle tools.

3) The ns and nf blocks do not need to be written immediately after the G73 block. The system can automatically search for and execute the ns-nf block. After the G73 instruction is completed, the block following G73 will be executed.

【Example 2-12】The part is shown in Figure 2-62. The blank is ϕ40. Use the G73 and G70 instructions to write the rough and finish machining program for the part. The machining program is shown in Table 2-26.

Table 2-26 The machining program

O245;	Program No.
G40 G97 G99 M03 S600 F0.2;	Cancel tool nose radius compensation, setting constant spindle speed control, feed rate 0.2mm/r, spindle CW at 600r/min
T0101;	Replace T01 external turning tool, substitute tool compensation No. 01
G00 Z5.0;	Rapid positioned to the starting point of the composite cycle for rough machining of a fixed shape (X45.0, Z5.0)
X45.0;	
G73 U20.0 W0.0 R15.0;	Define G73 coarse turning cycle, the back cutting amount in X direction is 20mm, the total cutting amount in Z direction is 0mm, and the cycle is 15 times
G73 P10 Q20 U0.5 W0.0;	The finishing route is specified by N10~N20, with X finishing margin of 0.5mm and Z finishing margin of 0.0mm
N10 G42 G01 X0 F0.1;	Sequence of finishing shape
Z0;	
G01 X20 C1.0;	
G01 Z-10.0;	
G01 X28.0;	
G03 X29.478 Z-33.01 R19.0;	
G02 X38.0 Z-45.0 R19.0;	
G01 Z-55;	
N20 G40 G01 X40;	
M05;	Spindle stop
M00;	Program stop
M03 S1000 T0101 F0.1;	Spindle CW at 1000r/min, Change T01 external turning tool, substitute tool compensation No. 01, feed rate 0.1mm/r

Continued Table 2-26

G00 Z5. 0;	Rapid positioned to the starting point of the composite cycle of rough machining with fixed shape and the tool tip radins is set up to compensate right (*X*45. 0, *Z*5. 0)
X45. 0;	
G70 P10 Q20;	Calling finishing cycle in turning
G00 X100. 0;	Rapid positioned to the tool change point, cancel the tool tip radius compensation (*X*100. 0, *Z*100. 0)
Z100. 0;	
M30;	End of program and reset

2. 4. 3 Operation

The commonly used tools for machining arcs are external eccentric, sharp and arc turning tools. The sharp tool setting is required to ensure the Z position of the tool. It is commonly used to complete the tool setting tool, as shown in Figure 2-65.

The ideal point of the arc turning tool is the apex of the arc, as shown in Figure 2-66. The process of arc turning tool setting is basically the same as that of the cutter. The difference is that when determining the Z value, the tool inputs Z0 and for point measurement, the arc turning tool inputs the arc radius Z_R of the tool and then points to determine the Z value position, as shown in Figure 2-67.

Figure 2-65 Tool setting instrument

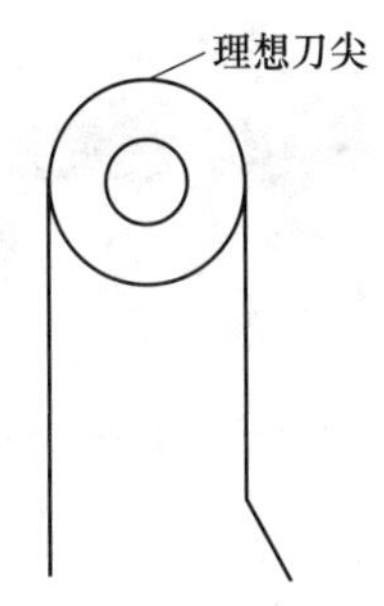

Figure 2-66 Arc turning tool tip

OFFSET/GEOMETRY O0001 N0000

NO.	X.	Z.	R	T
G 01	-148.360	0.000	0.000	0
G 02	0.000	0.000	0.000	0
G 03	0.000	0.000	0.000	0
G 04	0.000	0.000	0.000	0
G 05	0.000	0.000	0.000	0
G 06	0.000	0.000	0.000	0
G 07	0.000	0.000	0.000	0
G 08	0.000	0.000	0.000	0

ACTUAL POSITION (RELATIVE)

U -24.180 W -332.291

>Z1.5 OS 50% T0101

MEM. *** *** *** 18:52:27

[NO.SRH] [MESUR] [INP.C.] [+INPUT] [INPUT]

Figure 2-67 Arc turning tool setting

When using a circular cutter to machine an arc, the arc radius compensation R of the tool is determined by the arc radius of the arc cutter. For example, if the arc radius of the arc cutter $R=2\text{mm}$, the tool compensation in the CNC system is $R=2.000$, and the tool tip orientation T is 8.

2.4.4 Measurement

For the general accuracy profile, the radius gauge can be used to detect the light gap method. As shown in Figure 2-68, the measuring surface of the radius gauge must be completely in close contact with the arc of the workpiece. When there is no gap in the middle of the arc, the arc degree of the workpiece is the number shown on the radius gauge. Because it is visual inspection, the accuracy is not very high, only qualitative measurement can be done. High-precision contours can be detected using a three-dimensional measuring machine. The three-dimensional measuring machine can detect the contour by scanning through the contour surface, as shown in Figure 2-69.

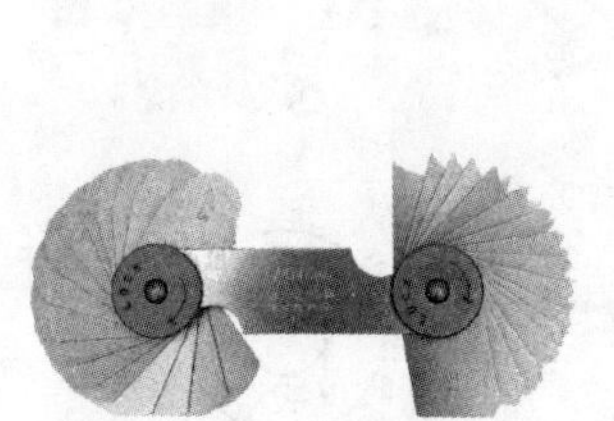

Figure 2-68 Radius gauge

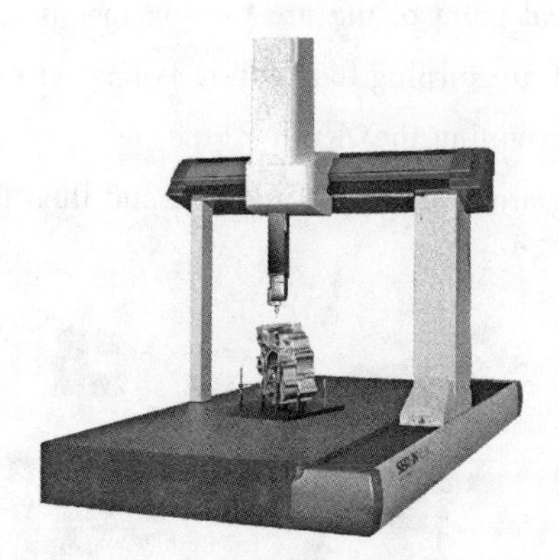

Figure 2-69 Coordinate measuring machine

Implemention

(1) Step 1 Drawing analysis:

The part shown in Figure 2-53 is a round pulley with two $R4$ arc grooves. Each arc groove has an $R0.5$ arc transition at the intersection of the outer circle, there is an $R2$ arc transition between the outer circle and the end face, the middle is a $\phi16$ through hole for mounting the motor shaft, and there is an M6 threaded hole on the outer circle of $\phi26$. The surface roughness is $Ra1.6$ and $Ra3.2$ respectively. This task is mainly to process the outer circle and arc groove of the part.

(2) Step 2 Making the machining plan:

1) Machining plan:

① The three-jaw self-centering chuck is used for mounting, and the part extends about 60mm from the chuck.

② Machining the outer contour of the part to the required size.

③ Cut off the workpiece.

2) Tool selection:

T01 90° external turning tool (Blade Angle 55°)

T02 *R*1.5 arc turning tool

T03 3mm grooving tool

The CNC machining tool card for the pulley part is shown as Table 2-27.

Table 2-27 CNC machining tool card

<table>
<tr><td colspan="2">Part name</td><td colspan="2">Belt pulley</td><td>Drawing No.</td><td colspan="3">2-53</td></tr>
<tr><td>S No.</td><td>Tool No.</td><td>Tool name</td><td>Quantity</td><td>Machined surface</td><td>Tool nose radius R/mm</td><td>Tool tip orientation T</td><td>Note</td></tr>
<tr><td>1</td><td>T01</td><td>90° external turning tool</td><td>1</td><td>Rough and finish turning the outer diameter</td><td>0.4</td><td>3</td><td></td></tr>
<tr><td>2</td><td>T02</td><td>R1.5 arc turning tool</td><td>1</td><td>Turning R4 arc groove</td><td>1.5</td><td>8</td><td></td></tr>
<tr><td>3</td><td>T03</td><td>Grooving tool</td><td>1</td><td>Cutting off the workpiece</td><td></td><td></td><td>3mm</td></tr>
<tr><td>Edit</td><td></td><td>Check</td><td></td><td>Approve</td><td>Date</td><td></td><td>Total 1 Page</td><td>Page1</td></tr>
</table>

3) Machining process:

Table 2-28 shows the CNC machining process of the pulley.

Table 2-28 CNC machining process card

<table>
<tr><td rowspan="2" colspan="2">Company name</td><td rowspan="2" colspan="2">Tianjin polytechnic college</td><td colspan="2">Part name</td><td colspan="2">Part drawing No.</td></tr>
<tr><td colspan="2">Belt pulley</td><td colspan="2">2-53</td></tr>
<tr><td colspan="2">Program No.</td><td>Fixture name</td><td>Machine tool</td><td colspan="2">CNC system</td><td colspan="2">Workshop</td></tr>
<tr><td colspan="2">0248</td><td>Three jaw chuck</td><td>CKA6140</td><td colspan="2">FANUC SERIES 0i</td><td colspan="2">CNC training center</td></tr>
<tr><td>Steps</td><td colspan="2">Contents</td><td>Tool</td><td>Spindle speed n/r · min^{-1}</td><td>Feed rate F/mm · r^{-1}</td><td>Back engagement a_p/mm</td><td>Note</td></tr>
<tr><td>1</td><td colspan="2">Extend the clamping parts 58mm and align them</td><td></td><td></td><td></td><td></td><td rowspan="4">Manual</td></tr>
<tr><td>2</td><td colspan="2">Setting external turning tool</td><td>T01</td><td></td><td></td><td></td></tr>
<tr><td>3</td><td colspan="2">Setting grooving tool</td><td>T02</td><td></td><td></td><td></td></tr>
<tr><td>4</td><td colspan="2">Setting threading tool</td><td>T03</td><td></td><td></td><td></td></tr>
<tr><td>5</td><td colspan="2">Rough turning belt pulley, leave machining allowances of 0.5mm</td><td>T01</td><td>800</td><td>0.2</td><td>1.5</td><td></td></tr>
<tr><td>6</td><td colspan="2">Finish turning belt pulley to ensure the size</td><td>T01</td><td>1000</td><td>0.1</td><td>0.25</td><td></td></tr>
</table>

Continued Table 2-28

Steps	Contents			Tool	Spindle speed n/r · min^{-1}	Feed rate F/mm · r^{-1}	Back engagement a_p/mm	Note
7	Rough turning R4 arc groove leave machining allowances of 0. 2mm			T02	800	0. 1	1	R1. 5mm
8	Finish turning R4 arc groove to ensure the size			T02	1200	0. 08	0. 1	
9	Cutting off the workpiece, leave 0. 5mm margin for the length			T03	600	0. 1	3	3mm
Edit		Check		Approve	Date		Total 1 Pages	Page1

(3) Step 3 Programming:

The part program is shown in Table 2-29.

Table 2-29 The part program

O2410;	Program No.
G40 G97 G99 M03 S600 F0. 2;	Cancel tool nose radius compensation, setting constant spindle speed control, feed rate 0. 2mm/r, spindle CW at 600r/min
T0101;	Replace T01 external turning tool, substitute tool compensation No. 01
M08	Cutting fluid drive
G42 G00 Z5. 0;	Rapid positioning to point (X40. 0, Z5. 0), call tool nose radius compensation
X40. 0;	
G71 U1. 5 R0. 5;	Define rough turning cycle, cutting depth of 1. 5mm, cutting back of 0. 5mm
G71 P10 Q20 U0. 5W0. 05;	The finishing line is specified by N10~N20, with a finishing margin of 0. 5mm in X direction and 0. 05mm in Z direction
N10 G00 X0. ;	Sequence of finishing shape
G01 Z0 F0. 1;	
X26. 0 C1. 0;	
Z-10. 0;	
X32. 0;	
G03 X36. 0 Z-12 R2. 0;	
G01 Z-46. 0;	
N20 G01 X40. 0;	
G40 G00 X100. 0;	Rapid positioning to point (X100. 0, Z100. 0), cancel tool nose radius compensation
Z100. 0;	
M05;	The spindle stop
M09	Cutting fluid clearance
M01;	Optional stop

Continued Table 2-29

M03 S1000 T0101 F0. 1;	Spindle CW at 1000r/min, change T01 external turning tool, substitute tool compensation No. 01, feed rate 0. 1mm/r
G42 G00 Z5. 0;	Rapid positioning to point (X40. 0, Z5. 0), call tool nose radius compensation
X40. 0;	
G70 P10 Q20;	Calling finishing cycle in turning
G40 G00 X100. 0;	Rapid positioning to point (X100. 0, Z100. 0), cancel tool nose radius compensation
Z100. 0;	
M05;	Spindle stop
M09	Cutting fluid clearance
M01;	Optional stop
M03 S800 T0202 F0. 1;	Spindle CW at 800r/min, change T02 R1. 5 arc turning tool, substitute tool compensation No. 02, feed rate 0. 1mm/r
M08;	Cutting fluid drive
G42 G00 Z5. 0;	Rapid positioning to point (*X*40. 0, *Z*-10. 0), call tool nose radius compensation
X45. 0;	
Z-10. 0;	
G73 U6. 0 R6;	Calling pattern repeating
G73 P30 Q40 U0. 2W0. 0;	
N30 G42 G01 X36. 0;	Sequence of finishing shape
G01 Z-14. 5;	
G03 X35. 0 Z-15. 0 R0. 5;	
G01 X32. 0;	
G02 Z-23. 0 R4. 0;	
G01 X35. 0;	
G03 X36. 0 Z-23. 5 R0. 5;	
G01 Z-28. 5;	
G03 X35. 0 Z-29. 0 R0. 5;	
G01 X32. 0;	
G02 Z-37. 0 R4. 0;	
G01 X35. 0;	
G03 X36. 0 Z-37. 5 R0. 5;	
G01 Z-42. 0;	
N40 G01 X40. 0;	
G40 G00 X100. 0;	Rapid return to the point (*X*100. 0, *Z*100. 0), cancel tool nose radius compensation
Z100. 0;	

Continued Table 2-29

M05;	Spindle stop
M09	Cutting fluid clearance
M01;	Optional stop
M03 S1000 T0202 F0. 08;	Spindle CW at 1000r/min, change T02 R1. 5 arc turning tool, substitute tool compensation No. 02, feed rate 0. 08mm/r
G42 G00 Z5. 0;	Rapid return to the point (*X*40. 0, *Z*-10. 0), call tool nose radius compensation
X40. 0;	
Z-10. 0;	
G70 P30 Q40. ;	Calling finishing cycle in turning
G40 G00 X100. 0;	Rapid return to the change point (*X*100. 0, *Z*100. 0), cancel tool nose radius compensation
Z100. 0;	
M05;	Spindle stop
M01;	Optional stop
M03 S600 T0303 F0. 08;	Spindle CW at 600r/min, change T03 grooving tool, substitute tool compensation No. 03, feed rate 0. 1mm/r
G00 X45. 0;	The quick point of the tool is located to the cutting point
G00 Z-45. 5;	
X41. 0;	
G01 X0. 0 F0. 05;	Parting off the workpiece
G00 X100. 0;	Rapid return to the change point (*X*100. 0, *Z*100. 0)
Z100. 0;	
M30;	End of program and reset

(4) Step 4 Part machining:

1) Select the tool according to the tool table and install the tool.

2) Setting tool, pay attention to input the tool nose radius and Direction of imaginary tool nose.

3) Input the program and simulation, ensure the correctness of the program.

4) Machining the parts, and the processing effect is shown in Figure 2-70.

Figure 2-70 The finished part

(5) Step 5　Dimension Measurement:

1) Measure the $\phi 26$, $\phi 36$, 10, 42, $R4$ slot width with vernier calipers.

2) Detect $R4$ arc size with radius gauge.

3) Roughness detection using roughness comparison board.

Assessment

The assessment table is shown in Table 2-30.

Table 2-30　The assessment table

Item	No.	Standard		Partition	Score
Machining plan (10%)	1	Machining steps	Meet the requirements of CNC turning	3	
	2	Dimension	Calculate dimensions	3	
	3	Tool selection	The cutter selection is reasonable and meets the machining requirements	4	
Programming (30%)	4	Program number	No program no point	2	
	5	Program segment number	No program segment number no point	2	
	6	Cutting parameters	Unreasonable choice of cutting parameters no point	4	
	7	Origin and coordinates	Mark the origin and coordinates of the program, otherwise there is no point	2	
	8	Program contents	Deduct 2 points for each segment that does not meet the requirements of program logic and format	20	
			Deduct 5 points if the procedure content is not in accordance with the process		
			Deduct 5 points in case of dangerous directive		
Operation (20%)	9	Program input	Deduct 2 points for each wrong section	10	
	10	Machining operation 30min	No point for setting error of workpiece coordinate system	10	
			No points for misoperation		
			Overtime no point		
Dimension measurement (30%)	11	External profile	$\phi 26$ out of tolerance without score	5	
	12		$\phi 36$ out of tolerance without score	4	
	13	Length	10 ± 0.2 out of tolerance without score	5	
	14		42 ± 0.3 out of tolerance without score	4	
	15	$R4$ groove width	8 ± 0.2 out of tolerance without score	5	
	16		8 ± 0.2 out of tolerance without score	5	
	17	Contour	$R4$	4	
	18		$R2$	4	
	19	Surface roughness	$Ra1.6$ over tolerance no points	4	

Continued Table 2-30

Item	No.	Standard	Partition	Score
Vocational ability (10%)	20	Learning ability	2	
	21	Communication skills	2	
	22	Team cooperation	2	
	23	Safe operation and civilized production	4	
Total				

Exercises

(1) Answer the following questions:

1) Briefly describe the method of dividing processes according to the principle of process concentration.

2) Briefly describe the functions and usage of G73 directive.

3) How to select G41 and G42 directives?

4) How to determine the machining direction of G02 and G03?

5) How to ensure the radius tolerance during machining?

(2) Synchronous Training:

The parts are shown in Figure 2-71 and Figure 2-72. The blank ϕ40mm. The material is hard aluminum alloy LY15. Please programming and machining.

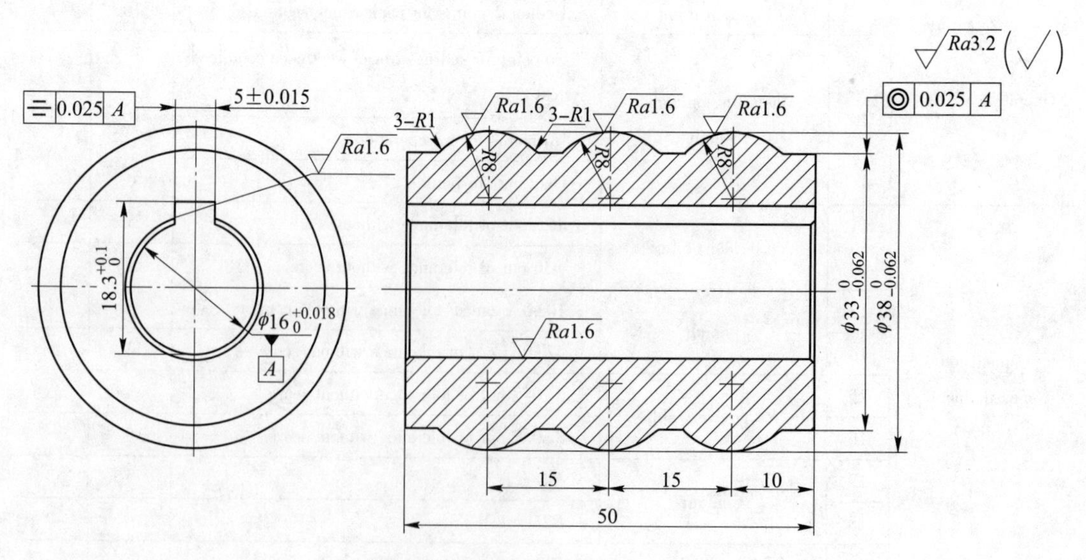

Figure 2-71 Synchronous training 1

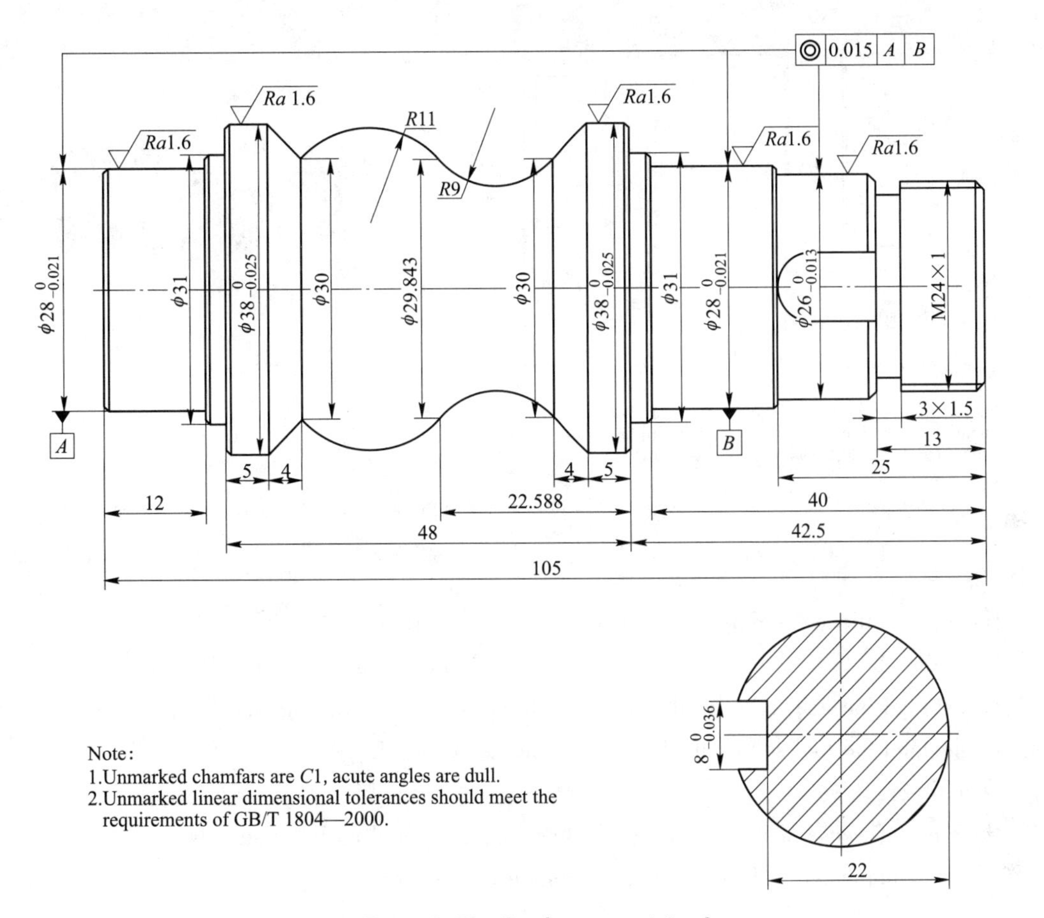

Figure 2-72 Synchronous training 2

Task 2. 5 CNC Turning of Threaded Parts

Introduction

As shown in Figure 2-73, the part is made of hard aluminum alloy LY15, and the blank is φ40× 75mm. It is produced in one piece. It is coordinated with task 2. 6. This task needs to prepare the program, machine the part using the CNC lathe, and then measure the part.

Competences

(1) Knowledge:

1) Understand the thread turning technology and thread parameters.

2) Explain G92 and G76 instructions.

3) Know the types and structures of thread measuring tools.

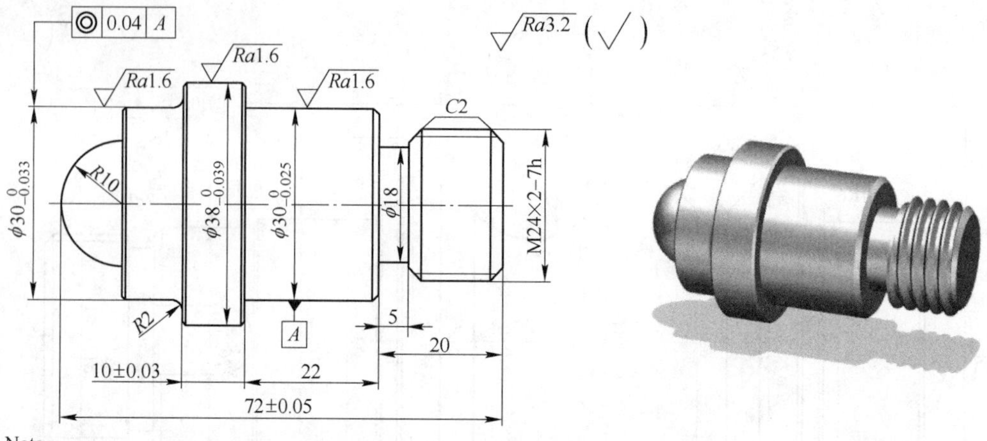

Note:
1. Unmarked chamfers are $C0.5$, acute angles are dull.
2. Unmarked linear dimensional tolerances should meet the requirements of GB/T 1804—2000.

Figure 2-73 The part drawing

(2) Skill:

1) Have the ability to complete the aligment of thread turning tool.

2) Have the ability to use G92 and G76 in the machining program of threaded parts.

3) Understand the inspection methods of thread gauge and thread micrometer.

4) Understand the tool setting method for turning another side.

(3) Quality:

1) Have a high sense of responsibility, dedication, unity and cooperation spirit.

2) Correctly implement safety technical operation procedures.

Relevant Knowledge

2.5.1 Process Preparation

2.5.1.1 The Use of Thread

Thread is a widely used removable fixed connection, which has the advantages of simple structure, reliable connection and convenient disassembly. Thread is mainly used in connection, fastening and transmission, as is shown in Figure 2-74.

2.5.1.2 Thread Types

There are two types of threads: external threads and internal threads. The thread used as the connection is called the connecting thread, and the thread which can be the driving is called the driving thread. According to the thread direction, it can be divided into left-hand thread and right-hand thread. The most frequently used thread in daily life is right-hand thread. The helix of thread includes single line, double line and multi line, and the connecting thread is generally single line.

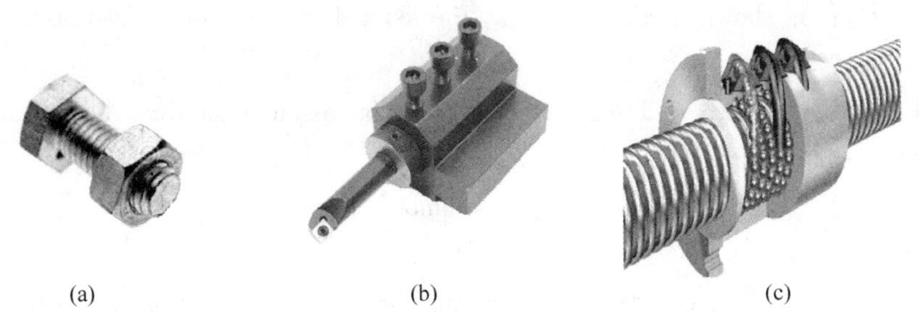
(a) (b) (c)

Figure 2-74 Applications of thread

(a) Connection; (b) Fastening; (c) Transmission

The thread types are common triangle thread, pipe thread, trapezoid thread, rectangle thread and serrated thread. The first two are mainly used for connection and the last three are mainly used for transmission, as is shown in Figure 2-75.

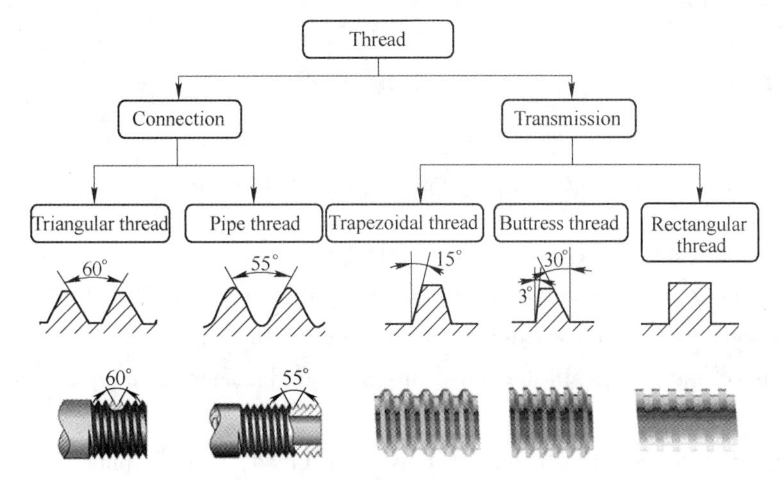

Figure 2-75 Types of thread

2.5.1.3 Main Elements of Thread

(1) Major diameter (d, D): The top diameter d of external thread and the bottom diameter D of internal thread are the maximum diameter of thread.

(2) Minor diameter (d_1, D_1): The bottom diameter d_1 of external thread and the top diameter D_1 of internal thread are the minor diameter of thread Figure 2-76.

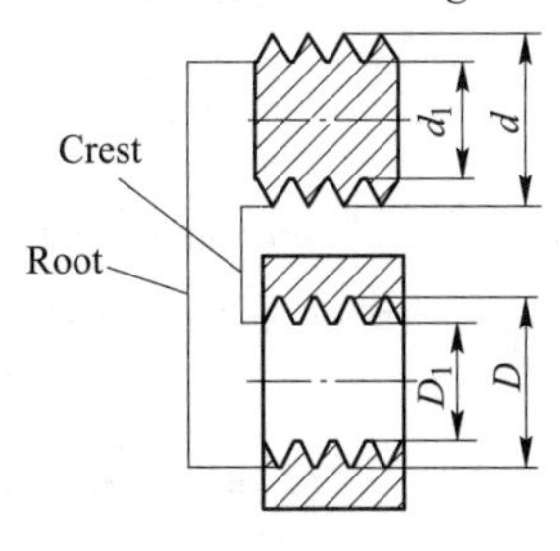

Figure 2-76 Major diameter and minor diameter of thread

(3) Pitch (P): As shown in Figure 2-77, the axial distance between two adjacent thread is pitch.

(4) Lead (L): As shown in Figure 2-78, the axial distance between two adjacent threads formed along the same helix is lead.

$$\text{Lead} = \text{pitch} \times \text{numbers of lines}$$

(5) Tooth angle (α)

As shown in Figure 2-79, on the section passing through the axial direction of the thread, the profile shape of the thread is called the thread profile, and the angle between the adjacent two flanks is the profile angle.

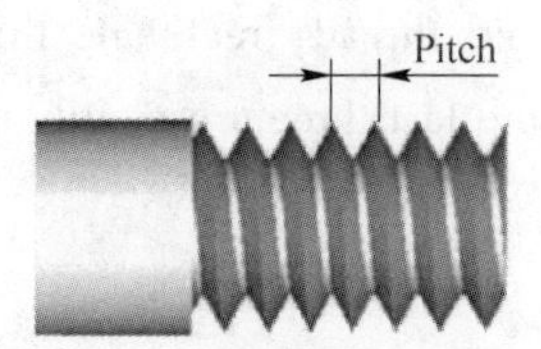

Figure 2-77 Pitch

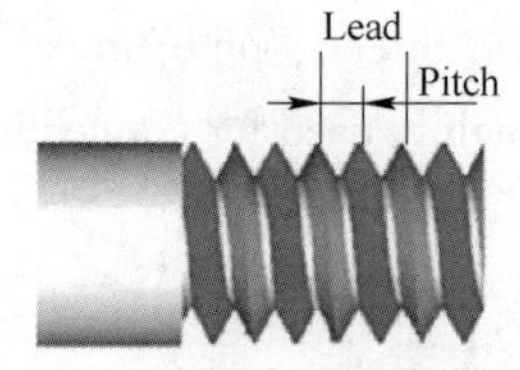

Figure 2-78 Lead

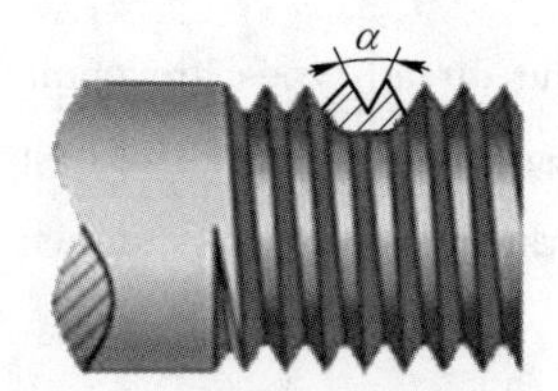

Figure 2-79 Pitch angle

2.5.1.4 External Thread Machining Dimension Calculation

(1) Actual Turning Outer Cylindrical Surface Diameter

When turning the triangular external thread, the size of the thread will swell due to the extrusion of the tool, so the diameter of the external cylinder before turning the thread should be smaller than the diameter of the thread. In the actual turning, the diameter of the cylindrical surface:

$$d_{actual} = d - 0.1P$$

Among them, d is the major diameter of the external thread, P is the pitch.

(2) Thread profile height.

According to the national standard of common thread, the profile height of triangular thread:

$$h_{teeth} = 0.65P$$

(3) Minor diameter of thread

$$d_1 = d - 2h_{teeth} = d - 1.3P$$

2.5.1.5 Thread Machining Method

(1) There are two ways of thread feed in CNC lathe, straight feed and oblique feed, as shown in Figure 2-80. When the pitch is $P<3$mm, the straight method is used. When the pitch is $P \geqslant 3$mm, the oblique method is used.

The times of cutting and the amount of back engagement in thread machining directly affect the quality of thread machining, and the way of decreasing back engagement distribution should be followed in turning. As shown in Figure 2-81.

(2) Layered cutting method. In the process of thread machining, the cutting force is large, so layered cutting is needed. Table 2-31 shows the common feed times and back engagement. In

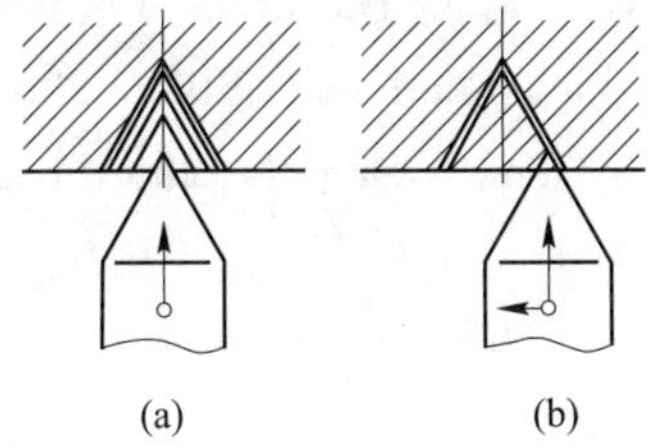

Figure 2-80 Thread feed mode

(a) The straight method; (b) The oblique method

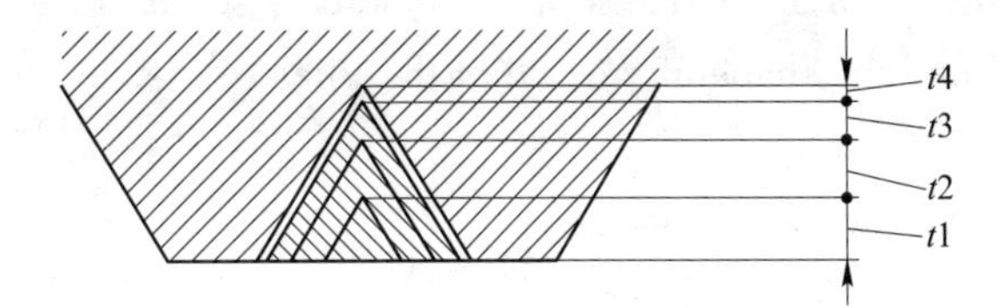

Figure 2-81 Distribution mode of back engagement

order to prevent excessive cutting force, the number of cuts can be increased properly. However, in order to improve the surface roughness of the thread, the cutting depth of the last tool should not be less than 0. 1mm when the carbide thread turning tool is used.

Table 2-31 Recommended cutting times and cutting depth

Pitch/mm		1. 0	1. 5	2. 0	2. 5
Depth of threading/mm		0. 65	0. 975	1. 3	1. 625
Depth of cutting/mm		1. 3	1. 95	2. 6	3. 25
Cutting condition /mm	1	0. 7	0. 8	0. 9	1. 0
	2	0. 5	0. 65	0. 7	0. 8
	3	0. 1	0. 4	0. 6	0. 6
	4		0. 1	0. 3	0. 5
	5			0. 1	0. 25
	6				0. 1

(3) Speed up section and speed down section of thread machining. Since there is an acceleration process at the beginning of turning thread, there is a deceleration process before the end. Therefore, when turning the thread, both ends must be equipped with enough speed up feed section and speed down retreat section. As shown in Figure 2-82, δ_1 is the distance between speed up feed section and δ_2 is the distance between speed up feed section and speed down retreat section.

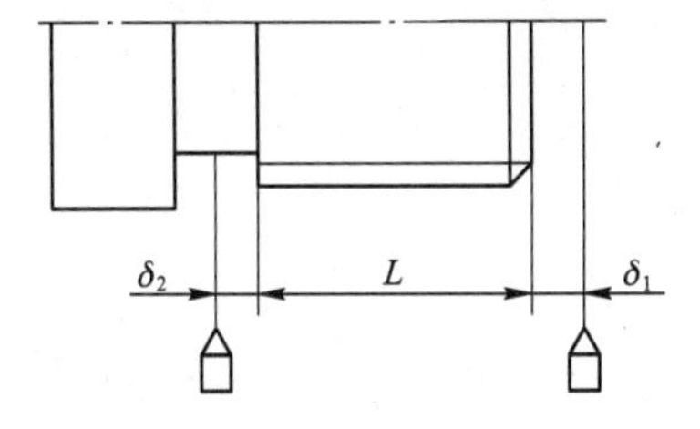

Figure 2-82 Speed up feed section and speed down retreat section

The values of δ_1 and δ_2 are related to the workpiece pitch and spindle speed. Generally, δ_1 is set as $1-2P$, and δ_2 is set as $0.5P$. In actual production, the value of δ_1 is generally set as 2~5mm, and the value of large pitch and high-precision thread is set as large value. The value of δ_2 is generally about half of the width of undercut, setting 1~3mm. If there is no undercut at the end of the thread, the shape of the end of the thread is related to the CNC system. Generally, the thread is undercut at 45°.

(4) Cutting parameters.

1) Spindle speed n.

When CNC lathe processes thread, spindle speed is affected by many factors such as CNC system, thread lead, cutting tools, workpiece size and material. Different CNC systems have different recommended spindle speed ranges. The operator should carefully consult the manual and select according to the specific situation. When most CNC lathe turning thread, the recommended spindle speed formula is as follows:

$$n \leqslant 1200/P - K$$

Among them P——Thread pitch (mm);

K——Insurance factor, generally taken as 80;

n——Spindle speed (r/min).

2) The cutting depth or back cutting engagement a_p is shown in Table 2-31.

3) Feed rate F (mm/r):

① The feed rate of single thread is equal to the pitch, i. e. $F=P$;

② The feed rate of multi thread is equal to the lead, i. e. $F=L$.

2.5.2 Programming

2.5.2.1 Thread Cutting Cycle G92

(1) Function.

It is used for circular machining of small pitch threaded parts, and the circular route is shown in Figure 2-83.

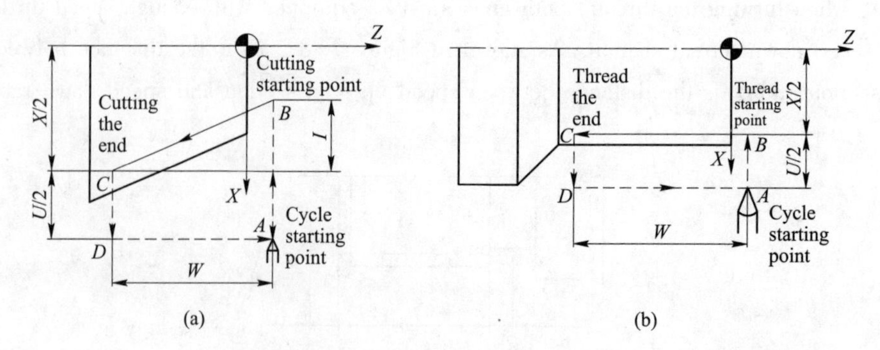

Figure 2-83 The route of G92 code

(a) Cylindrical thread cutting cycle route; (b) Tapered thread cutting cycle route

(2) Format.

G92 X(U)_Z(W)_I(R)_F_;

Among them X, Z——Absolute coordinate of thread end point;

U, W——Coordinate of the end point of thread relative to the starting point;

F——Thread lead;

I(R)——The difference between the start radius and the end radius of the taper thread. I(R) is negative when the end radius of taper thread is greater than the start radius. I(R) is positive when the end radius of taper thread is less than the start radius. Cylindrical thread I = 0, I can be omitted.

(3) Note.

1) When turning the thread, it is not allowed to use the code of constant linear speed control. It is required to use the code of G97. The speed of rough turning and finishing turning spindle is the same, or the disordered teeth will occur.

2) When turning the thread, the feed speed multiplier and the spindle speed multiplier are invalid (fixed to 100%).

3) Due to the influence of machine tool structure and CNC system, the rotation speed of main chuck is limited.

4) For taper thread, when the angle is below 45°, the thread lead is specified in the Z-axis direction; when the angle is 45° ~ 90°, the thread lead is specified in the X-axis direction.

【Example 2-13】 As shown in Figure 2-84, the external diameter of thread has been turned to the size, the 4×2 undercut has been processed. It is required to compile a machining program for M24×2 threads with G92.

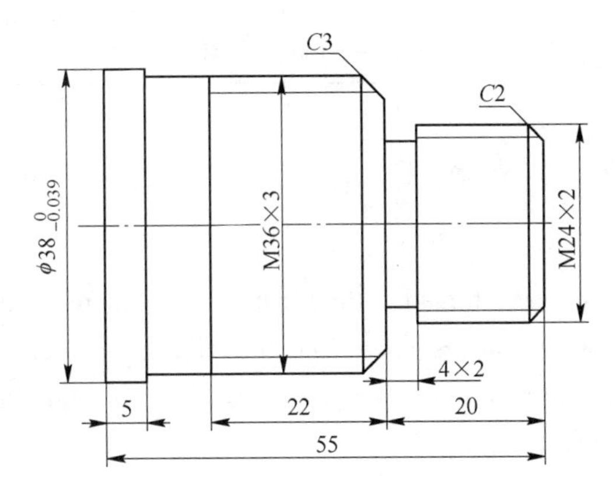

Figure 2-84 Drawing of threaded parts

(1) Dimension calculation:

The diameter of cylindrical surface in actual turning $d_{actual} = d - 0.1P = 24 - 0.1 \times 2 = 23.8$mm.

Actual thread profile height $d_{teeth} = 0.65P = 1.3$mm.

Actual minor diameter $d_{small} = D - 1.3P = 21.4$mm.

The δ_1 of the speed-increasing speed is 5mm, and the δ_2 of the speed-retracting speed is 2mm.

(2) Cutting parameters:

Spindle speed n is set as 400r/min. Feed rate F is 2mm.

From Table 2-31, it can be seen that there are five cutting times for thread cutting, and it is 0.9mm, 0.7mm, 0.6mm, 0.3mm and 0.1mm respectively.

(3) The program:

The M24×2 thread program is shown in Table 2-32.

Table 2-32 The M24×2 thread program

O1411;	Program No.
G40 G97 G99 M03 S400;	Clockwise rotation of main chuck, speed 400r/min
T0404;	Change T04 thread turning tool, substitute tool compensation No. 04
G00 Z5.0;	Tool moves to start point of thread cutting cycle
G00 X25.0;	
G92 X23.1 Z-18.0 F2.0;	The first cutting depth 0.9mm, and the feed rate is 2mm
X22.4;	The second cutting depth is 0.7mm
X21.8;	The third cutting depth is 0.6mm
X21.5;	The fourth cutting depth is 0.3mm
X21.4;	The fifth cutting depth is 0.1mm
X21.4;	Finish threading without cutting depth
G00 X100.0;	Fast move back to tool changing point
Z100.0;	
M30;	End of program and return to start

2.5.2.2 Multiple Thread Cutting Cycle G76

(1) Function.

It is used for multiple automatic thread cutting cycles. It is more simple than G92 instruction, only need to make relevant parameters once, the thread can be processed. It is often used to process threads without undercut and large pitch threads. The compound cycle route of G76 thread cutting is shown in Figure 2-85.

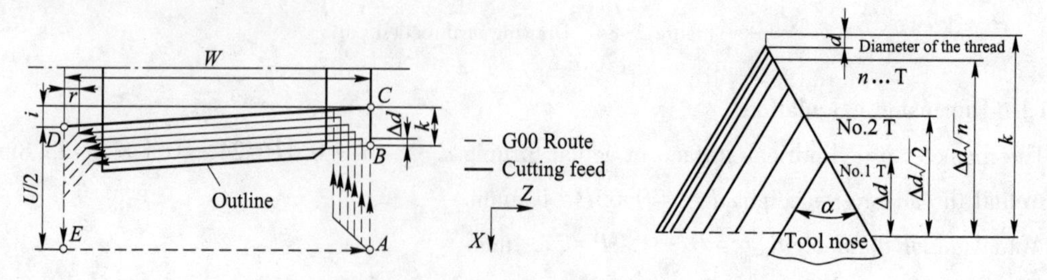

Figure 2-85 The route of G76 cycle code

(2) Format.

G76 P(m)(R)(α)Q(Δdmin)R(d);

G76 X(U)Z(W)R(i)P(k)Q(Δd)F(L);

Among them m——The number of times of fine turning repetition (01-99 unit: number of times);

R——Chamfer amount of thread tail, expressed by two integers between 00 and 99 (unit: 0.1×L, l is pitch);

α——Tip angle, that is, the tooth angle;

Δdmin——Minimum turning depth, specified with radius value (μm);

D——Finishing allowance, cutting depth of thread alarm car, specified with radius value (mm);

X(U), Z(W)——Absolute or incremental coordinates of the thread end point (in mm, X(U) is the diameter value);

i——Taper value of thread, radius difference between two ends of thread, cylindrical thread, I=0;

k——Thread height, specified by radius value (μm);

Δd——The first turning depth, specified with radius value (μm);

L——Lead, single head is pitch (mm).

【Example 2-14】 As shown in Figure 2-86, the outer diameter has been turned to the size requirement. It is required to compile a machining program for M36×3 threads with G76.

(1) Dimension calculation.

Diameter of cylindrical surface in actual turning $d_{actual}=d-0.1P=36-0.1\times3=35.7$mm.

Actual thread profile height $d_{teeth}=0.65P=0.65\times3=1.95$mm.

Actual diameter of thread $d_{small}=D-1.3P=36-1.3\times3=32.1$mm.

Speed up feed section $\delta_1=5$mm.

(2) Thread parameters.

Repeat the fine turning twice, $m=02$.

There is no chamfer at the end of thread, r is taken as 00.

The sharp angle of triangular thread cutter is 60°, $\alpha=60$.

The minimum turning depth δdmin is 0.05mm, $Q=50\mu$m.

Leave 0.1mm finishing allowance, $R=0.1$mm.

Calculate the end point coordinate of thread according to the part drawing ($X32.1$, $Z-42.0$).

Cylindrical thread I is 0, which can be omitted.

The thread height K is 1.95mm, P is 1950μm.

The first turning depth Δd is 0.45mm, Q is 450μm.

The pitch of single head thread is 3, $F=3$.

(3) The program.

The M36×3 thread program is shown in Table 2-33.

Table 2-33 The M36×3 thread program

O1412;	Programming No.
G40 G97 G99 M03 S400;	Clockwise rotation of main chuck, speed 400r/min
T0404;	Change T04 thread turning tool, substitute tool compensation No. 04
G00 Z-15.0;	Fast point positioning of tool to the start of cycle of thread cutting
G00 X37.0;	
G76 P020060 Q50 R0.1;	G76 threading cycle
G76 X32.1 Z-42.0 P1950 Q450 F3.0;	
G00 X100.0;	Move the tool back to changing tool point
Z100.0;	
M30;	End of program and return to start

2.5.3 Operation

2.5.3.1 Setting of Thread Tools

When thread processing, common tools have the external thread turning tool shown in Figure 2-86 and the external thread turning tool shown in Figure 2-87. Taking indexable external thread cutters as an example, the indexable thread cutters of CNC machine clamps have high precision in manufacturing the tool bar, and generally only need to hold the tool bar against the side of the tool holder.

Figure 2-86 Mechanical clamping thread turning tool

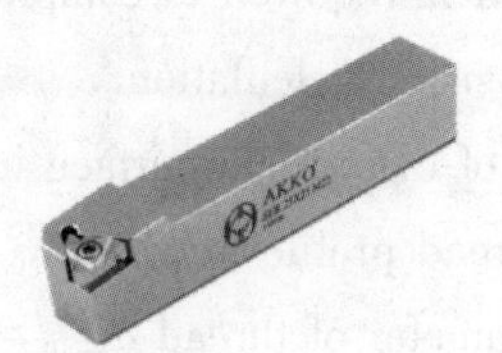

Figure 2-87 Indexable thread turning tool

Note: The rigidity of the thread knife extended too long will reduce the quality of the part, as shown in Figure 2-88.

Figure 2-88 Setting of the indexable external thread turning tool

2.5.3.2 Tool Setting Operation and Tool Compensation Input of External Threading Cutter

(1) Install workpiece.

(2) Install thread cutter.

(3) Switch the working mode of machine tool to handwheel mode, input code M03 S400 cycle start in MDI to make the spindle rotate forward.

(4) Operate the hand wheel to make the external thread turning tool close to the workpiece.

(5) External thread tool to Z-axis.

Align the thread tip with the end face of the workpiece, as shown in Figure 2-89, and then click [OFS/SET] to enter [OFFS] function in [GEOM], move the "cursor" to the "compensation" position where the tool is, input "Z0", and then click [MEASUR] to complete the Z-axis tool setting.

Figure 2-89 External thread tool to Z-axis

(6) External thread tool to X-axis. Use vernier calipers to measure the cylindrical surface of the part, record the measured value (such as X39.32), then click the spindle forward rotation on the operation panel, shake the hand wheel to slightly contact the thread tool tip with the cylindrical surface of the workpiece, when there is a "bright line" on the cylindrical surface, enter the [GEOM] function in [OFFS] again, corresponding to the tool compensation number, input "X39.32", and then click [MEASUR], as shown in Figure 2-90, Figure 2-91.

Figure 2-90 Diameter measurement

Figure 2-91 External thread tool to X axis

2.5.4 Measurement

2.5.4.1 Thread Measuring Tools

(1) Pitch gauge.

Pitch gauge is mainly used for the inspection of pitch and profile angle of low-precision screw workpiece, as is shown in Figure 2-92. All working surfaces of the thread template shall be free of rust, bruise, burr and other defects affecting the use or appearance quality. The connection between the sample plate and the guard plate should be able to make the sample plate rotate smoothly around the axis without jamming or loosening.

Figure 2-92 Pitch gauge

(2) Threading gauge.

There are two kinds of thread gauges: plug gauge and ring gauge, as is shown in Figure 2-93. Thread gauges have a through end and a stop end. The through end is indicated by letter "T". The stop is indicated by the letter "Z". The thread go gauge has a complete profile, and the thread length is equal to the screw in length of the thread to be measured. The thread stop gauge has a truncated profile, and the thread length is shorter than the thread length of the go gauge. Thread gauge is suitable for mass production and inspection of parts.

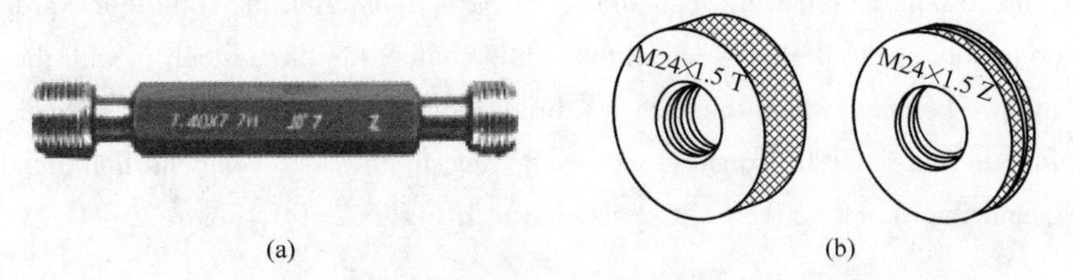

Figure 2-93 Threading gauge

(a) Threading plug gauge; (b) Threading ring gauge

(3) Thread micrometer.

The scale line and reading mode of the thread micrometer can refer to the outside micrometer. The difference is that the thread micrometer has two sets of measuring heads: 60° and 55°, which are suitable for different thread angles and different pitches. The measuring heads can be replaced according to the needs of measurement. They are suitable for the processing and testing of single thread and have strong universality, as shown in Figure 2-94.

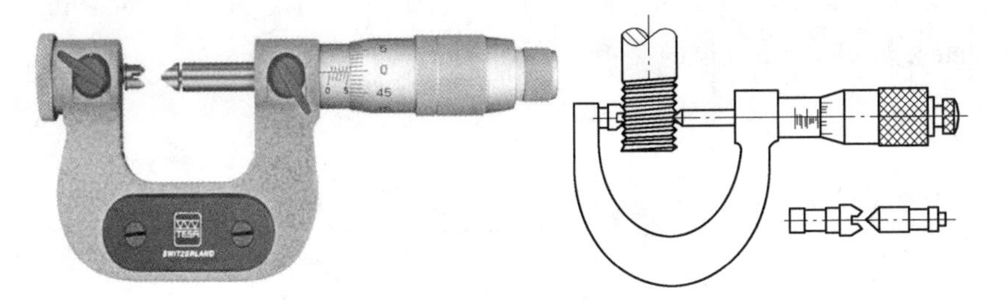

Figure 2-94 Thread micrometer

2.5.4.2 The Method of Measuring Thread

(1) Pitch measurement.

For the thread with general accuracy requirements, the pitch is usually measured with vernier caliper and pitch gauge. The measurement methods are as follows:

1) When measuring the pitch, as shown in Figure 2-95, use the pitch gauge as a template and clamp it on the thread part to be measured. If it is not sealed, replace it with another one until it is sealed. At this time, the number recorded on the pitch gauge is the pitch of the thread part to be measured. The pitch gauge should be clamped on the thread profile by using the working part length of the thread as much as possible to make the measurement result more correct.

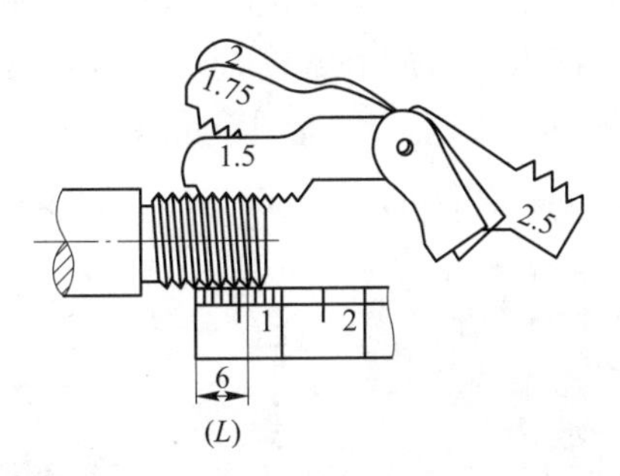

Figure 2-95 Measuring method of pitch gauge

2) When measuring the profile angle, put the parts together with the matching pitch gauge, and then check the contact between them. If there is uneven gap and light transmission, it means that the profile angle of the thread to be measured is not accurate. This measurement method can only roughly judge the profile angle tolerance.

(2) Measurement of pitch diameter.

Calculation formula of pitch diameter of common triangle thread $d=d-0.65P$.

1) Select the corresponding side head according to the thread profile angle, install it on the thread micrometer and calibrate it.

2) Select 2、3 times of measurement to ensure that the error value is within the allowable range. (for pitch diameter tolerance of thread, please refer to mechanical design manual).

3) Use the thread micrometer to measure, and the reading method is the same as that of the outside micrometer, as shown in Figure 2-96.

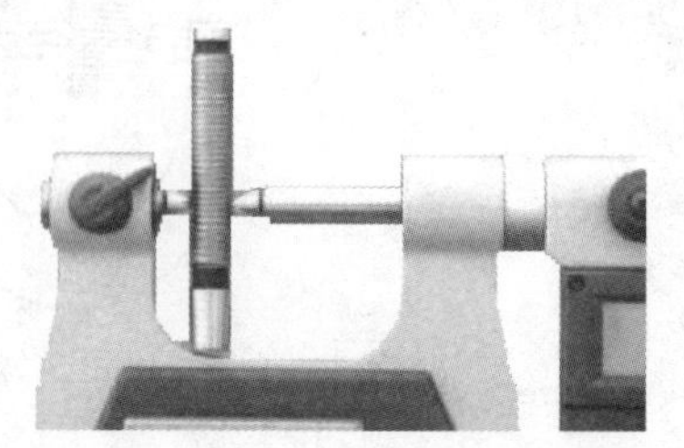

Figure 2-96 Measurement method of thread micrometer

Caculation formula of pitch diameter for ordinary triangular thread: $d_{middle} = d - 0.65P$

Among them d——Nomindl diameter;

P——Pitch.

(3) Comprehensive measurement.

The comprehensive measurement is mainly to measure the thread with thread gauge, which is suitable for the thread detection of mass production and has the advantage of high measurement efficiency.

During the measurement, if the thread to be measured can freely screw through with the thread go gauge, and cannot screw in with the thread stop gauge or no more than 2 pitches, it indicates that the effective pitch diameter of the thread to be measured does not exceed the pitch diameter of its maximum solid profile. The applied pitch diameter does not exceed the pitch diameter of the minimum solid profile, and the thread tested is qualified, as shown in Figure 2-97.

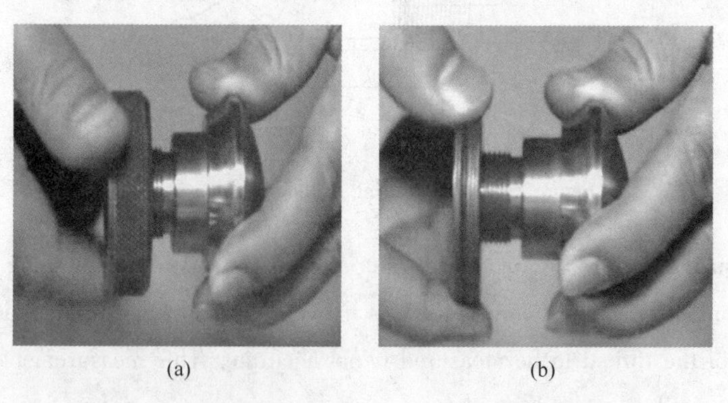

(a) (b)

Figure 2-97 Measurement method of thread ring gauge

(a) General gauge measurement; (b) Stop gauge measurement

Task Implemention

(1) Step 1 Drawing analysis:

The surface of the part to be machined includes M24×2 thread、$\phi 30^{0}_{-0.025}$、$\phi 38^{0}_{-0.039}$、$\phi 30^{0}_{-0.033}$

outer circle surface, $R10$ arc, $R2$ arc, 5mm thread undercut, $C2$ chamfer, etc. , and the surface roughness is $Ra3.2$.

(2) Step 2 Making the machining plan:

1) Machining plan

① Use three-jaw self centering chuck to install the clamp, the parts extend out of the chuck about 58mm, and aligned.

② For tool setting, set the center of the right end face of the part as the programming origin.

③ Rough turning $\phi30^{0}_{-0.025}$、$\phi38^{0}_{-0.039}$ outer cylindrical surface and M24×2 large diameter cylindrical surface at the right end of parts.

④ The outer contour of the right end of the finish turning part shall meet the dimensional requirements.

⑤ Turn 5mm undercut.

⑥ Turn M24×2 thread.

⑦ Turn around the $\phi30^{0}_{-0.025}$mm stepped shaft at the right end of the clamping part, and align the dial indicator to ensure the coaxiality of both ends is 0.04mm.

⑧ Rough turning $R10$ ball head and $\phi30^{0}_{-0.033}$mm cylindrical surface.

⑨ Finish turning the left end of the part to the dimension requirements to ensure the length and dimension of the part.

2) Tool selection

T01 90° cemented carbide external circle deviation tool (tool tip angle 80°).

T02 undercut tool (insert width 3mm).

T03 CNC thread cutter (60° blade angle).

The machining tool card for threaded shaft parts, as shown in Table 2-34.

Table 2-34 CNC machining tool card

Part name		Threaded shaft		Drawing No.	2-73		
S No.	Tool No.	Tool name	Quantity	Machined surface	Tool nose radius R/mm	Tool tip orientation T	Note
1	T01	90° external turning tool	1	Rough and finish turning the outer diameter	0.4	3	
2	T02	Grooving tool	1	Grooving			3mm
3	T03	60° threading tool	1	Turning thread			
Edit		Check		Approve	Date	Total 1 Page	Page1

3) Machining process

The machining card of threaded shaft is shown in Table 2-35.

Table 2-35 CNC machining process card

Company name	Tianjin polytechnic college		Part name		Drawing N	
			Threaded shaft		2-37	
Program No.	Fixture name	Machine tool	CNC system		Workshop	
0131	Three jaw chuck	CKA6140	FANUC SERIES Oi		CNC training center	
Steps	Contents	Tool	Spindle speed n/r · min^{-1}	Feed rate F/mm · r^{-1}	Back engagement a_p/mm	Note
1	Extend the clamping parts 58mm and align them					Manual
2	Setting external turning tool	T01				
3	Setting grooving tool	T02				
4	Setting threading tool	T03				
5	Rough turning right end outer diameter, leaving 0. 5mm allowance	T01	600	0. 2	1. 5	0251
6	Finish turning the right outer diameter	T01	1200	0. 1	0. 25	
7	Grooving	T02	400	0. 05	3	
8	Threading	T03	400	2	$0.05<a_p<0.4$	
9	Turn to another side and clamp					Manual
10	External cirular effset knife	T01				0252
11	Rough turning left end outer diameter, leaving 0. 5mm allowance	T01	600	0. 2	1. 5	
12	Finish turning the left outer diameter	T01	1200	0. 1	0. 25	
Edit	Check	Approve	Date		Total 1 Pages	Page1

(3) Step 3 Programming:

The program right end of threaded shaft is shown in Table 2-36.

Table 2-36 The Program for right end of threaded shaft

0251;	Program No.
G40 G97 G99 M03 S600 F0. 2;	Cancel the tool radius compensation, cancel the constant rotation speed of the spindle, set the feed rate per revolution, the spindle rotates forward, the rotation speed is 600r/min, and the feed rate is 0. 2mm/r
T0101;	Replace T01 external turning tool, substitute tool compensation No. 01
M08;	Turn coolant on
G42 G00 Z5. 0;	Set the right compensation of tool tip radius, fast feed to the starting point of rough turning cycle
X41. 0;	
G71 U1. 5 R0. 5;	Set rough turning cycle of outer circle
G71 P10 Q20 U0. 5 W0. 05;	

Continued Table 2-36

N10 G00 X0. 0;	Outer profile of the right end of the finished part
G01 Z0. 0;	
X23. 8 C2. 0;	
Z-20. 0;	
X29. 987 C0. 5;	
Z-42. 0;	
X37. 981 C0. 5;	
N20 Z-54. 0;	
G40 G00 X150. 0;	Cancel tip radius compensation and return to tool change point
Z150. 0;	
M05;	Spindle stop, program stop
M00;	
M03 S1200 F0. 1;	The main chuck rotates forward, the rotating speed is 1200r/min, and the fine turning feed rate is set to 0. 1mm/r
T0101;	Substitute tool compensation No. 01
G42 G00 X41. 0;	Set the right compensation of tool tip radius, fast feed to the starting point of finishing cycle
Z5. 0;	
G70 P10 Q20;	Set the outer finishing circle
G40 G00 X150. 0;	Cancel tip radius compensation and return to tool change point
Z150. 0;	
M05;	Spindle stop, program stop
M00;	
M03 S400 F0. 05;	The main chuck rotates forward, the rotation speed is 400r/min, and the feed rate of groove cutting is set to 0. 05mm/r
T0202;	Replace T02 grooving tool, substitute tool compensation No. 02
G00 X26. 0;	Machining 5×3 groove and *C*2 left chamfer
Z-20. 0;	
G01 X18. 2;	
X26. 0;	
W2. 0;	
X18. 2;	
X26. 0;	
W2. 0;	
X23. 8;	
U-4. 0 W-2. 0;	
X18. 0;	
Z-20. 0;	

Continued Table 2-36

U0. 5 W0. 25;	Tool back up
G00 X150. 0;	Return to the change point
Z150. 0;	
M05;	Spindle stop, program stop
M00;	
M03 S400 T0303;	The spindle rotates forward, the rotating speed is 400r/min, and change T03 thread turning tool, substitute tool compensation No. 03
G00 X26. 0;	Fast feed to M24×2 thread cycle starting point
Z4. 0;	
G76 P020060 Q50 R0. 1;	Set G76 thread cycle compound cycle to machine M24×2 thread
G76 X21. 4 Z-17. 0 P1300 Q400 F2. 0;	
G00 X150. 0;	Return to tool change point
Z150. 0;	
M09;	Turn coolant off
M30;	End of program and return to program header

The program for left end of threaded shaft part is shown in Table 2-37.

Table 2-37 The program for left end of threaded shaft part

O252;	Program No.
G40 G97 G99 M03 S600 F0. 2;	Cancel the tool radius compensation, cancel the constant rotation speed of the spindle, set the feed rate per revolution, the spindle rotates forward, the rotation speed is 600r/min, and the feed rate is 0. 2mm/r
T0105;	Change T01 external turning tool, substitute tool compensation No. 05
M08;	Turn coolant on
G42 G00 Z8. 0;	Set the right compensation of tool tip radius, fast feed to the starting point of rough turning cycle
X41. 0;	
G71 U1. 5 R0. 5;	Set rough outer turning cycle
G71 P10 Q20 U0. 5 W0. 05;	
N10 G00 X00;	Finish turn the left outer profile
G01 Z0. 0;	
G03 X20. 0 Z-10. 0 R10. 0;	
G01 X29. 983 C0. 5;	
X33. 983 R2. 0;	
X38. 983 C1. 0;	
N20 W-1. 0;	

Continued Table 2-37

G40 G00 X150.0;	Cancel tip radius compensation and return to tool change point
Z150.0;	
M05;	Spindle stop, program stop
M00;	
M03 S1200 F0.1;	The main shaft rotates forward, the rotating speed is 1200r/min, and the fine turning feed is set to 0.1mm/r
T0105;	Substitute tool compensation No. 05
G42 G00 X41.0;	Set the right compensation of tool tip radius, fast feed to the starting point of finishing cycle
Z5.0;	
G70 P10 Q20;	Set the outer finishing cycle
G40 G00 X150.0;	Cancel tip radius compensation and return to tool change point
Z150.0;	
M09;	Turn coolant off
M30;	End of program and return to program start

(4) Step 4　Part machining:

1) Select a reasonable blank according to the drawing requirements and install the blank.

2) Correctly install the cutting tools required for parts processing.

3) Carry out tool setting operation and establish workpiece coordinate system.

4) Refer to Table 2-36 for the right end program of the machine tool input parts.

5) Finish the machining of the outer circle outline, groove and thread at the right end of the part according to the requirements of the drawing, as shown in Figure 2-98.

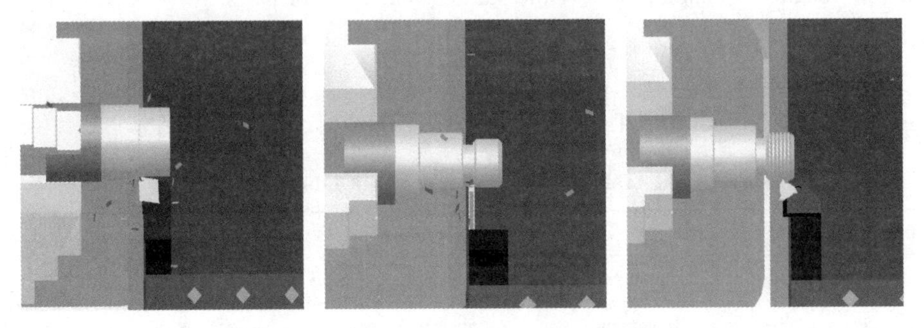

Figure 2-98　Right end machining of parts

6) Turn around and clamp, and align.

7) Set the external turning tool. After the parts turn around, the X-axis does not need to be aligned, the Z-axis is aligned, the end face is turned first, the X-axis is extended forward to withdraw the tool, and the total length of the parts (or the length direction dimension of the machined end face and the Z-axis face to be aligned) is measured with a vernier caliper. Calcula-

tion: measured value of Z-axis of tool setting = actual length of measured part - length dimension of this part in the drawing. Input the measured value to complete the tool setting. With this method, the effective value of the measured value of Z-axis after calculation is positive.

8) Refer to Table 2-37 for the left end program of machine tool input parts.

9) Finish the machining of the outer circle contour at the left end of the parts according to the requirements of the drawing, as shown in Figure 2-99.

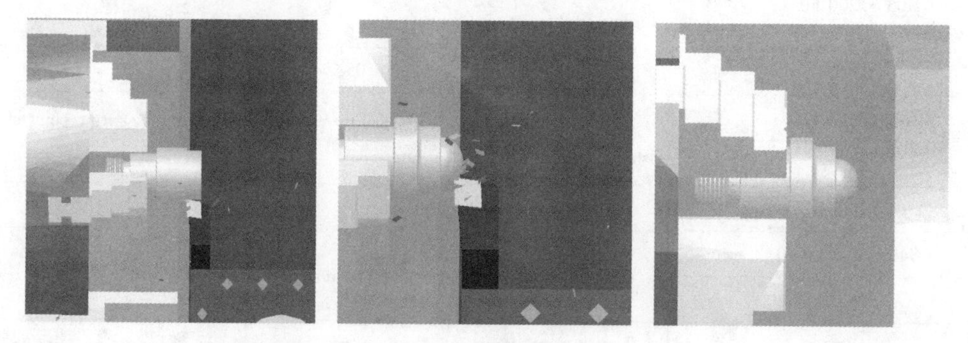

Figure 2-99 Left end machining of parts

(5) Step 5 Dimension Measurement:

1) Use 25~50mm outside micrometer to measure the outside diameter of $\phi30^{0}_{-0.025}$、$\phi38^{0}_{-0.039}$、$\phi30^{0}_{-0.033}$.

2) Use vernier caliper to measure the length of 10mm and 72mm.

3) Use vernier caliper to measure the size of 5mm groove.

4) Use thread gauge to measure M24×2 thread size.

5) Use R10 arc template to measure R10 arc.

6) Use the roughness template to test the surface roughness of parts.

Assessment

The assessment table is shown in Table 2-38.

Table 2-38 The assessment table

Item	No.	Standard		Partition	Score
Machining plan (10%)	1	Machining steps	Meet the requirements of CNC turning	3	
	2	Dimension	Calculate dimensions	3	
	3	Tool selection	The cutter selection is reasonable and meets the machining requirements	4	
Programming (30%)	4	Program No.	No program no point	2	
	5	Program segment number	No program segment number section no point	2	
	6	Cutting parameters	Unreasonable selection of cutting parameters no point	4	

Continued Table 2-38

Item	No.	Standard		Partition	Score
Programming (30%)	7	Origin and coordinates	Mark the origin and coordinates of the program, otherwise no point	2	
	8	Program contents	Deduct 2 points for each segment that does not meet the requirements of program logic and format	20	
			Deduct 5 points if the procedure content is not in accordance with the process		
			Deduct 5 points in case of dangerous codes		
Operation (20%)	9	Program input	Deduct 2 points for each wrong section	10	
	10	Machining operation 60minutes	No point for setting error of workpiece coordinate system	10	
			No points for misoperation		
			Overtime no point		
Dimension measurement (30%)	11	External profile	$\phi38^{0}_{-0.025}$ over tolerance 0.01, deduct 3 points	3	
	12		$\phi30^{0}_{-0.033}$ over tolerance 0.01, deduct 3 points	3	
	13		$\phi38^{0}_{-0.039}$ over tolerance 0.01, deduct 3 points	3	
	14		$\phi18$ over tolerance no points	2	
	15	Thread	M24×2 over tolerance no point	5	
	16	Radius10	$R10$ over tolerance no point	2	
	17	Radius2	$R2$ over tolerance no point	2	
	18	Length	72±0.05	3	
	19		10±0.03	2	
	20	Chamfer	2×$C2$	2	
	21	Chamfer	4×$C0.5$	2	
	22	Surface roughness	$Ra3.2$ over tolerance no points	1	
Vocational ability (10%)	23	Learning ability		2	
	24	Communication ability		2	
	25	The team cooperation ability		2	
	26	Safe operation and civilized production		4	
Total					

Exercises

(1) Answer the following questions:

1) Compare the similarities and differences of G92 and G76 instructions in thread processing.

2) How to select the spindle speed for machining threads with large lead?

3) If the go gauge and stop gauge of the processed external thread can't be screwed in, how to correct it?

4) If the external thread can't be screwed in with the go gauge and the stop gauge, what is the cause?

(2) Synchronous Training:

1) As is shown in Figure 2-100. The blank is a $\phi 35$ aluminum bar. Please programming and machining.

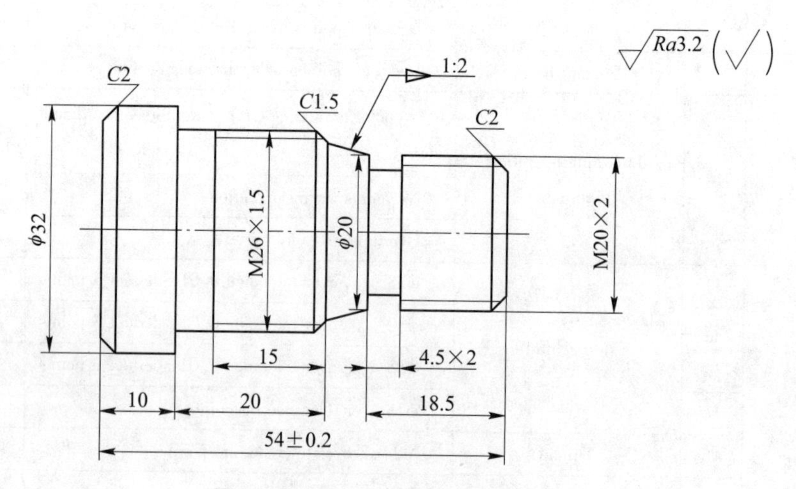

Figure 2-100 Synchronous training 1

2) As is shown in Figure 2-101. The blank is a $\phi 50 \times 69$ aluminum alloy. Please programming and machining.

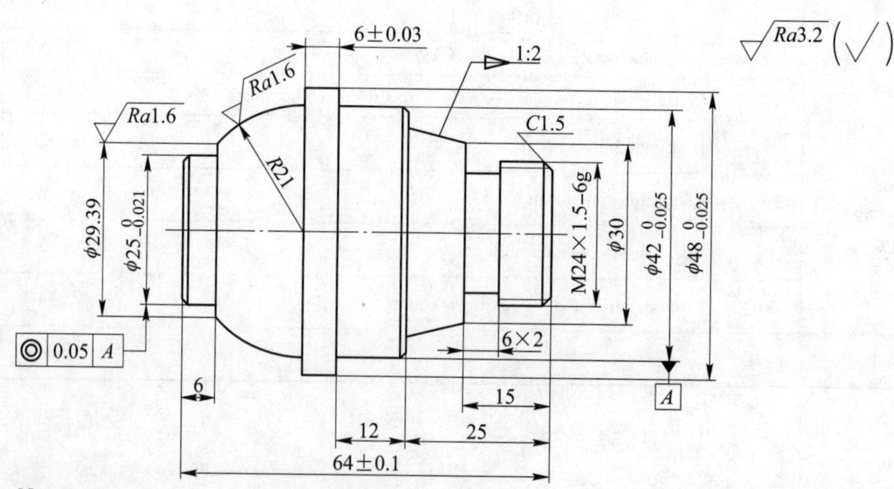

Note:
1.Unmarked chamfers are C0.5, acute angles are dull.
2.Unmarked linear dimensional tolerances should meet the requirements of GB/T 1804—2000.

Figure 2-101 Synchronous training 2

Task 2. 6 CNC Turning of Sleeve Parts

Introduction

As shown in Figure 2－102, the part is made of hard aluminum alloy LY15, and the blank is ϕ40mm×45mm. It is produced in one piece. It is coordinated with task 2. 5. This task needs to prepare the program, machine the part using the CNC lathe, and then measure the part.

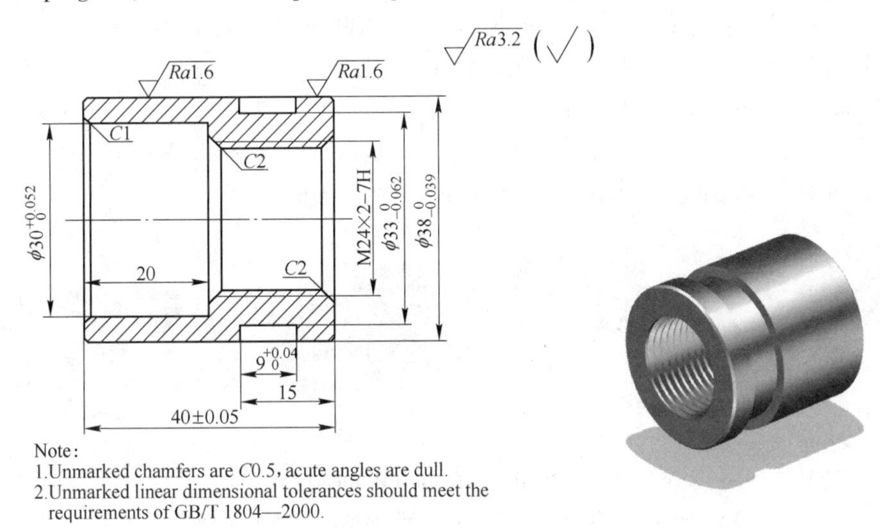

Figure 2－102 The part drawing

Competences

(1) Knowledge:

1) Familiar with the boring process technology and the selection of related cutting tools and measuring tools in lathes.

2) Familiar with the internal thread processing technology of lathes and the selection of related cutting tools and measuring tools.

3) Explain the application of G71 and G76 instructions in the processing of sleeve parts.

(2) Skill:

1) Have the ability to analyze the processing technology of sleeve parts.

2) Have the ability to program boring and turning threaded holes.

3) Have the ability to use lathe tailstock to drill holes correctly.

4) Understand the inspection methods of inner hole and inner thread.

5) Have the ability to ensure good parts fitting.

(3) Quality:

1) Cultivate safety awareness and standardized operation awareness.

2) Cultivate the craftsman spirit of keeping improving according to the work pieces.

Relevant Knowledge

2.6.1 Process Preparation

2.6.1.1 Tool Selection for Sleeve Parts

The machining surface of the sleeve parts include both the outer surface and the inner surface. The outer surface machining tools are consistent with previous tasks. The inner surface machining tools are commonly used in boring tools, center drills, twist drills and accessories. When the inner surface has threads, internal threads cutters will be used.

(1) Common drilling tools. The center drill is shown in Figure 2-103 (a), the twist drill bit is shown in Figure 2-103 (b), the drill chuck for installation is shown in Figure 2-103 (c), the morse reducer for installation of taper shank is shown in Figure 2-103 (d).

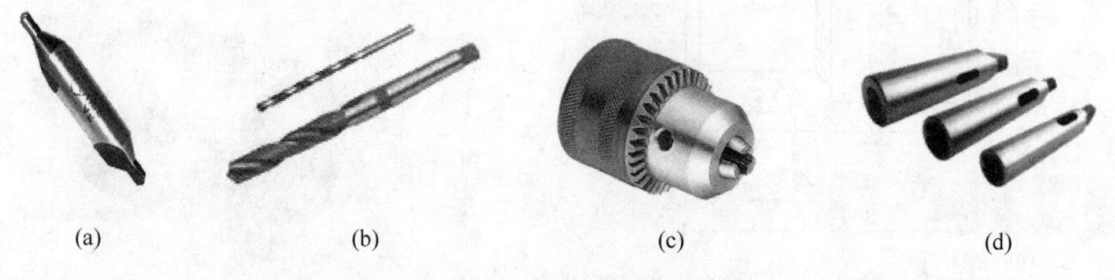

(a) (b) (c) (d)

Figure 2-103 Common drilling tools

(a) Center drill; (b) Twist drill; (c) Drill chuck; (d) Mo tailstock sleeve

(2) Boring tool. Boring tool is the most commonly used for making internal hole in lathe, which can guarantee the processing accuracy well, as is shown in Figure 2-104.

(3) Internal thread turning tool. The machining principle is the same as that of external thread turning tool, as is shown in Figure 2-105.

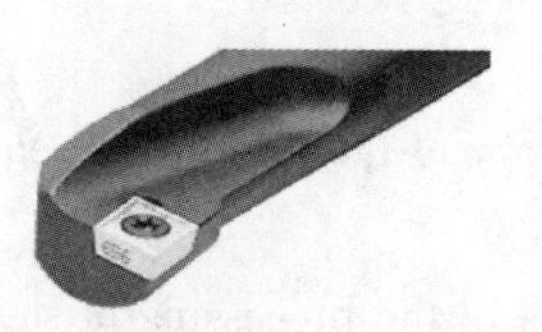

Figure 2-104 Boring tool

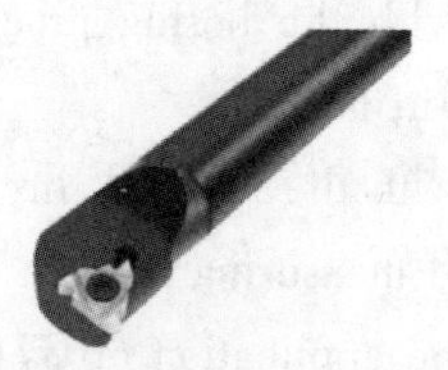

Figure 2-105 Internal thread turning tool

(4) Lathe tailstock. As shown in Figure 2-106, the lathe tailstock can move forward and backward in Z direction by shaking the hand wheel 1, the tailstock can be fixed by locking the tailstock fixed handle 2, the tailstock can be fixed by locking the spindle clamping handle 3.

Lathe tailstock mainly has the following three functions;

1) Eccentric shaft can be turned by tailstock.

2) Use tailstock to adjust lathe precision during installation.

3) The tailstock can be used to drill and center holes, and the long shaft can be machined

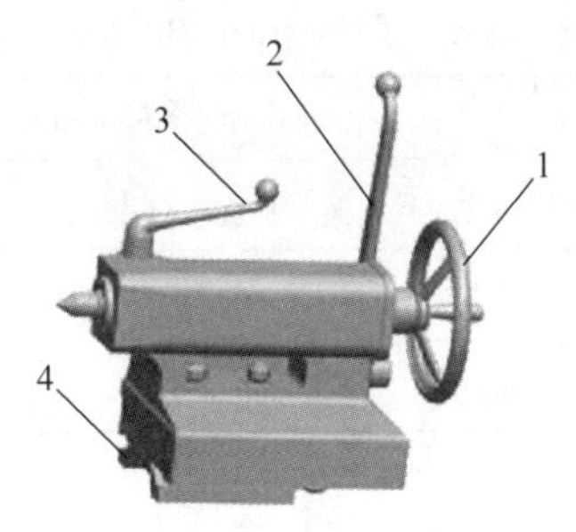

Figure 2-106 Lathe tailstock

1—Main spindle wheel; 2—Tail stock locate handle; 3—Main spindle clamp handle; 4—Tail stock bottom plate

against the other end.

2.6.1.2 Selection of Cutting Parameters for Internal Surface Machining

When machining the inner surface, chip removal is difficult, the tool bar extends long, the tool head is weak, so the rigidity is low and vibration is easy to occur. Therefore, the cutting amount on the inner surface is lower than that on the outer surface.

2.6.1.3 Arrangement of Machining Steps for Inner Surface

To machine the inner surface on solid material, first turn the end face, then drill the center hole (the center hole may not be drilled when the accuracy of the bottom hole is not high), then use a suitable drill bit to drill manually, use a suitable boring tool to machine the inner contour surface, and finally turn the internal thread.

2.6.2 Programming

2.6.2.1 Tool Compensation

(1) T value of tool tip position. The tool tip position T of internal turning tool is taken as 2.

(2) Tool compensation codes. For front tool holder, use G41 code to compensate the inner tool.

2.6.2.2 Stock Removal in Turning G71

(1) Format:

G71 U(Δd)R(e);

G71 P(ns)Q(nf)U(Δu)W(Δw);

(2) Parameter description is the same as before.

(3) Note: when machining the inner surface, Δu is negative.

2.6.2.3 The Change of the Point the Tool Move On and Back

To ensure the safety of the actual processing, it is necessary to change the feed and withdraw points of turning tools, as is shown in Table 2-39.

Table 2-39　The change of the point the tool move on and back

No.	Contents	External profile programming	Internal profile programming
1	Feed point (point before G71/G70) *X* value	Workpiece blank diameter	Hole diameter for drilling
2	Feed programming, *X*, *Z* sequence	First *Z*-axis, then *X*-axis	First *X*-axis, then *Z*-axis
3	Tool back programming, *X*, *Z* sequence	First *X*-axis, then *Z*-axis	First *Z*-axis, then *X*-axis

(1) When programming the outer contour, the *X* value of feed point is the diameter of workpiece blank, while when programming the inner hole, the position of machining feed point is the hole diameter.

(2) When programming the outer contour, it is recommended to run the *Z*-axis first, then the X axis. When programming the inner contour, the *X*-axis first, then the *Z*-axis.

(3) When programming the outer contour, it is recommended to input *X* value first, then *Z* value. For programming the inner contour, it is suggested to input *Z* direction first, and then feed in *X* direction.

Note: when the G70 command is used to program the inner hole, the tool advance and retreat modes are also changed according to G71 programming.

【Example 2-15】The part is shown in Figure 2-107. The outer surface has been machined to the size requirements. It is required to program for the internal contour. The reference processing program is shown in Table 2-40.

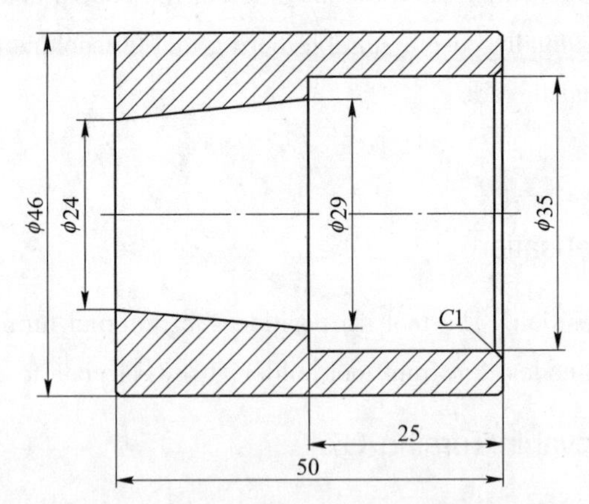

Figure 2-107　Example for the inner surface programming

Table 2-40　The inner surface programming

O266;	Program No.
G40 G97 G99 M03 S400 F0. 1;	Cancel the cutter radius compensation, cancel the constant rotation speed of the spindle, set the feed rate per revolution, the spindle rotates forward, the rotation speed is 400r/min, and the feed rate is 0. 1mm/r
T0202;	Replace T02 boring tool, substitute tool compensation No. 02
M08;	Open cutting fluid

Continued Table 2-40

G00 X20. ;	Fast feed to the starting point of rough turning cycle in X direction (drilling diameter position)
Z5. ;	Fast feed to the starting point of rough turning cycle in Z direction
G71 U1. 5 R0. 5;	Set the rough turning cycle of the inner hole, with the cutting depth of 1. 5mm and the tool withdrawal of 0. 5mm
G71 P30 Q40 U−0. 5 W0. 05;	The finishing route is specified by N10 ~ N20, with a finishing allowance of 0. 5mm in X direction and 0. 05mm in Z direction
N30 G00 X37. 0;	Left end finishing inner profile
G01 Z0. ;	
X35. 0 C1. 5. ;	
Z−25. 0;	
X29. 0;	
X24. 0 Z−50. 0;	
N40 X20. ;	
G00 Z150. ;	Quickly back to tool changing point
X150. ;	
M09;	Close cutting fluid
M05;	Spindle stop
M00;	Program pause
M03 S1200 F0. 08;	The main spindle rotates forward, the rotating speed is 1200r/min, and the set feed rate is 0. 08mm/r
T0202;	Substitute tool compensation No. 02
M08;	Open cutting fluid
G00 X20. ;	Fast feed to start of finishing cycle
Z5. ;	
G70 P30 Q40;	Finishing compound cycle
G00 Z150. ;	Quickly back to tool changing point
X150. ;	
M30;	The program ends and the cursor returns to the program start

2. 6. 2. 4 Multiple Thread Cutting Cycle G76

(1) G76 code is preferred for internal thread machining.

G76 code uses oblique cutting method to cut thread, that is to say, the thread tool always uses one cutting edge to cut, which reduces cutting resistance, improves the tool is life and thread finishing quality. In addition, the code has high machining efficiency, so G76 code is preferred for machining internal thread.

(2) Format.

$$G76\ P(m)(r)(\alpha)Q(\Delta dmin)R(d);$$
$$G76\ X(U)Z(W)R(i)P(k)Q(\Delta d)F(L);$$

The parameter description is the same as above.

(3) Calculation of internal thread dimension.

1) The diameter of the inner cylindrical surface actually turned:

Plastic material: $D_{actual} = D - P$

Brittle material: $D_{actual} = D - 1.05P$

2) Thread profile height: $h_{teech} = 0.65P$

3) X-direction end coordinate of internal thread is the nominal diameter of the thread: $D_{bottom} = D$

【Example 2-16】The part is shown in Figure 2-108. The outer contour and the threaded bottom hole have been turned to the required size. It is required to program for M28×1.5 internal thread.

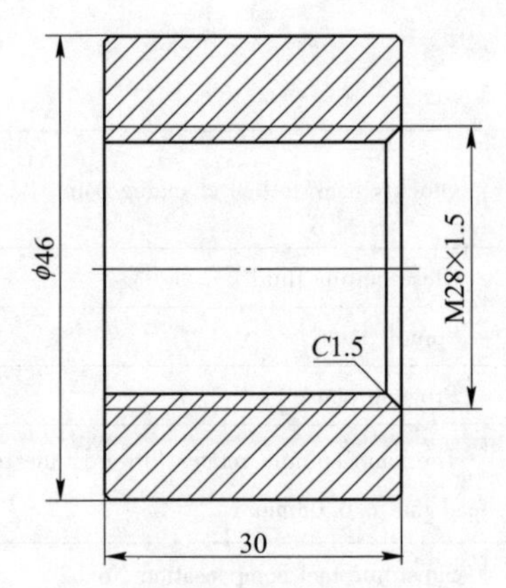

Figure 2-108 Example for internal thread programming

(1) Dimension calculation.

Diameter of the inner cylindrical surface in actual turning: $D_{actual} = D - P = 28 - 1.5 = 26.5$mm

Actual thread profile height: $d_{teech} = 0.65P = 0.65 \times 1.5 = 0.975$mm

Actual diameter of thread: $D_{bottom} = D = 28$mm

(2) Thread parameters.

Repeat the fine turning twice, $m = 02$.

There is no chamfer at the end of thread, $R = 00$.

The sharp angle of triangular thread tool is 60°, $\alpha = 60°$.

The minimum turning depth δdmin is 0.03mm, $Q = 30\mu$m.

Leave 0.05mm finishing allowance, $R = 0.05$mm.

Calculate the end point coordinate of thread according to the part drawing ($X28.0$, $Z-31.0$).

Cylindrical thread i is 0, which can be omitted.

The thread height K is 0.975mm, P is 975μm.

The first turning depth ΔD is 0.4mm, Q is 400μm.

The pitch of single head thread is 1.5, $F=1.5$.

(3) The program.

The M28×1.5 thread program is shown in Table 2-41.

Table 2-41 The M28×1.5 thread program

0263;	Program No.
G40 G97 G99 M03 S400;	The spindle is turning at a speed of 400r/min
T0404;	Replace the T04 internal thread turning tool, substitute tool compensation No. 04
M08;	Open cutting fluid
G00 X25.0;	Fast feed to start of machining cycle
Z5.;	
G76 P020060 Q30 R0.05;	Compound cycle of thread cutting
G76 X28. Z-31. P975 Q400 F1.5.;	
G00 Z150.;	Quickly back to tool changing point
X150.;	
M30;	The program ends and the cursor returns to the program start

2.6.3 Operation

2.6.3.1 Install the Drill and Drilling

(1) The installation methods.

1) Straight shank twist drill installation: generally, a straight shank twist drill is clamped with a drill chuck, and the taper shank of the drill chuck is inserted into the taper hole of the tailstock of the lathe.

2) Installation of taper shank twist drills: Taper shank twist drills can be inserted into the taper hole of the tailstock of the lathe directly or with a Morse transition taper sleeve (reducing sleeve).

(2) Drilling operations.

1) Machining process: first flat end surface, then drill the center hole, finally drill. When the accuracy is not high, the center hole may not be drilled.

2) Precautions for drilling.

① The end face must be leveled before drilling.

② Align the drill axis with the workpiece rotation axis.

③ When the drill bit is about to penetrate through the end face of the workpiece and the drilled through hole, the feed rate should be small to prevent the drill bit from breaking.

④ When drilling small and deep holes, first use the center drill to drill the center hole to avoid skewing the hole.

⑤ When drilling deep holes, the chips are not easy to discharge, and the drill must be pulled out frequently to clear the chips.

⑥ When drilling steel, sufficient cutting fluid must be poured to cool the drill bit. Cutting fluid can be omitted when drilling cast iron.

2.6.3.2 Tool Setting of Boring Tool

(1) Installation of boring tool.

1) The tool bar should be made substantially parallel to the axis of the workpiece.

2) The extended length of the boring tool holder should be as short as possible, and generally a tool head width longer than the hole depth can be used.

3) The tool tip is at the same height or slightly higher than the center of rotation of the main shaft to prevent the tool bar from bending under the action of cutting force to produce a "blade."

(2) Tool setting operation of boring tool.

1) Trial cut the inner hole of the workpiece, as shown in Figure 2-109.

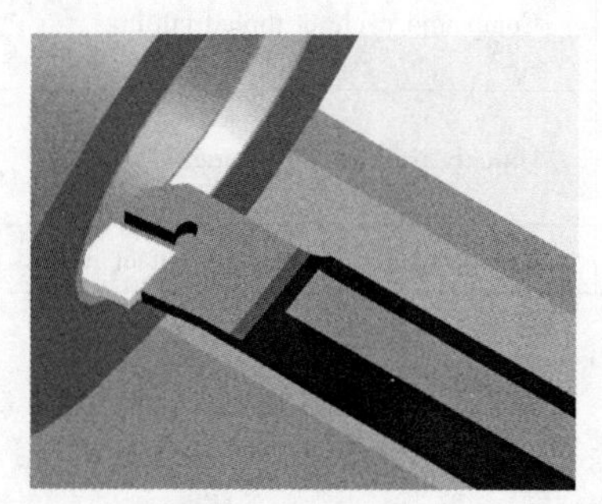

Figure 2-109 Turning the inner surface of the workpiece

2) After the tool is retracted in the Z direction, the spindle is completely stopped, and the inner hole diameter of the trial cutting is measured.

3) Set the X direction to complement and correct.

4) Trial cutting the end face.

5) Set the Z direction correction.

2.6.3.3 Tool Setting of Internal Thread Turning Tool

(1) Installation of internal thread turning tool.

1) The tool bar is substantially parallel to the axis of the workpiece.

2) The tool tip is higher or slightly higher than the center of rotation of the main shaft to prevent the tool bar from bending under the action of cutting force to produce a "blade."

3) The extension length of the internally threaded tool bar should be as short as possible to increase the rigidity of the tool bar and prevent vibration.

(2) Tool setting operation of internal thread tool.

1) Move the tool to the intersection of the inner hole and the end face, as shown in Figure 2-110.

2) Set the Z direction correction.

3) Set the X-direction correction: At the X-direction tool parameter, enter the hole diameter of

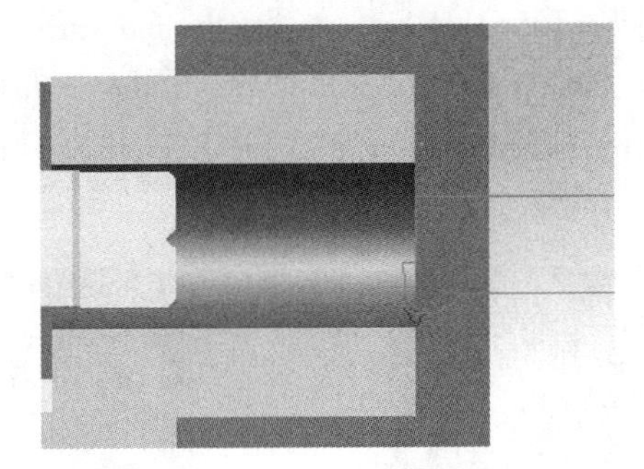

Figure 2-110　Internal thread tool setting

the boring tool after turning to complete the internal thread tool setting.

2. 6. 4　Measurement

2. 6. 4. 1　Inside Diameter Dial Indicator

(1) Structure.

The inner diameter dial indicator is a combination kit, which consists of a clock face dial indicator and a lever measuring frame. It is a measuring tool for relative measurement of inner hole, as shown in Figure 2-111.

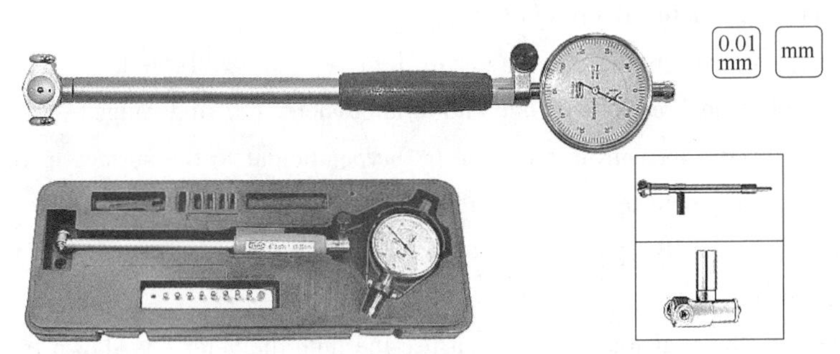

Figure 2-111　Inner diameter dial indicator

(2) Indication.

The dial is engraved with 100 equal divisions, and the scale value is 0. 01mm. When the pointer rotates for one turn, the small pointer will rotate for a small space, and the scale value of the revolution indicator is 1mm.

When you turn the bezel by hand, the dial also turns to align the pointer with any scribe line. This function is convenient for the operator to calibrate the meter.

(3) Correct the indicator.

1) Select the dial indicator and check or modify the metering label.

2) Press the dial indicator gently by hand, and observe whether the pointer rotates flexibly, as shown in Figure 2-112.

3) Check the inner diameter lever measuring frame: there is no dusk inside the measuring

frame. The contact can be ejected flexibly by pressing the measuring head.

4) Install the dial indicator, insert the dial indicator into the shaft hole of the measuring frame, compress the small pointer of the dial indicator at about 0.5mm, and fix the meter head, as shown in Figure 2-113.

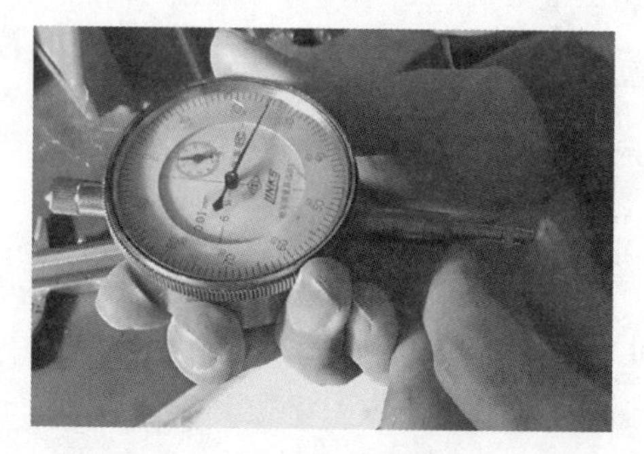

Figure 2-112 Inspection of dial indicator

Figure 2-113 Fixed indicator

5) Select, install and adjust the replaceable probe according to the measurement range.

6) Press the movable probe by hand, and the pointer of dial indicator shall rotate smoothly and flexibly.

7) Use vernier caliper to measure roughly, the length of replaceable probe should be about 0.5mm longer than the length of zero calibration.

8) Adjust the calibrated outside micrometer to the size to be calibrated.

9) Put the movable side head against one end of micrometer (or ring gauge) and put the replaceable side head into it. The measuring rod must be perpendicular to the surface of the tested part.

10) Gently shake the gauge to stop the probe at the minimum position.

11) Turn the dial so that the pointer points to zero and tighten the screw.

(4) Measurement method.

Use the inside diameter dial indicator to measure the hole diameter, as shown in Figure 2-114. During measurement, the measuring claw shall be in full contact with the inner hole to be measured.

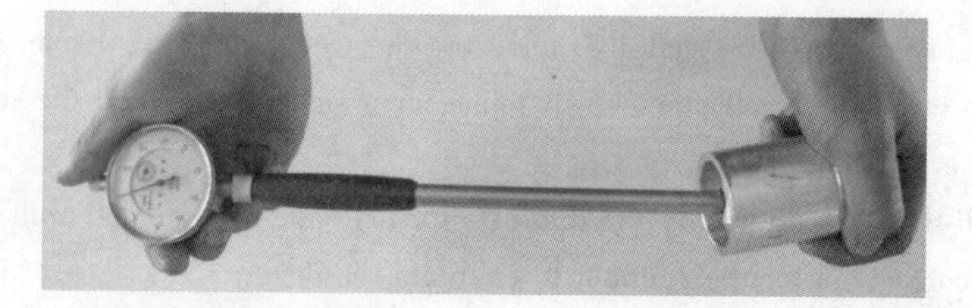

Figure 2-114 Measuring workpiece with inner diameter dial indicator

2.6.4.2 Inside Micrometer

(1) Application. The inside micrometer is mainly used to measure the apertures with high

accuracy and the width of grooves.

(2) Structure. The inside micrometer is composed of movable measuring claw, guide tube, fixed measuring claw, differential tube and ratchet knob, as shown in Figure 2-115.

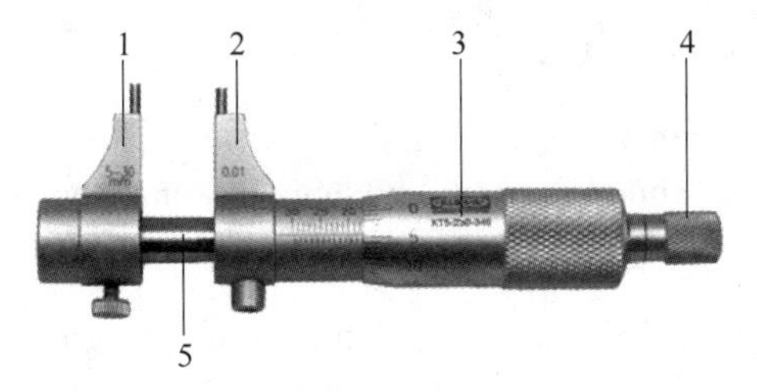

Figure 2-115 The structure of inside micrometer

1—Movable measuring; 2—Fixed measuring claw; 3—Differential tube; 4—Ratchet knob; 5—Guide tube

(3) Measurement methods. When measuring, first calibrate the zero position, then put the inside micrometer into the hole to be measured, the contact tightness is appropriate, and read out the correct diameter value.

(4) The readings and precautions are basically the same as the outside micrometer.

2.6.4.3 Thread Gauge

Gauges for measuring internal threads are generally called thread plug gauges, as shown in Figure 2-116. The end used for passing is the through end, which is indicated by the letter "T"; The end used for restricting the passing is indicated by the letter "Z" .

Figure 2-116 Thread gauge

2.6.4.4 The Method of Using the Thread Plug Gauge

(1) Use a thread gauge to screw on the measured thread. If the thread plug gauge can pass, it means that the effective diameter of the measured thread does not exceed the diameter of its largest solid tooth shape.

(2) Use the thread stop gauge to screw with the measured thread. The number of turns does not exceed two pitches, that is, the thread stop gauge is not fully pass through. The thread diameter is acceptable.

Implemention

(1) Step 1 Drawing analysis:

This part has a total length of (40±0.05) mm and is a typical set of parts. It has both outer and inner surfaces. The outer circle $\phi38^{0}_{-0.039}$ is divided into left and right sections by a groove with a diameter $\phi33^{0}_{-0.062}$ of 9mm on the bottom surface. Internal thread and surface roughness are $Ra3.2$. The 9mm wide groove on the outer surface of this part is required to match with the drawing of sub module 5.

(2) Step 2 Making the machining plan:

1) Machining plan:

① Three-jaw self centering chuck is used for clamping, and the parts extend about 35mm out of the chuck.

② Machine the outer contour $\phi38^{0}_{-0.039}$ of the left end of the part to the required size.

③ Machine the inner contour $\phi30^{+0.052}_{0}$ of the left end of the part to the size requirements. The left end finished workpiece is shown in Figure 2-117.

④ Turn around the part for clamping, and the length of the parts for clamping is about 20mm. The workpiece clamping is shown in Figure 2-118.

Figure 2-117 Left end of finished part

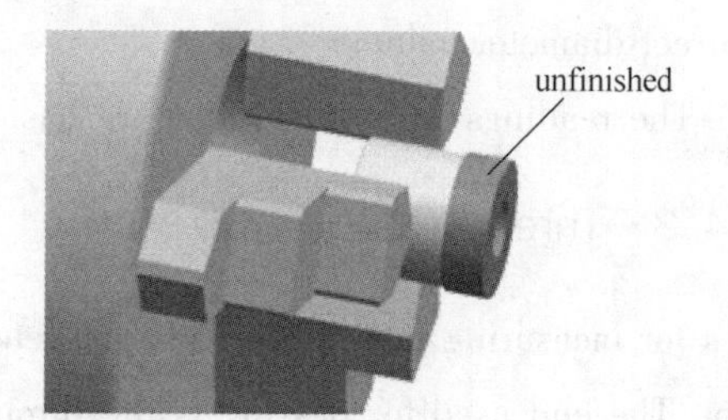

Figure 2-118 Schematic diagram of workpiece clamping

⑤ Align the workpiece.

⑥ Machine the outer contour $\phi38^{0}_{-0.039}$ of the right end of the part to the required size, and ensure the total length of the workpiece (40±0.05)mm.

⑦ Machine the groove with a bottom diameter $\phi33^{0}_{-0.062}$ and the width of 9mm.

⑧ Machine the bottom hole of M24×2-7H internal thread at the right end of the part to the size requirements.

⑨ Machine internal thread.

2) Tool selection:

This part is produced in a single piece. In order to save tool changing time and reduce processing costs, the same external turning tool is used for the end face, outer rough machining and outer finishing during contour machining. The inner hole and inner hole step roughing and finishing The same inner hole turning tool is also used for machining. The CNC machining tool card is shown in Table 2-42.

T01 carbide eccentric external turning tool blade with 55° tip angle.

T02 grooving tool, blade width 4mm.

T03 boring tool, tool diameter 16mm.

T04 internal thread tool, tool diameter 16mm.

Table 2-42 CNC machining tool card

Part name		Sleeve part		Drawing No.	2-102		
S No.	Tool No.	Tool name	Quantity	Machined surface	Tool nose radius R/mm	Tool tip orientation T	Note
1	Manual	Center drill	1	Drilling the center			A3
2	Manual	Twist drill	1	Drilling			ϕ20
3	T01	90° external turning tool	1	Rough and finish turning the outer diameter	0. 4	3	
4	T02	Boring tool	1	Rough and finish the inner dimension	0. 4	2	ϕ16
5	T03	Grooving tool	1	Grooving	0. 4	0	4mm
6	T04	Inner threading tool	1	Turn internal thread	0	0	ϕ16
Edit		Check		Approve		Date		Total 1 Page	Page1

3) Machining process:

CNC machining process card of the left end is shown in Table 2-43.

Table 2-43 CNC machining process card of the left end

Company name	Tianjin polytechnic college			Part name		Drawing No.	
				Sleeve part		2-102	
Program No.	Fixture name		Machine tool	CNC system		Workshop	
O61	Three jaw chuck		CKA6140	FANUC SERIES 0i		CNC training center	
Steps	Contents		Tool	Spindle speed n/r · min^{-1}	Feed rate F/mm · r^{-1}	Back engagement a_p/mm	Note
1	Extend the clamping parts 35mm and align them						Manual
2	Drill center		Manual	1200	0. 05		A3
3	Drill the hole		Manual	600	0. 05		ϕ20
4	Rough turning left end outer diameter, leaving 0. 5mm allowance		T01	800	0. 2	2	
5	Finish the outer diameter of the left end to reach the required size		T01	1500	0. 1	0. 25	
6	Rough turning the inner diameter of the left end, leaving 0. 5mm allowance		T02	400	0. 1	1. 5	
7	Finish the inner contour of the left end to reach the required size		T02	1200	0. 08	0. 25	
Edit		Check		Approve		Date		Total 2 Pages	Page1

CNC machining process card of right end is shown in Table 2-44.

Table 2-44 CNC machining process card of right end

Company name	Tianjin polytechnic college				Part name		Drawing No.	
					Sleeve part		2-102	
Program No.	Fixture name		Machine tool		CNC system		Workshop	
O62	Three jaw chuck		CKA6140		FANUC SERIES 0i		CNC training center	
Steps	Contents			Tool	Spindle speed n/r · min^{-1}	Feed rate F/mm · r^{-1}	Back engagement a_p/mm	Note
1	Turn to another side and clamp							Manual
2	Rough turning the outer diameter of the left end, leaving 0.5mm allowance			T01	800	0.2	2	
3	Finish the outer diameter of the left end to reach the required size			T01	1500	0.1	0.25	
4	Grooving with width of 9mm			T03	400	0.05	4	
5	Rough turning the inner diameter of the right end, leaving 0.5mm allowance			T02	400	0.1	1.5	
6	Finish the inner diameter of the right end to reach the required size			T02	1200	0.08	0.25	
7	Turning internal thread			T04	400	2	0.3	
Edit	Check		Approve		Date		Total 2 Pages	Page2

(3) Step 3 Programming:

The program for the left end is shown in Table 2-45.

Table 2-45 The program for the left end

O61;	Program No.
G40 G97 G99 M03 S800 F0.2;	Cancel the cutter radius compensation, cancel the constant rotation speed of the spindle, set the feed rate per revolution, the spindle rotates forward, the rotation speed is 800r/min, and the feed rate is 0.2mm/r
T0101;	Replace T01 external turning tool, substitute tool compensation No. 01
M08;	Cutting fluid drive
G00 Z5.0;	Fast feed to the starting point of rough turning cycle in Z direction
X42.0;	Fast feed to the starting point of rough turning cycle in X direction (blank diameter plus 2mm position)
G71 U1.5 R0.5;	Set the rough turning cycle of the outer circle, the cutting depth is 1.5mm, and the tool withdrawal is 0.5mm

Continued Table 2-45

G71 P10 Q20 U0. 5 W0. 05;	The finishing route is specified by N10 ~ N20, with a finishing allowance of 0. 5mm in X direction and 0. 05mm in Z direction
N10 G00 X36. 981;	Left end finishing outer profile
G01 Z0. ;	
X37. 981 C0. 5;	
Z-30. ;	
N20 X42. ;	
G00 X150. ;	Quickly back to tool change point
Z150. ;	
M09;	Close cutting fluid
M05;	Spindle speed off
M00;	Program pause
M03 S1500 F0. 1;	The main spindle rotates forward, the rotating speed is 1500r/min, and the feed rate is 0. 1mm/r
T0101;	Substitute tool compensation No. 01
M08;	Open cutting fluid
G00 Z5. ;	Fast feed to start of finishing cycle
X42. ;	
G70 P10 Q20;	Finishing codes cycle
G00 X150. ;	Quickly back to tool changing point
Z150. ;	
M09;	Close cutting fluid
M05;	Spindle stop
M00;	Program pause
M03 S400 F0. 1;	The main spindle rotates forward, the rotation speed is 400r/min, and the set feed rate is 0. 1mm/r
T0202;	Change T02 boring tool, substitute tool compensation No. 02
M08;	Open cutting fluid
G00 X20. ;	Fast feed to the starting point of rough turning cycle in X direction (drilling diameter position)
Z5. ;	Fast feed to the starting point of rough turning cycle in Z direction
G71 U1. 5 R0. 5;	Set the rough turning cycle of the inner hole, with the cutting depth of 1. 5mm and the tool withdrawal of 0. 5mm
G71 P30 Q40 U-0. 5 W0. 05;	The finishing route is specified by N30 ~ N40, with a finishing allowance of 0. 5mm in X direction and 0. 05mm in Z direction

Continued Table 2-45

N30 G00 X32. 026;	Left end finishing inner profile
G01 Z0. ;	
X30. 026 C1. ;	
Z-20. ;	
X26. ;	
X21. W-2. 5;	
N40 X20. ;	
G00 Z150. ;	Quickly back to tool changing point
X150. ;	
M09;	Close cutting fluid
M05;	Spindle stop
M00;	Program pause
M03 S1200 F0. 08;	The main spindle rotates forward, the rotating speed is 1200r/min, and the set feed rate is 0. 08mm/r
T0202;	Substitute tool compensation No. 02
M08;	Open cutting fluid
G00 X20. ;	Fast feed to start of finishing cycle
Z5. ;	
G70 P30 Q40;	Finishing compound cycle
G00 Z150. ;	Quickly back to tool changing point
X150. ;	
M30;	The cursor returns to the program start at the end of the program

The program for the right end is shown in Table 2-46.

Table 2-46 The program for the right end

O62;	Program No.
G40 G97 G99 M03 S800 F0. 2;	Cancel the cutter radius compensation, cancel the constant rotation speed of the spindle, set the feed rate per revolution, the spindle rotates forward, the rotation speed is 800r/min, and the feed rate is 0. 2mm/r
T0101;	Replace T01 external turning tool, substitute tool compensation No. 01
M08;	Open cutting fluid
G00 Z5. 0;	Fast feed to the starting point of rough turning cycle in *Z* direction
X42. 0;	Fast feed to the starting point of rough turning cycle in *X* direction (blank diameter plus 2mm position)
G71 U1. 5 R0. 5;	Set the rough turning cycle of the outer circle, the cutting depth is 1. 5mm, and the tool withdrawal is 0. 5mm

Continued Table 2-46

G71 P10 Q20 U0. 5 W0. 05;	The finishing route is specified by N10 ~ N20, with a finishing allowance of 0. 5mm in X direction and 0. 05mm in Z direction
N10 G00 X36. 981;	Left end finishing outer profile
G01 Z0. ;	
X37. 981 C0. 5;	
Z-10. 5;	
N20 X42. ;	
G00 X150. ;	Quickly back to tool changing point
Z150. ;	
M09;	Close cutting fluid
M05;	Spindle stop
M00;	Program pause
M03 S1500 F0. 1;	The main spindle rotates forward, the rotating speed is 1500r/min, and the set feed rate is 0. 1mm/r
T0101;	Substitute tool compensation No. 01
M08;	Open cutting fluid
G00 Z5. ;	Fast feed to start of finishing cycle
X42. ;	
G70 P10 Q20;	Finishing compound cycle
G00 X150. ;	Quickly back to tool changing point
Z150. ;	
M09;	Close cutting fluid
M05;	Spindle stop
M00;	Program pause
M03 S400 F0. 05;	Clock wise rotation of main spindle, rotation speed 400r/min, feed rate 0. 05mm/r
T0303;	Change T03 grooving tool, substitute tool compensation No. 03
M08;	Open cutting fluid
G00 Z5. ;	Tool fast positioning to feed point
X40. ;	
G01 Z-15. F20;	Tool positioning to grooving point
X33. 4 F0. 05;	Grooving, diameter with one side 0. 2 machining allowance
X38. 5 F5;	Tool take back
W3. 5;	Tool move right
X33. 4 F0. 05;	Grooving, diameter with one side 0. 2 machining allowance
X38. 5 F5;	Tool take back
W1. 52;	Shift the tool to the right to ensure the size

Continued Table 2-46

X32. 969 F0. 05;	Grooving to the size
Z-15. ;	Tool move left
X38. 5;	Tool take back
G00 X150. ;	Quickly back to tool changing point
Z150. ;	
M09;	Close cutting fluid
M05;	Spindle stop
M00;	Program pause
M03 S400 F0. 1;	The main spindle rotates forward, the rotation speed is 400r/min, and the set feed rate is 0. 1mm/r
T0202;	Change T02 boring tool, substitute tool compensation No. 02
M08;	Open cutting fluid
G00 X20. ;	Fast feed to the starting point of rough turning cycle in *X* direction (drilling diameter position)
Z5. ;	Fast feed to the starting point of rough turning cycle in *Z* direction
G71 U1. 5 R0. 5;	Set the rough turning cycle of the inner hole, with the cutting depth of 1. 5mm and the tool withdrawal of 0. 5mm
G71 P30 Q40 U-0. 5 W0. 05;	The finishing route is specified by N30 ~ N40, with a finishing allowance of 0. 5mm in *X* direction and 0. 05mm in *Z* direction
N30 G00 X26. ;	Left end finishing inner profile
G01 Z0. ;	
X24 C2. ;	
Z-21. ;	
N40 X20. ;	
G00 Z150. ;	Quickly back to tool changing point
X150. ;	
M09;	Close cutting fluid
M05;	Main spindle stop
M00;	Program pause
M03 S1200 F0. 08;	The main spindle rotates forward, the rotating speed is 1200r/min, and the set feed rate is 0. 08mm/r
T0202;	Substitute tool compensation No. 02
M08;	Open cutting fluid
G00 X20. ;	Fast feed to start of finishing cycle
Z5. ;	
G70 P30 Q40;	Finishing compound cycle

Continued Table 2-46

G00 Z150. ;	Quickly back to tool changing point
X150. ;	
M09;	Close cutting fluid
M05;	Main spindle stop
M00;	Program pause
M03 S400;	Clock wise rotation of main spindle, speed of 400r/min
T0404;	Change the T04 internal thread turning tool, substitute tool compensation No. 04
M08;	Open cutting fluid
G00 X20. ;	Fast feed to start of machining cycle
Z5. ;	
G76 P020060 Q30 R0. 05;	Compound cycle of thread cutting
G76 X24. Z-21. P1300 Q400 F2. ;	
G00 Z150. ;	Quickly back to tool changing point
X150. ;	
M30;	The cursor returns to the program start at the end of the program

(4) Step 4 Part machining:

1) This part is a turning machining part, so when machining, it should first consider whether it can be clamped during turning, whether it has a pressure gauge alignment position, whether it is convenient to measure, and whether there are obvious traces in two directions. Therefore, when machining the outer circle at the left end of the workpiece, we choose to process it to the middle part of the outer circle groove of the part, and guarantee the machining dimension of the part according to the previous training content.

2) When machining the inner hole, it is necessary to note that the U value in the second line of G71 rough machining compound cycle command is set to a negative value. At the same time, if it is necessary to input the value at the [OFFSET/WEAR] position of the machine tool, carry out semi finishing to ensure the dimensional accuracy of the parts, and also need to input the negative value to reserve the margin; when programming, it is also necessary to pay attention to the operation sequence of feed and return to ensure the safety of rapid positioning.

3) When measuring the inner hole, because the inner diameter dial indicator is a relative measurement, the accuracy of the outer diameter micrometer used to check is also crucial. It is necessary to calibrate the accuracy of the outer diameter micrometer first, and pay attention to the reading of the outer diameter micrometer. During the measurement, it is necessary to find points for several times in the axial and radial direction.

4) When machining the right end of the workpiece, pay attention to how to determine the *Z* value of the workpiece coordinate system when turning around. This step is the key to ensure the total

length of the workpiece.

5) During turning machining, the X coordinate system parameters of the tool do not need to be determined again because the distance between the tool and the workpiece rotation center (spindle center) does not change.

6) When machining the outer circular groove, pay attention to the width of the groove tool blade, and program after careful calculation.

7) When machining the bottom hole of the inner hole thread, whether the thread is qualified mainly depends on the pitch diameter. Therefore, the bottom hole size is not the main size, so it is not necessary to carry out semi finishing. At the same time, it is not necessary to measure carefully. It can be measured by vernier caliper.

8) G76 code adopts oblique cutting method to cut thread, that is, the thread tool always uses one cutting edge to cut, which reduces cutting resistance, improves tool life and thread finishing quality. In addition, the instruction has high machining efficiency, so G76 programming instruction is preferred to machining the internal thread.

The finished part is shown in Figure 2-119.

Figure 2-119 The finished part

(5) Step 5 Dimension measurement:

1) This task is the final task. The vernier caliper is only allowed to be used when the tool determines the workpiece coordinate system, and the vernier caliper is not used to measure the final workpiece.

2) Outer diameters: use 25~50mm outer diameter micrometer to measure the outer diameter of both ends $\phi38^{0}_{-0.039}$ and groove diameter $\phi33^{0}_{-0.062}$.

3) Inner diameter part: use the inner diameter dial indicator to measure the inner hole size $\phi30^{+0.052}_{0}$.

4) Length dimensions: use 25~50mm outside micrometer to measure the length dimension (40±0.05)mm of the workpiece; use 5~30mm inside micrometer to measure the width dimension of the groove $9^{+0.04}_{0}$; use 0~25 depth micrometer to measure the depth dimension of the inner hole of the workpiece 20mm.

5) Internal thread: use thread plug gauge to check whether the internal hole thread M24×2-7H is qualified.

6) Surface roughness: use roughness comparison sample block to compare whether the overall

roughness of workpiece $Ra3.2$ is qualified.

(6) Step 6 Part assembly:

Task 2.6 part "thread sleeve" can be assembled with task 2.5 part "thread shaft", and the assembly drawing is shown in Figure 2-120. It can be seen that the outer circle of the threaded shaft is matched with the inner hole of the threaded shaft. The M24×2 outer thread is matched with the inner thread. The fit clearance is required to be (2±0.03) mm, and the total length is guaranteed to be (72±0.02) mm.

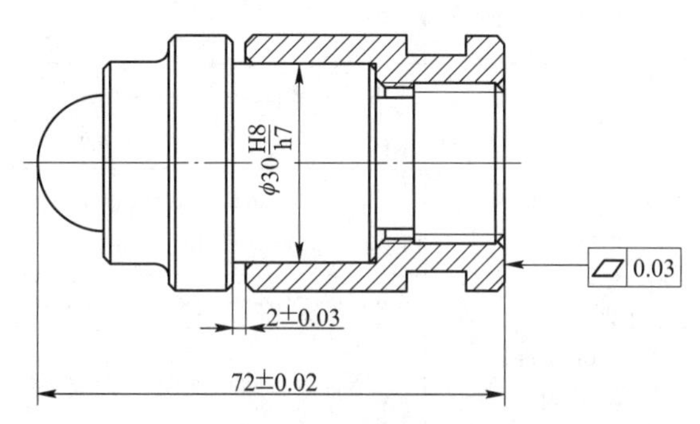

Figure 2-120 Assembly drawing

Method of ensure fitting clearance: under the condition of ensuring the dimension of threaded shaft (due to the teaching task, task 2.5 part shall be completed first), the depth dimension 20mm of threaded sleeve inner hole of task 2.6 part shall be processed on the premise of ensuring the requirement of fitting clearance (2±0.03) mm first.

When assembling, the chips on the internal and external threads shall be cleaned first to prevent the chips from screwing into the thread, which will lead to the thread unable to be removed.

When assembling parts for the first time, it is necessary to keep the axes of two workpieces coincident, one hand is used to stabilize the workpieces, the other hand is used to try to rotate the workpieces smoothly, and it is not allowed to use brute force.

There are two matching parts of thread and cylindrical surface in the assembly assembly assembly. If it is impossible to screw in during assembly, the cylindrical surface and thread of the two parts shall be tested to see if they are qualified. After the dimensional accuracy is modified, adjust the length of the workpiece to ensure the fitting clearance of the parts.

Assessment

The assessment table is shown in Table 2-47.

Table 2-47 The assessment table

Item	No.	Standard		Partition	Score
Machining plan (10%)	1	Machining steps	Meet the requirements of CNC turning process	3	
	2	Dimension	Calculate dimensions of relevant parts	3	
	3	Tool selection	The selection and installation of cutting tools are reasonable and meet the machining requirements	4	

Continued Table 2-47

Item	No.	Standard		Partition	Score
Programming (30%)	4	Program No.	No program number no point	1	
	5	Segment number	No segment number no point	1	
	6	Cutting condition	No point for unreasonable selection of cutting parameters	4	
	7	Origin and coordinates	Mark the origin and coordinates of the part, otherwise there is no point	2	
	8	Program contents	Deduct 2 points for each segment that does not meet the requirements of program logic and format	22	
			Deduct 5 points if the procedure content is not in accordance with the process		
			Deduct 5 points in case of dangerous code		
Operation (20%)	9	Program input and search	Deduct 2 points for each wrong section	10	
	10	Operation 30 minutes	No point for setting error of workpiece coordinate system	10	
			No points for misoperation		
			No scores for overtime		
Dimension measurement (30%)	11	Outer profile	No point if left outer profile over $\phi 38^{0}_{-0.039}$	3	
	12		No point if right outer profile over $\phi 38^{0}_{-0.039}$	2	
	13	Outer groove	No point if grooving size over $\phi 33^{0}_{-0.062}$	2	
	14		No point if groove width over $9^{+0.04}_{0}$	1	
	15	Inner hole	No point if left inner profile over $\phi 30^{+0.052}_{0}$	5	
	16	Inner thread	M24×2-7H	5	
	17	Length	No point if length over (40±0.05) mm	2.5	
	18	Chamfer	5	2.5	
	19	Roughness *Ra*3.2	Deduct 0.5 points for each unqualified outer contour and end face	1.5	
	20		Inner contour, no point for degradation	1.5	
	21	Fitting size	Fit clearance (2±0.03) mm out of tolerance full thread	4	
Vocational ability (10%)	22	Learning ability		2	
	23	Communication ability		2	
	24	The team Cooperation		2	
	25	Safe operation and civilized production		4	
Total		100			

Exercises

(1) Answer the following questions:

1) Briefly describe the difference between inner hole programming and outer contour programming.

2) Briefly describe the use of internal micrometer.

3) Briefly describe the processing method of matching dimension.

(2) Synchronous training:

Write the programs and machine the workpiece as shown in Figure 2-121 and Figure 2-122, and complete the assembly, as shown in Figure 2-123 and Figure 2-124.

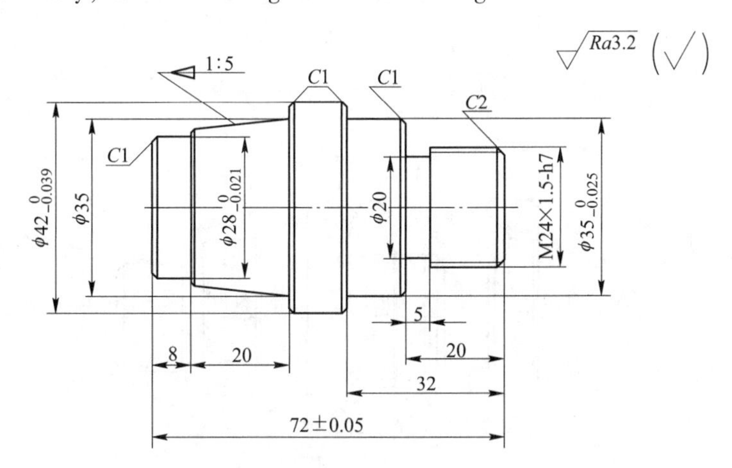

Note:
1.Unmarked chamfers are C0.5, acute angles are dull.
2.Unmarked linear dimensional tolerances should meet the requirements of GB/T 1804—2000.

Figure 2-121 Part 1

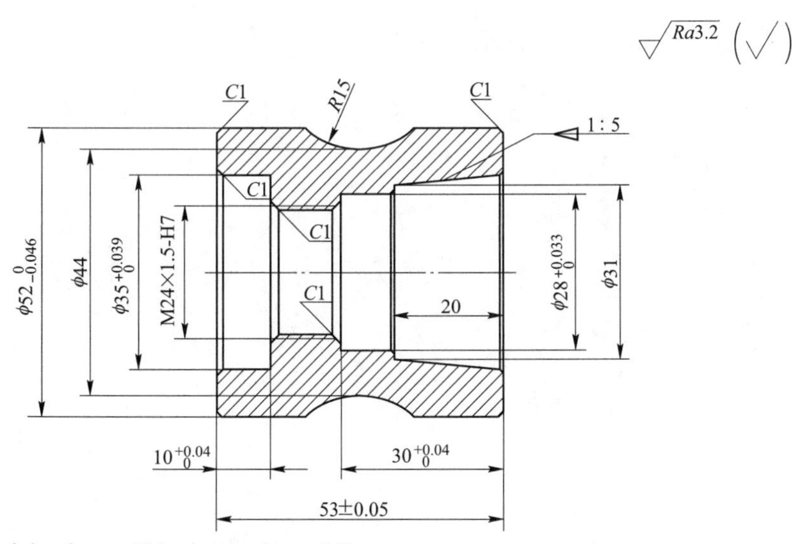

Note:
1.Unmarked chamfers are C0.5, acute angles are dull.
2.Unmarked linear dimensional tolerances should meet the requirements of GB/T 1804—2000.

Figure 2-122 Part 2

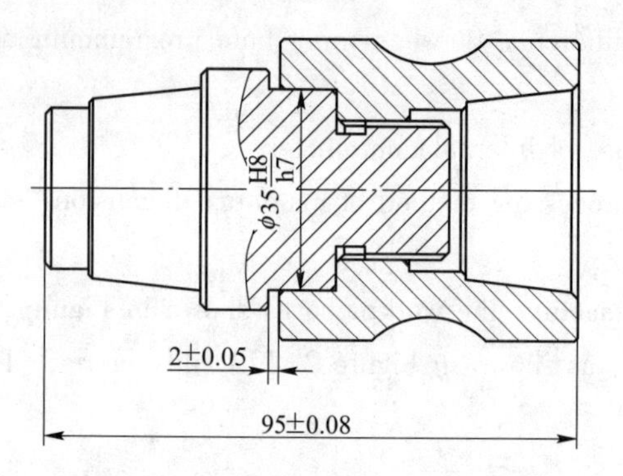

Figure 2-123　Assembly drawing 1

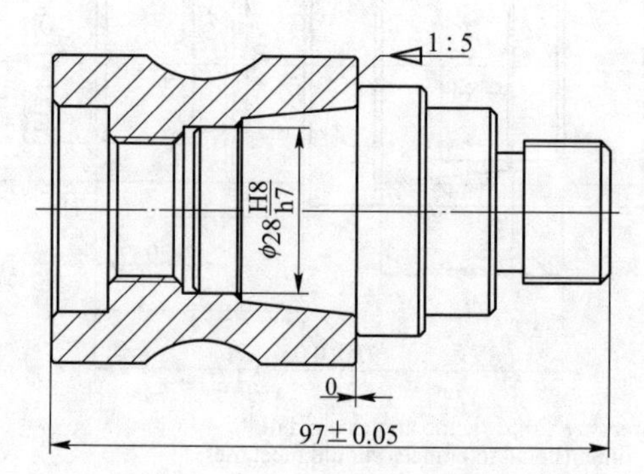

Figure 2-124　Assembly drawing 2

Appendix

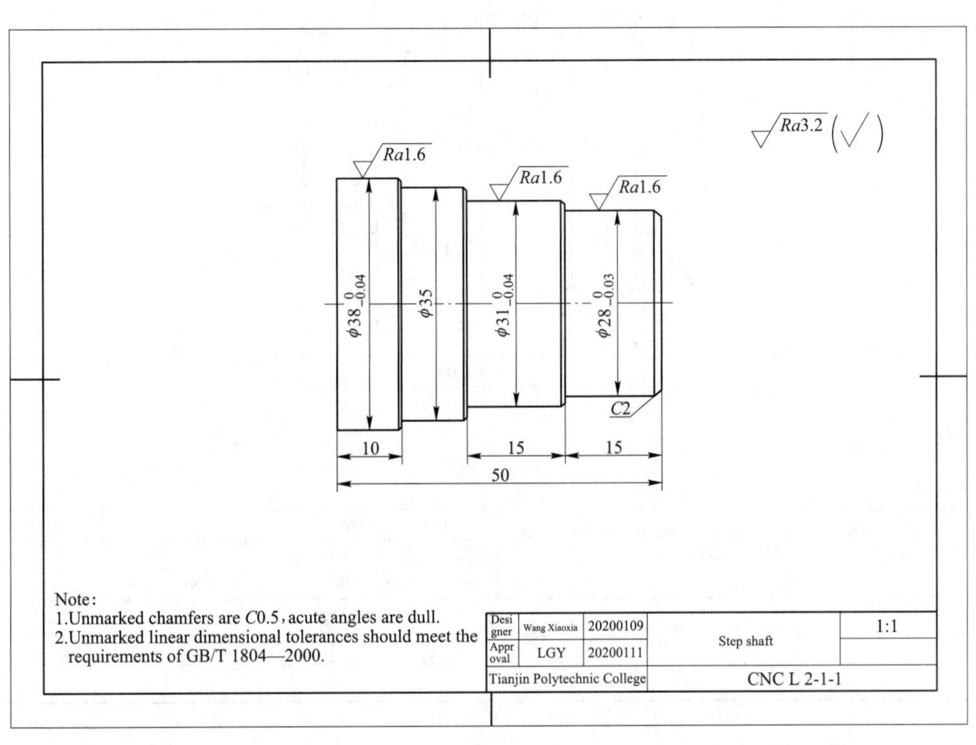

Ra3.2 (√)
Ra1.6
Ra1.6
Ra1.6
φ38 0 −0.04
φ35
φ31 0 −0.04
φ28 0 −0.03
C2
10
15
15
50
Note:
1.Unmarked chamfers are C0.5, acute angles are dull.
2.Unmarked linear dimensional tolerances should meet the requirements of GB/T 1804—2000.
Designer
Wang Xiaoxia
20200109
Step shaft
1:1
Approval
LGY
20200111
Tianjin Polytechnic College
CNC L 2-1-1

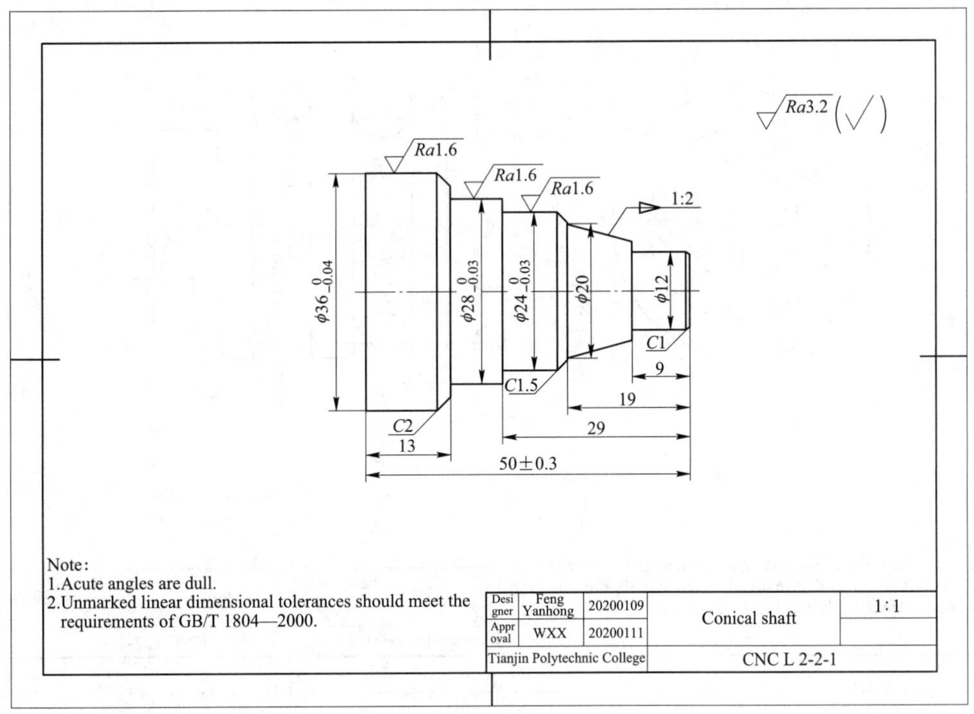

Ra3.2 (√)
Ra1.6
Ra1.6
Ra1.6
1:2
φ36 0 −0.04
φ28 0 −0.03
φ24 0 −0.03
φ20
φ12
C1
9
C1.5
19
29
C2
13
50±0.3
Note:
1.Acute angles are dull.
2.Unmarked linear dimensional tolerances should meet the requirements of GB/T 1804—2000.
Designer
Feng Yanhong
20200109
Conical shaft
1:1
Approval
WXX
20200111
Tianjin Polytechnic College
CNC L 2-2-1

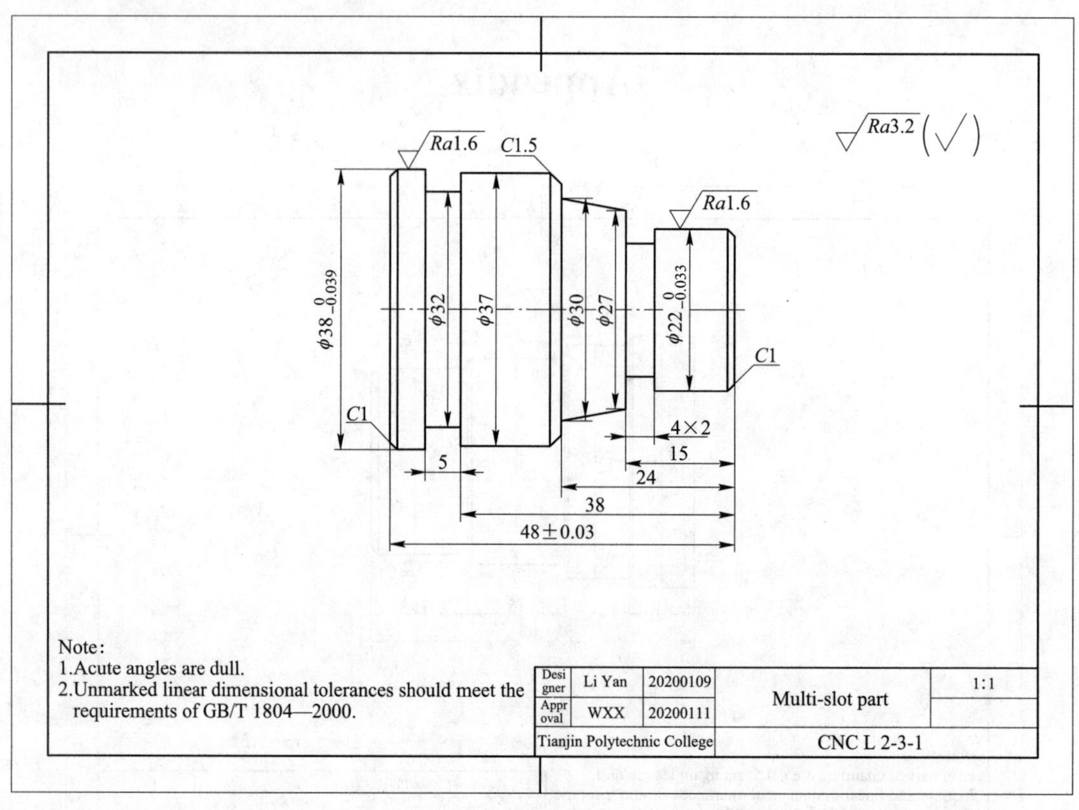

Note:
1.Acute angles are dull.
2.Unmarked linear dimensional tolerances should meet the requirements of GB/T 1804—2000.

Desi gner	Li Yan	20200109	Multi-slot part		1:1
Appr oval	WXX	20200111			
Tianjin Polytechnic College			CNC L 2-3-1		

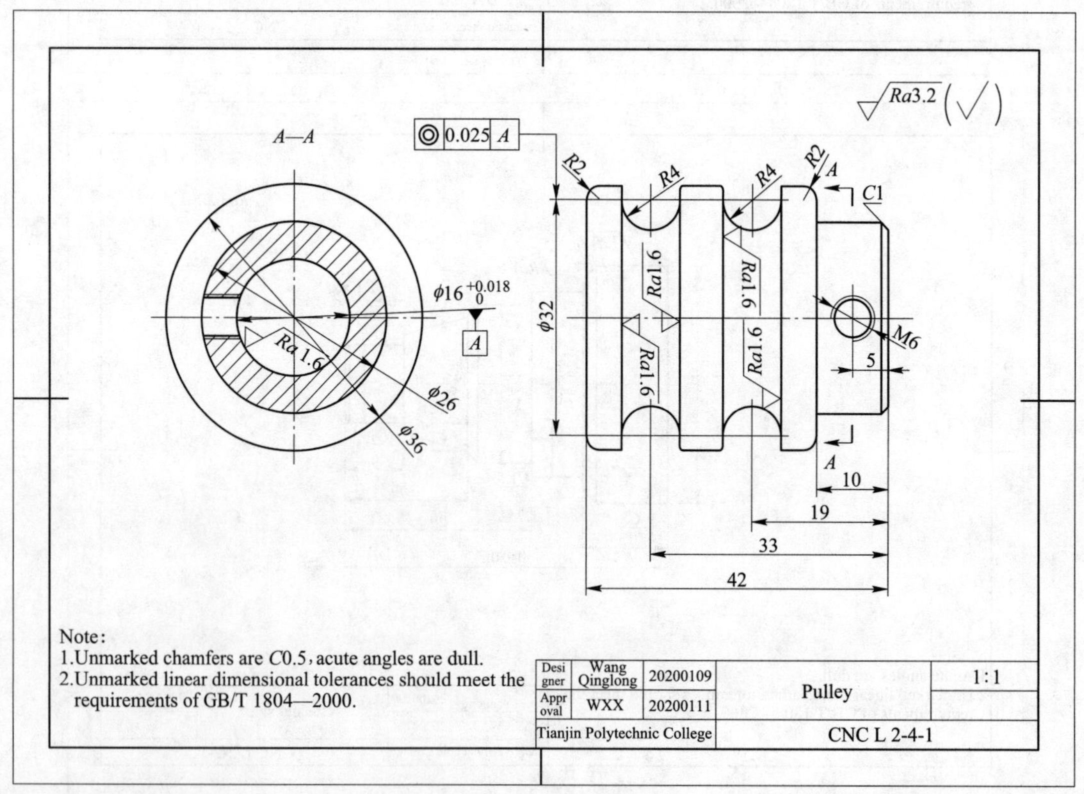

Note:
1.Unmarked chamfers are C0.5, acute angles are dull.
2.Unmarked linear dimensional tolerances should meet the requirements of GB/T 1804—2000.

Desi gner	Wang Qinglong	20200109	Pulley		1:1
Appr oval	WXX	20200111			
Tianjin Polytechnic College			CNC L 2-4-1		

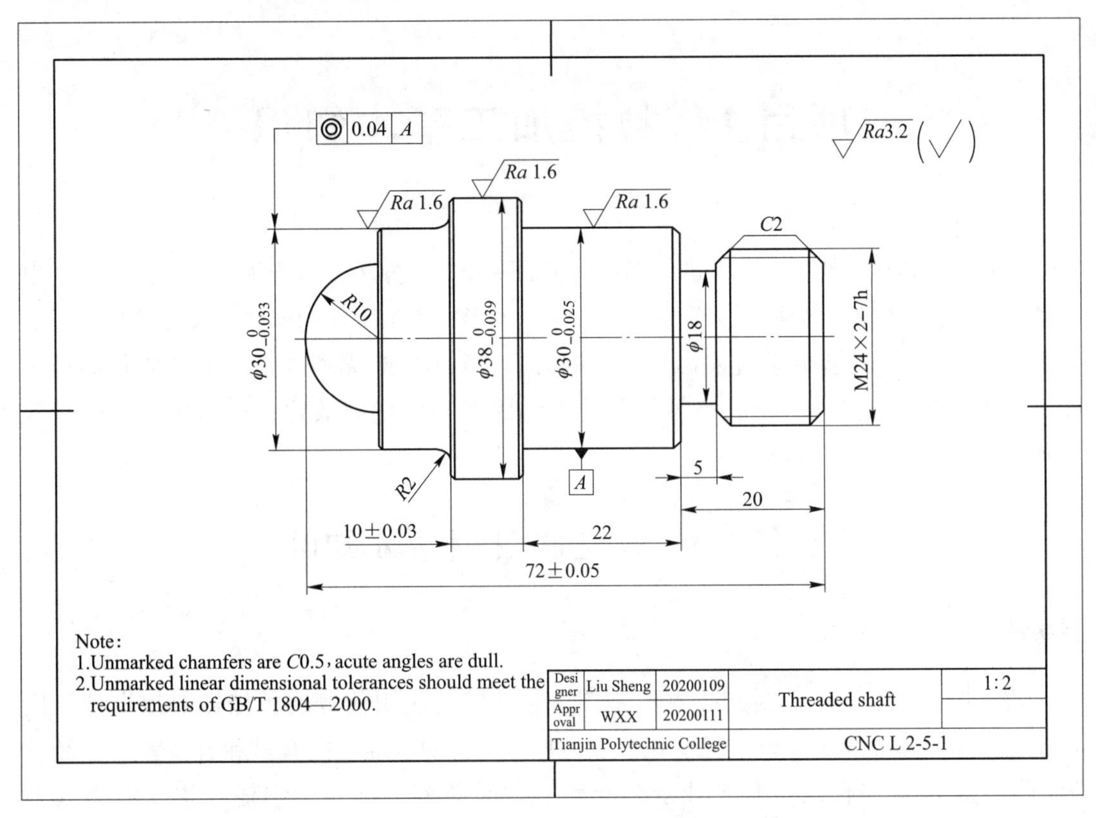
0.04 A
Ra3.2
Ra 1.6
Ra 1.6
Ra 1.6
C2
R10
φ30 0 -0.033
φ38 0 -0.039
φ30 0 -0.025
φ18
M24×2-7h
R2
A
5
20
10±0.03
22
72±0.05
Note:
1.Unmarked chamfers are C0.5, acute angles are dull.
2.Unmarked linear dimensional tolerances should meet the requirements of GB/T 1804—2000.
Desi gner
Liu Sheng
20200109
Threaded shaft
1:2
Appr oval
WXX
20200111
Tianjin Polytechnic College
CNC L 2-5-1

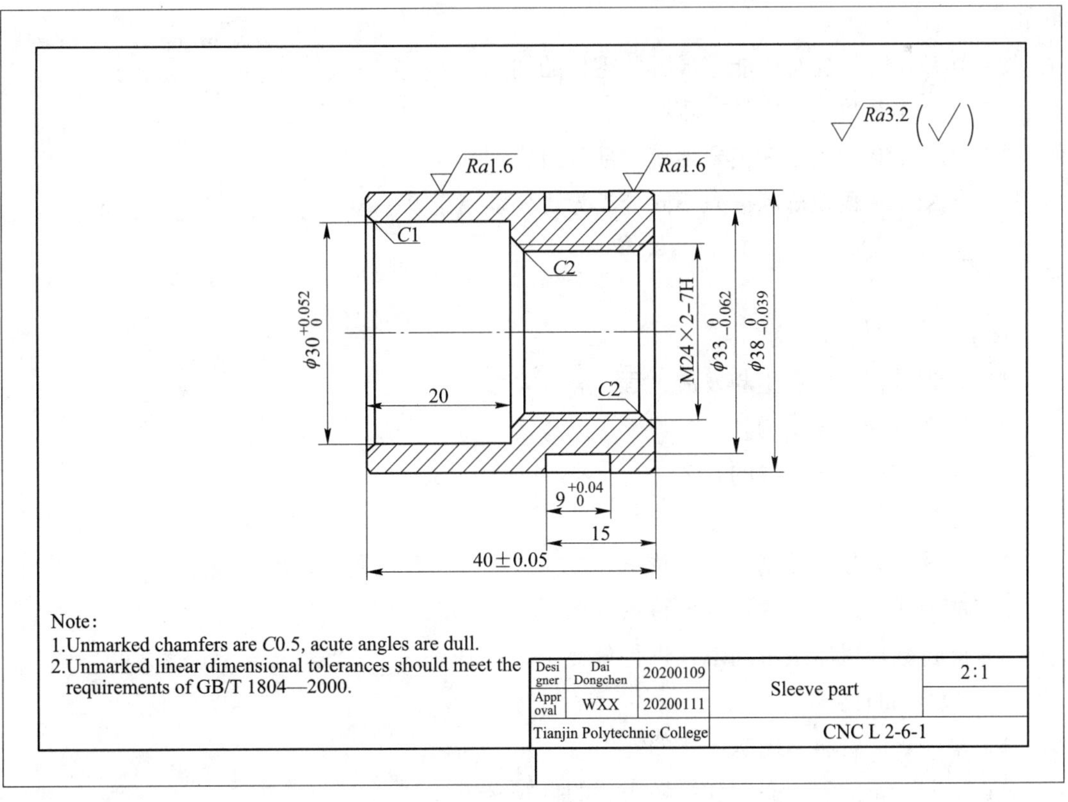
Ra3.2
Ra1.6
Ra1.6
C1
C2
C2
φ30 +0.052 0
20
M24×2-7H
φ33 0 -0.062
φ38 0 -0.039
9 +0.04 0
15
40±0.05
Note:
1.Unmarked chamfers are C0.5, acute angles are dull.
2.Unmarked linear dimensional tolerances should meet the requirements of GB/T 1804—2000.
Desi gner
Dai Dongchen
20200109
Sleeve part
2:1
Appr oval
WXX
20200111
Tianjin Polytechnic College
CNC L 2-6-1

项目1　数控加工基本知识

数控加工是指在数控机床上进行零件加工的一种工艺方法，与传统机床加工的工艺规程基本一致，不同的是数控加工是以数字和字母的形式表示工件的形状和尺寸等技术要求和加工工艺要求，由控制系统发出指令控制刀具进行加工。数控加工是解决零件品种多变、形状复杂、批量小、精度高等问题的关键因素和实现高效化、自动化加工的有效途径。

任务1.1　数控加工技术基本知识

任务描述

数控加工技术是现代制造技术的基础，它的广泛应用使普通机械被数控机械所代替，使全球制造业发生了根本性变化。数控加工技术的水准、拥有量和普及程度已经成为衡量一个国家综合国力和工业现代化水平的重要标志之一。本任务主要解决以下问题：

（1）CK6140数控车床能完成哪些零件的加工？

（2）CK6140数控车床具有哪些特点？

（3）CK6140数控车床主要由哪些部分组成？

（4）CK6140数控车床按照不同分类方式分别属于哪一类？

任务目标

（1）知识目标：

1）认识数控机床产生与发展过程。

2）熟知数控车床结构。

3）认识数控机床发展的趋势。

（2）技能目标：

1）能够区分数字控制、数控技术、数控系统、计算机数控系统等概念。

2）能说明数控加工的过程、特点及应用。

3）能将数控机床按不同形式进行分类。

（3）素质目标：

1）培养查阅资料、团队协作的能力。

2）培养端正的学习态度和流畅表达的良好习惯。

相关知识

1.1.1 数控概念、产生与发展

1.1.1.1 基本概念

（1）数字控制。数字控制简称数控，是一种借助数字、字符或者其他符号对某一工作过程进行编程控制的自动化方法。

（2）数控技术。数控技术是一种用数字信息对机械运动和工作过程进行控制的技术。它通常控制位置、角度、速度等机械量和与机械量流向有关的开关量，已经成为制造业实现自动化、柔性化、集成化生产的基础技术。

（3）数控系统。数控系统是一种根据计算机存储器中的控制程序，执行部分或全部数值控制功能，并配有接口电路和伺服驱动装置的专用计算机系统。

（4）计算机数控系统。计算机数控系统是利用一个专用的可存储程序的计算机，执行一些或全部的基本数字控制功能的数控系统。

1.1.1.2 数控机床的产生与发展

数控机床是指采用数字控制技术对机床的加工过程进行自动控制的一类机床。国际信息处理联盟（IFIP）第五技术委员会对数控机床定义如下：数控机床是一个装有程序控制系统的机床，该系统能够逻辑性地处理具有号码或其他符号编码指令规定的程序。定义中所说的程序控制系统即数控系统。

1952 年，第一台数控机床（三坐标立式数控铣床）在美国问世，成为世界机械工业史上一件划时代的事件。中国自 1958 年开始研制数控机床，1966 年成功生产第一台数控机床。

数控机床伴随着电子技术和计算机技术的发展而发展，经历了以下几个阶段：

1952~1959 年为第一代数控机床，电子器件为电子管；

1959~1964 年为第二代数控机床，电子器件为晶体管；

1965~1970 年为第三代数控机床，电子器件为中小规模的集成电路；

1971~1974 年为第四代数控机床，电子器件为大规模集成电路的小型通用电子计算机控制的系统（CNC）；

1974 年后为第五代数控机床，电子器件为微型电子计算机控制的系统（MNC）。

前三代为 NC 硬接线技术，后两代为 CNC 软接线技术。目前正在发展第六代数控机床。

1.1.2 数控加工过程、特点及应用

1.1.2.1 数控加工过程

数控机床加工零件时，首先要把加工零件需要的工艺信息以程序的形式记录下来，存

储在某些载体上，输入到数控系统中，由数控装置处理程序，发出控制信号指挥伺服系统驱动伺服电动机，协调机床的动作，使其产生主运动和进给运动的一系列机床运动，完成零件的加工。数控机床加工零件过程如图 1-1 所示。

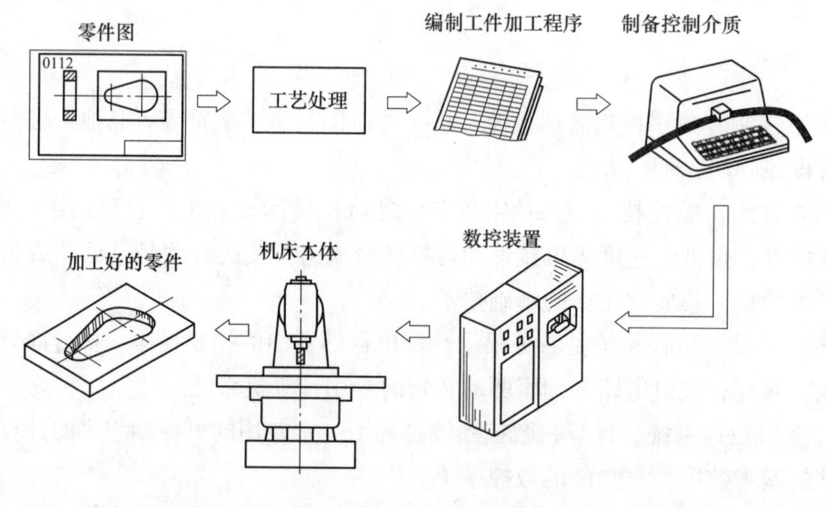

图 1-1　数控加工过程

1.1.2.2　数控加工特点

数控加工有下列优点：

（1）大量减少工装数量，加工形状复杂的零件不需要复杂的工装。如果要改变零件的形状和尺寸，只需要修改零件加工的程序，适用于新产品研制和改型。

（2）加工质量稳定、加工精度高、重复精度高。

（3）多品种、小批量生产情况下，生产效率较高，能减少生产准备、机床调整和工序检查的时间。

（4）可加工使用常规方法难于加工的复杂型面，甚至能加工一些无法观测的部位。

数控加工的缺点是机床设备价格较贵，必须由高级技术人员来编程、安装、操作和维护，要求维修人员具有较高水平。

1.1.2.3　数控加工的应用

随着电子技术及计算机技术的发展，数控机床不断更新换代，飞速发展。如今，数控机床已经广泛应用于宇宙飞船、船舶、机床、汽车、轨道交通、能源（火电、水电、热电、风力发电等）、冶金、轻工、纺织、电子、医疗器件、通用机械、工程机械等几乎所有的制造行业。

数控机床作为一种高度机电一体化的产品，技术含量高，成本高，从最经济的方面出发，数控机床适用于加工多品种小批量零件，结构较复杂，精度要求较高的零件，价格昂贵，不允许报废的关键零件和需要最短生产周期的急需零件。数控机床加工的典型零件包括回转体零件、模具类零件、箱体类零件及异形零件，如图 1-2 所示。

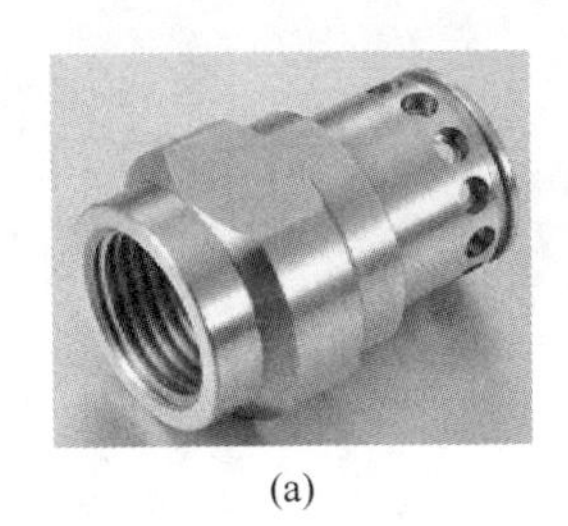
(a)
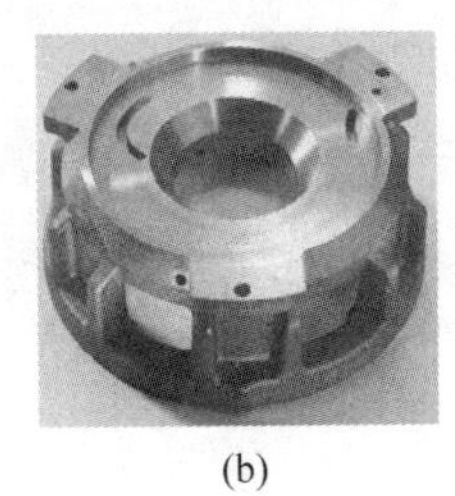
(b)
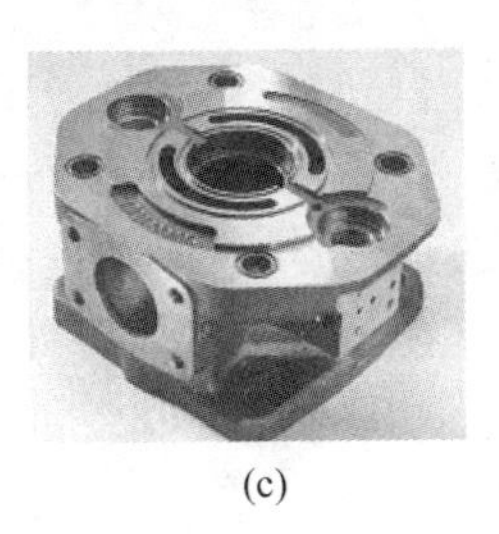
(c)
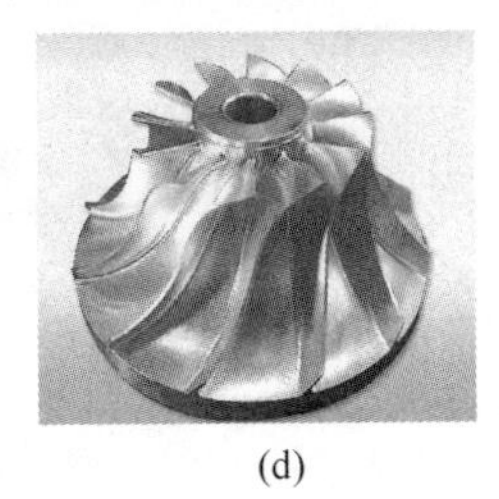
(d)

图 1-2　数控加工应用
（a）回转体零件；（b）模具类零件；（c）箱体类零件；（d）异形零件

1.1.3　数控机床分类

1.1.3.1　按控制运动方式分类

按照刀具运动方式，数控机床分为点位控制数控机床、直线控制数控机床和轮廓控制数控机床，如图 1-3 所示。

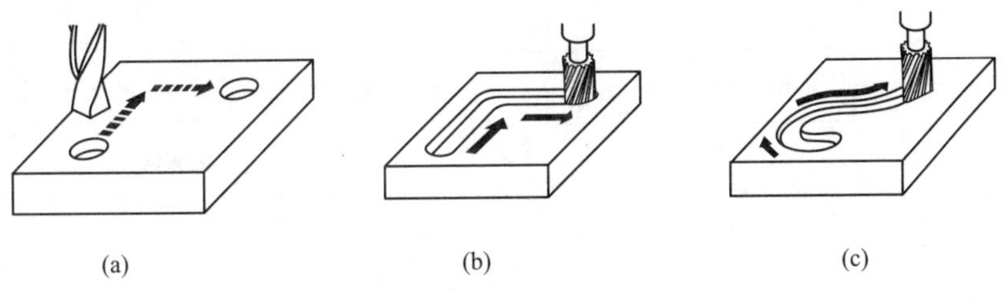
(a)　(b)　(c)

图 1-3　数控机床按控制运动分类
（a）点位控制；（b）直线控制；（c）轮廓控制

（1）点位控制数控机床。点位控制系统将刀具从一点移动到另一点。数控钻床是最典型的点位控制机床，钻头从一点移动到下一点的过程中，路径和进给无关紧要，不控制从起点到终点的运动轨迹。

（2）直线控制数控机床。直线控制数控机床的特点是除了控制机床的运动部件目标点的精确位置外，还能实现在起点和目标点之间沿直线或斜线的切削进给运动。典型的直线控制数控机床有早期的数控车床、数控铣床和数控镗床。

（3）轮廓控制数控机床。轮廓控制系统能够在至少两个坐标轴上同时调整工作台（或主轴）的进给。这种控制方式需要复杂的控制和驱动系统。数控车床、数控铣床、加工中心等数控机床常用轮廓控制运动方式。

1.1.3.2　按伺服系统的控制方式分类

数控机床可按照对伺服驱动被控量有无检测反馈装置分为开环控制和闭环控制。根据检测装置安装的部位不同，闭环控制又分为全闭环控制和半闭环控制两种，如图 1-4 所示。

（1）开环控制数控机床。开环控制系统不给控制单元提供位置反馈信息。开环控制系统的优点是成本低，缺点是很难检测位置误差。经济型数控机床一般采用开环控制。

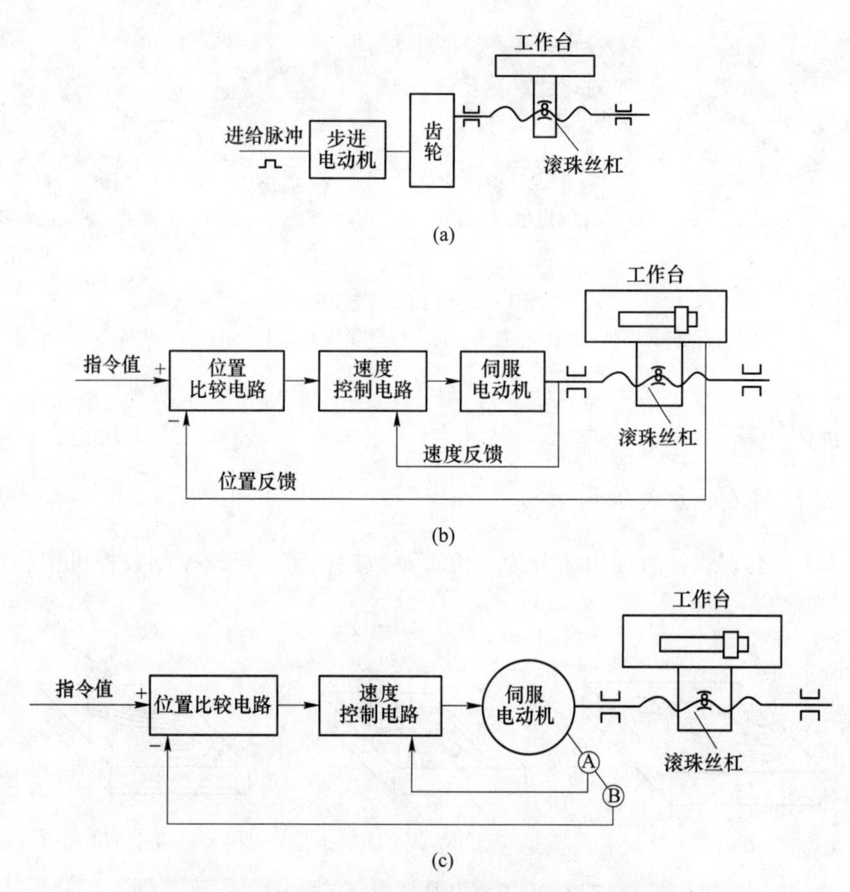

图 1-4 数控机床按伺服系统的控制方式分类

（a）开环控制数控机床；（b）全闭环控制数控机床；（c）半闭环控制数控机床

（2）全闭环控制数控机床。全闭环控制系统精度高，通常采用交流、直流或液压伺服电动机。精度要求较高的数控镗铣床、超精数控车床和数控加工中心等数控机床一般采用全闭环控制。

（3）半闭环控制数控机床。半闭环控制的数控机床将位置检测装置安装在驱动电动机或传动丝杠的端部，间接测量执行部件的实际位置和位移；其精度低于闭环系统，但测量装置结构简单，安装调试方便，中档数控机床一般采用半闭环控制。

1.1.3.3 按工艺用途分类

数控机床按工艺用途分为以下几类。

（1）切削类数控机床。切削类数控机床是指具有切削加工功能的数控机床。

1）普通型数控机床。最常用的普通型数控机床有数控车床、数控铣床、数控镗床、数控钻床、数控磨床、数控齿轮加工机床等，而且每种类型中又有很多品种，如图 1-5 所示。

2）加工中心。加工中心是在普通数控机床上加装刀库和自动换刀装置，工件经一次装夹后，可进行多道工序的集中加工，减少了工件装卸次数、更换刀具等辅助时间，机床的生产效率较高。常见的加工中心如图 1-6 所示。

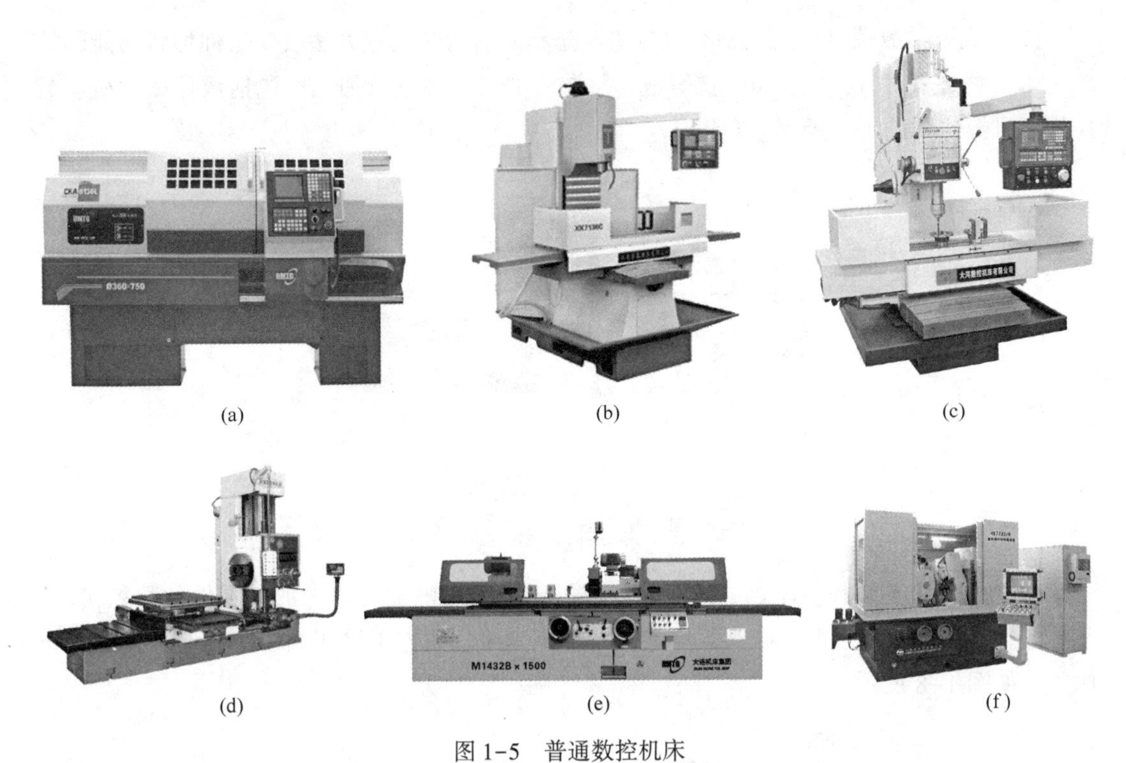

(a)　(b)　(c)

(d)　(e)　(f)

图 1-5　普通数控机床

(a) 数控车床；(b) 数控铣床；(c) 数控钻床；(d) 数控镗床；(e) 数控磨床；(f) 数控齿轮加工机床

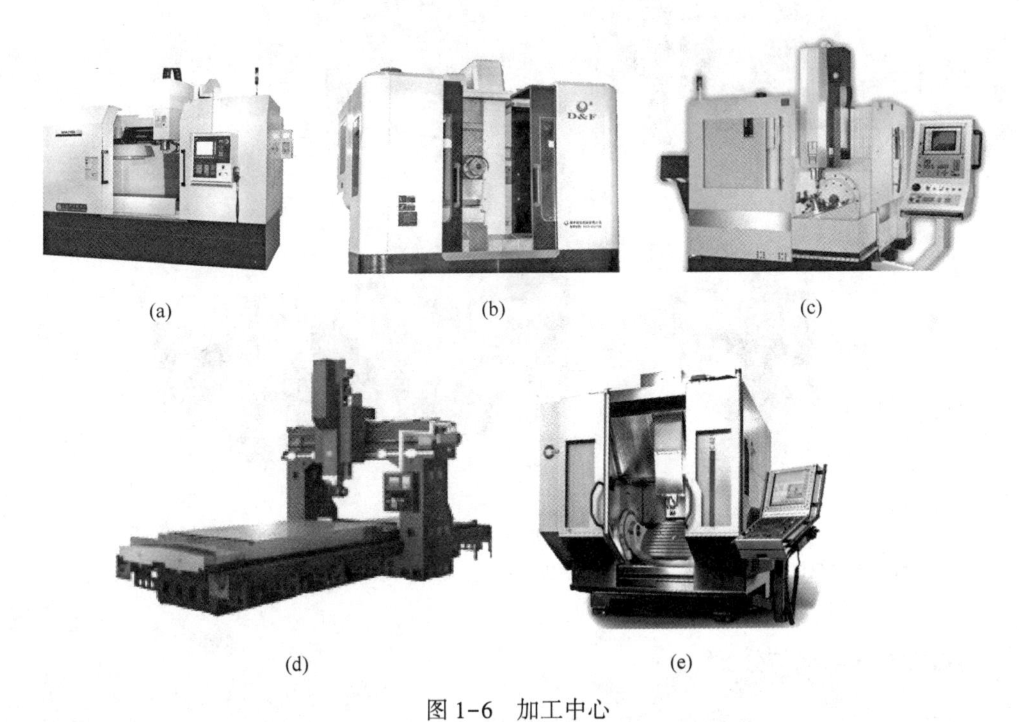

(a)　(b)　(c)

(d)　(e)

图 1-6　加工中心

(a) 立式加工中心；(b) 卧式加工中心；(c) 万能加工中心；(d) 龙门加工中心；(e) 五轴加工中心

（2）成型类数控机床。成型类数控机床是指具有通过物理方法改变工件形状功能的数控机床。它是采用挤、冲、压、拉等成型工艺方法对零件进行加工，包括数控压力机、数控折弯机等，如图 1-7 所示。

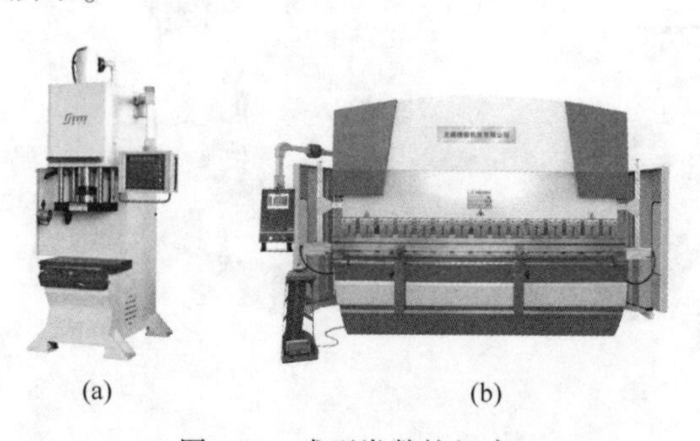

(a)　　(b)

图 1-7　成型类数控机床

（a）数控压力机；（b）数控折弯机

（3）电加工类数控机床。电加工类数控机床是采用电加工技术加工零件的数控机床，常见的有数控电火花成型机床、数控电火花切割机床、数控火焰切割机、数控激光加工机床等，如图 1-8 所示。

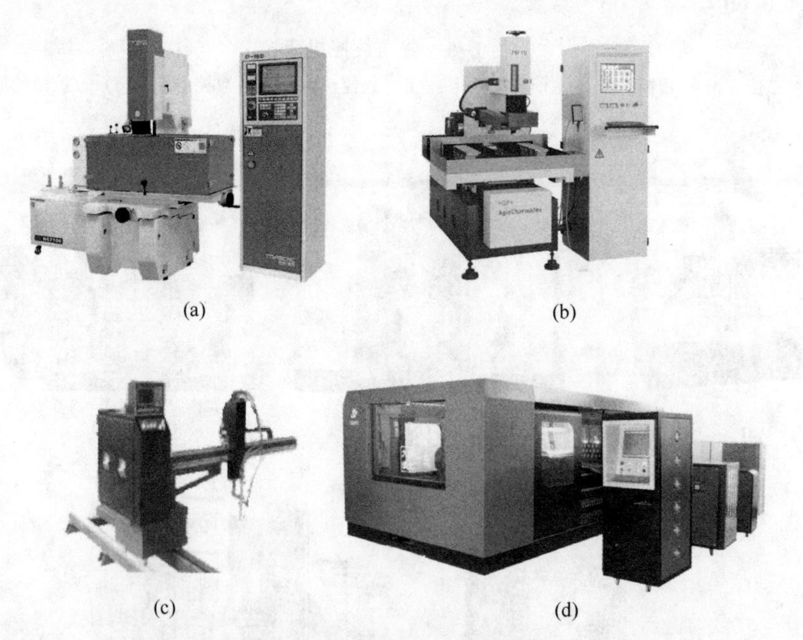

(a)　　(b)

(c)　　(d)

图 1-8　电加工类数控机床

（a）数控电火花机床；（b）数控电火花切割机床；（c）数控火焰切割机；（d）数控激光加工机床

1.1.4　数控车床组成与分类

数控车床是计算机数字控制车床。数控车床能完成车端面、圆柱面、车槽、切断、钻孔、车孔、车螺纹等加工操作。

1.1.4.1　数控车床组成部件

数控车床的主要部件有主轴箱、床身、导轨、转塔刀架、尾座、进给机构、操作面板和辅助系统等，如图 1-9 所示。

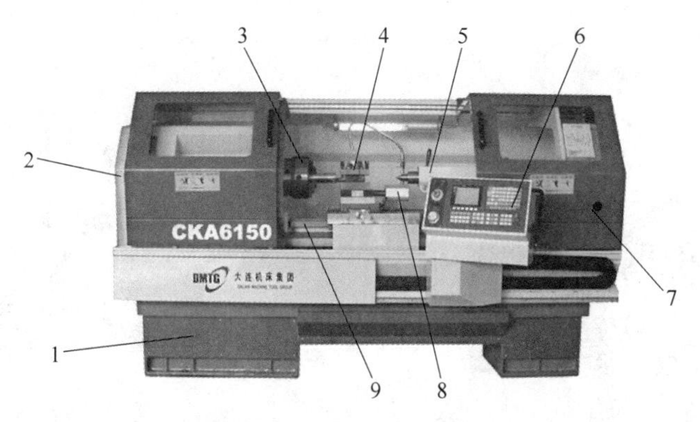

图 1-9　数控车床组成部件

1—床身；2—主轴箱；3—三爪卡盘；4—回转刀架；5—尾座；6—操作面板；7—防护罩；8—刀架电机；9—导轨

床身固定在底座上，其上安装着车床的各主要部件，并使它们在工作时保持准确的相对位置。

主轴箱固定于床身的最左边，主要用于支承并带动主轴，以实现机床的主运动。

刀架滑板由纵向（Z 方向）滑板和横向（X 方向）滑板组成。纵向滑板安装在床身导轨上，沿床身实现纵向（Z 方向）运动；横向滑板安装在纵向滑板上，沿纵向滑板上的导轨实现横向（X 方向）运动。刀架滑板可使安装在其上的刀具实现纵向和横向进给运动。

转塔刀架安装在机床的刀架滑板上，用于装夹各种刀具。加工时可根据加工要求自动换刀。

尾座安装在床身导轨上，可沿导轨进行纵向移动以调整位置。尾座主要用于安装顶尖，在加工中对工件进行辅助支承。尾座上也可安装钻头、铰刀等刀具进行孔加工。

防护罩安装在机床底座上，加工时保护操作者的安全和保护环境的清洁。

机床液压传动系统用来实现机床上的一些辅助运动，主要是实现机床主轴的变速、尾座套筒的移动及工件自动夹紧机构的动作。

机床的电气控制系统主要由数控系统（包括数控装置、伺服系统及可编程序控制器）和机床的强电气控制系统组成，它能够完成对机床的自动控制。

1.1.4.2　数控车床分类

（1）按照主轴位置分类。按照主轴位置不同，数控车床分为卧式数控车床和立式数控车床。卧式数控车床用于加工各种轴类、套筒类和盘类零件的回转表面，如内外圆柱面、圆锥面、螺纹面，如图 1-9 所示。立式数控车床用于回转直径较大的盘类零件车削加工，如图 1-10 所示。

（2）按照控制系统特征分类。按照控制系统特征不同，数控车床分为经济型数控车床、全功能型数控车床和车削中心。

1）经济型数控车床。经济型数控车床一般是在普通车床的基础上进行改进设计的，装备开环伺服系统，其控制部分采用单板机。此类车床结构简单，价格低廉，与其他数控车床相比，没有刀尖圆弧半径自动补偿和恒线速度切削等功能，如图 1-11 所示。

图 1-10　立式数控车床

图 1-11　经济型数控车床

2）全功能型数控车床。全功能型数控车床通常简称数控车床，又称为标准型数控车床，如图 1-9 所示。全功能型数控车床具有标准型系统，带有高分辨率的 CRT 显示器，具有图形仿真、刀具补偿，通信或网络接口、多轴联动等功能。全功能型数控车床采用闭环或半闭环控制的伺服系统，具有高刚度、高精度和高效率等特点。

(3) 车削中心。如图 1-12 所示，车削中心以全功能型数控车床为主体，配有刀库、自动换刀装置、分度装置、铣削装置和机械手等部件，以实现多工序复合加工。工件一次装夹后，在车削中心可完成回转类零件的车、铣、钻、铰、攻螺纹等多道加工工序。其效率和自动化程度高，但价格较贵。

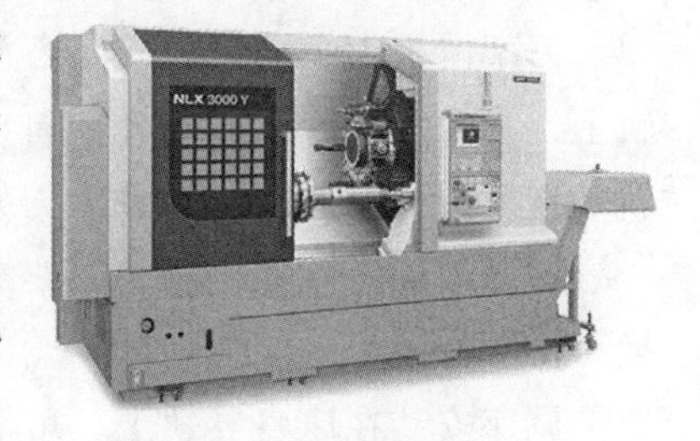

图 1-12　车削中心

1.1.5　数控机床的发展趋势

随着科学技术的发展，当今的数控机床正在不断采用最新技术成果，朝着高速度、高精度、多功能、智能化、自动化和可靠性最大化等方向发展，具体表现在以下几个方面。

1.1.5.1　高速度、高精度

(1) 数控机床系统采用位数及频率更高的处理器，以提高系统的基本运算速度，使得高速运算、模块化及多轴成组控制系统成为可能。

(2) 在采用全数字伺服系统的基础上，采用直线伺服电动机直接驱动机床工作台，实现“零传动”的直线伺服进给方式。

(3) 采用机床静、动摩擦的非线性补偿控制技术。

(4) 应用高速大功率电主轴。

(5) 配置高速、强功能的内装式可编程控制器。

1.1.5.2 多功能

（1）数控机床一机多能，最大限度地提高设备的利用率。

（2）采用“前台加工，后台编辑”技术。

（3）数控装备网络化。

1.1.5.3 智能化

（1）追求加工效率和加工质量方面的智能化，如加工过程的自适应控制和工艺参数自动生成。

（2）提高驱动性能及使用连接方便的智能化，如前馈控制、电机参数的自适应运算、自动识别负载、自动选定模型等。

（3）简化编程、操作，如智能化的自动编程、智能化的人机界面等。

（4）智能诊断、智能监控方面的内容，方便系统地诊断及维修等。

1.1.5.4 自动化

（1）CAD/CAM 图形交互式自动编程。

（2）CAD/CAPP/CAM 集成的全自动编程。

1.1.5.5 可靠性最大化

（1）数控系统将采用更高集成度的电路芯片。

（2）通过自动运行诊断、在线诊断、离线诊断等多种诊断程序，实现对系统内硬件、软件和各种外部设备进行故障诊断和报警。

任务实施

参观数控加工实训中心，结合所学，观察 CK6140 数控车床，完成表 1-1。

表 1-1 任务实施工作单

设备名称		CK6140 数控车床	根据现场观察结果，填写记录单
序号	工作内容		情况记录
1	加工内容	数控机床加工典型零件	
		数控机床适合加工零件	
		数控车床切削零件表面	
2	特点	数控加工优点	
		数控加工缺点	
3	结构组成	基本组成	
		各组成部分作用	
4	所属类别	按控制运动方式分类	
		按伺服系统控制方式分类	

续表

序号	工作内容		情况记录
4	所属类别	按工艺用途分类	
		按主轴位置分类	
		按控制系统特征分类	

任务评价

任务评价如表 1–2 所示。

表 1–2　任务评价表

项　目	序号	评 分 标 准	配分	得分
加工内容（15%）	1	数控机床加工典型零件	5	
	2	数控机床适合加工零件	5	
	3	数控车床切削零件表面	5	
特点（15%）	4	数控加工优点	12	
	5	数控加工缺点	3	
结构组成（30%）	6	基本组成	10	
	7	各组成部分作用	20	
所属类别（20%）	8	控制运动方式分类	4	
	9	伺服系统控制方式分类	4	
	10	工艺用途分类	4	
	11	主轴位置分类	4	
	12	控制系统特征分类	4	
职业能力（20%）	13	学习能力	4	
	14	表达沟通能力	4	
	15	团队合作	4	
	16	安全操作与文明生产	8	
合计				

同步思考与训练

（1）思考题：

1）简要说明数控加工的特点。

2）数控加工适用于加工什么样的零件？

3）数控机床按照工艺用途可分为哪些类？

（2）同步训练题：

1）观看 CK6140 数控车床加工过程，说明数控加工的过程。

2）观察 CKA6150 数控车床，说明该车床的结构组成和所属类别。

任务 1.2　数控机床基本操作与维护

任务描述

对于初学数控机床的人员，掌握一定的数控机床操作技巧是非常重要的。一方面可以避免发生机床碰撞事故，导致机床损坏；另一方面可以在较短的时间内，迅速提高操作者的数控机床操作技能，胜任本职工作。本任务针对实训设备，要求完成以下操作：

（1）启动 CK6140 数控车床。

（2）在手动方式、MDI 方式下实现主轴正转。

（3）在手动方式、手摇模式下实现刀具进给。

（4）完成任意刀号的换刀操作。

（5）完成 CK6140 数控车床日常维护。

任务目标

（1）知识目标：

1）认识数控车床机械结构。

2）识别数控车床系统面板、操作面板。

3）熟知数控车床安全操作规程。

（2）技能目标：

1）能按照操作规程，正确进行数控车床的启动与关闭操作。

2）能熟练进行数控车床的手动和 MDI 方式操作。

3）能按照要求，完成数控车床的日常维护和保养。

（3）素质目标：

1）培养学生独立思考的能力，注重细节的品质。

2）培养学生不轻言放弃懂得坚持，及有责任感。

相关知识

数控车床又称为 CNC 车床，即用计算机进行数字控制的车床，它具有加工灵活、通用性强、能适应产品的品种和规格频繁变化的特点，能够满足新产品的开发和多品种、小批量、生产自动化的要求，因此广泛应用于机械制造业，例如汽车制造厂、发动机制造厂等。

1.2.1　CK6140 数控车床机械结构

如图 1-13 所示，CK6140 数控车床是一台在计算机控制下工作的 2 轴数控车床，配置 4 工位刀架，主轴转速通过频率变化调整，机床主驱动电机在变频器控制下无级调速工作。适用于内外圆柱面、端面、任意锥面、弧面及各种螺纹等的车削加工，也可完成切断、切槽、

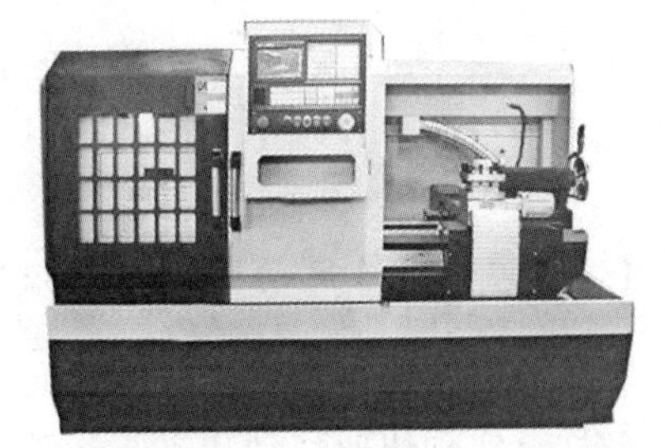

图 1-13　CK6140 卧式数控车床

钻孔、切孔、铰孔等工序的车削加工，加工效率高、精度高、使用方便、运行可靠。

1.2.1.1 床身

采用平床身结构，树脂砂造型铸造工艺，优质铸铁铸造。导轨采用中频淬火磨削和贴塑工艺，具有良好的耐磨性和精度保持性。

1.2.1.2 主轴箱

主轴结构如图 1-14 所示，采用前后端两点支承典型结构，具有很高的刚度和旋转精度。主轴孔直径为 ϕ60mm，主轴孔锥度为 MT6。

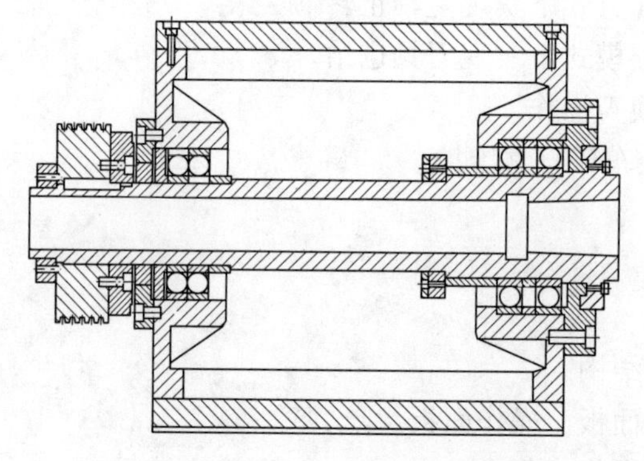

图 1-14 主轴结构示意图

数控车床主轴采用三角皮带传动形式，如图 1-15 所示。三角皮带传动可调整皮带张紧力获得最佳的传动性能。其优点是在数控车床变换速度运转时无噪音。

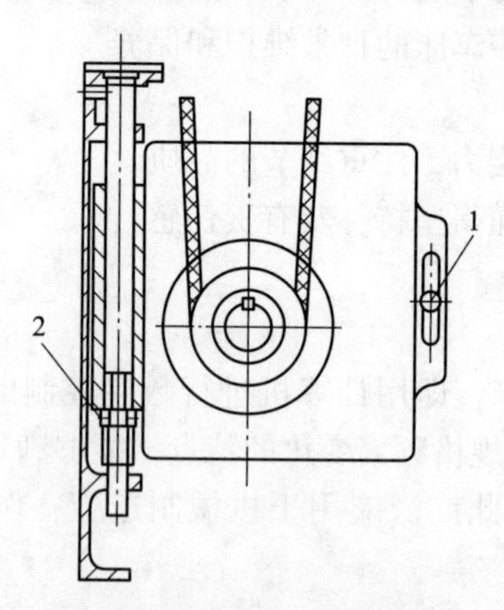

图 1-15 主轴电机皮带调节示意图

1—锁紧螺栓；2—主轴皮带调节螺栓

1.2.1.3 进给运动

数控车床进给系统分为纵向（X 轴）、横向（Z 轴）。进给运动均由伺服电机驱动滚珠丝杠，实现快速移动和进给运动，进给运动示意图如图 1-16 所示。

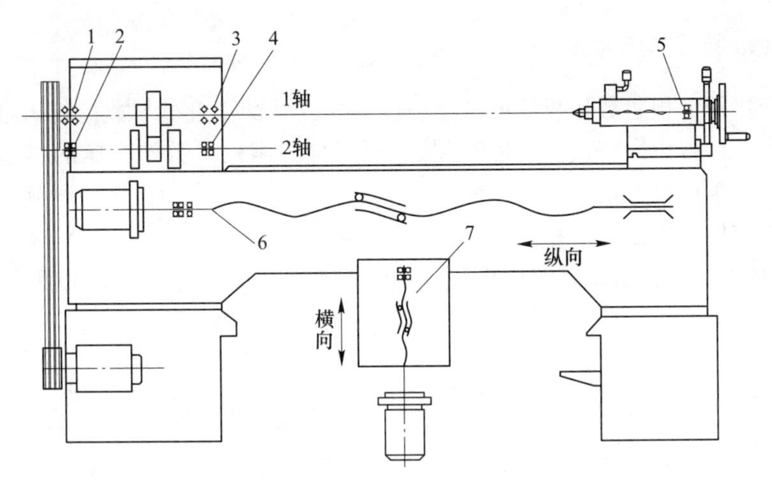

图 1-16　进给运动示意图

1~5—轴承；6—纵向进给丝杠；7—横向进给丝杠

纵向进给运动（X 轴）：X 轴是径向的，且平行于横滑座。

横向进给运动（Z 轴）：传递主要切削力的主轴为 Z 轴，且垂直于工件装夹面。若有多个主轴，则选择一个垂直于工件装夹面的主轴为 Z 轴。

1.2.1.4　数控车床主要技术参数

CK6140 数控车床主要技术参数见表 1-3。

表 1-3　CK6140×750 数控车床主要技术参数

项 目 名 称	参　数
机床上最大回转直径	ϕ400mm
最大回转托架	ϕ210mm
工件最大长度	750mm
主轴转速范围	80-750/400-1800r/min 或 80-1800r/min
主轴孔	ϕ52/ϕ82mm
主轴孔锥度	MT6
刀架工位	4
刀架最大尺寸	20×20（4 站）
尾座套筒锥度	MT4
尾架套筒最大范围	140mm
主电机功率	5.5kW
（Z）切断电机（供电）	6N · m
（X）切断电机（供电）	4N · m
包装尺寸	2300mm×1380mm×1720mm
净重	14000kg

1.2.2 CK6140 数控车床面板

数控机床操作面板是数控机床的重要组成部件，是操作人员与数控机床进行交互的工具，主要由显示装置、NC 键盘、MCP、状态灯等部分组成。数控车床的类型和数控系统的种类很多，各生产厂家设计的操作面板也不尽相同，但操作面板中各种旋钮、按钮和键盘的基本功能与使用方法基本相同。

CK6140 数控车床面板由系统面板和操作面板两部分组成，如图 1-17 和图 1-18 所示。

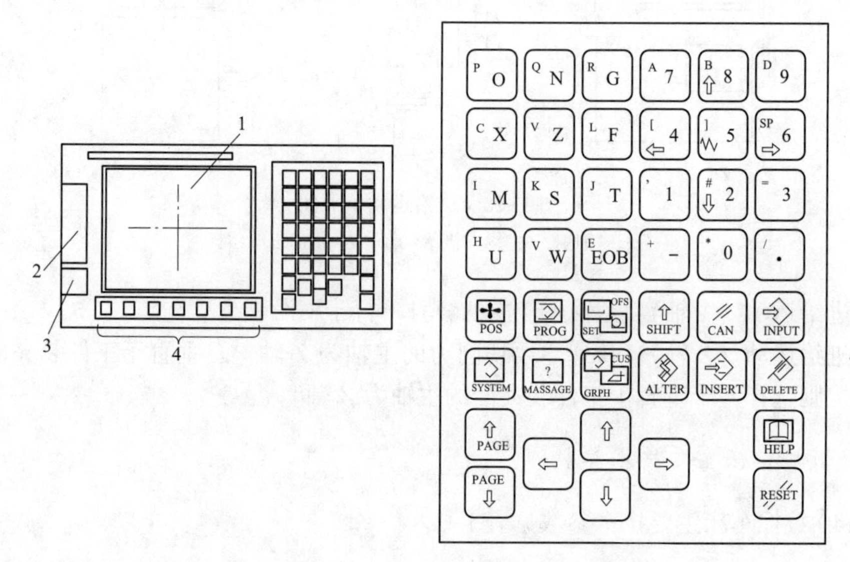

图 1-17 FANUC Series 0i-MODEL F 系统面板

1—液晶显示器（LCD）；2—PCMCIA 端口；3—USB 端口；4—软键

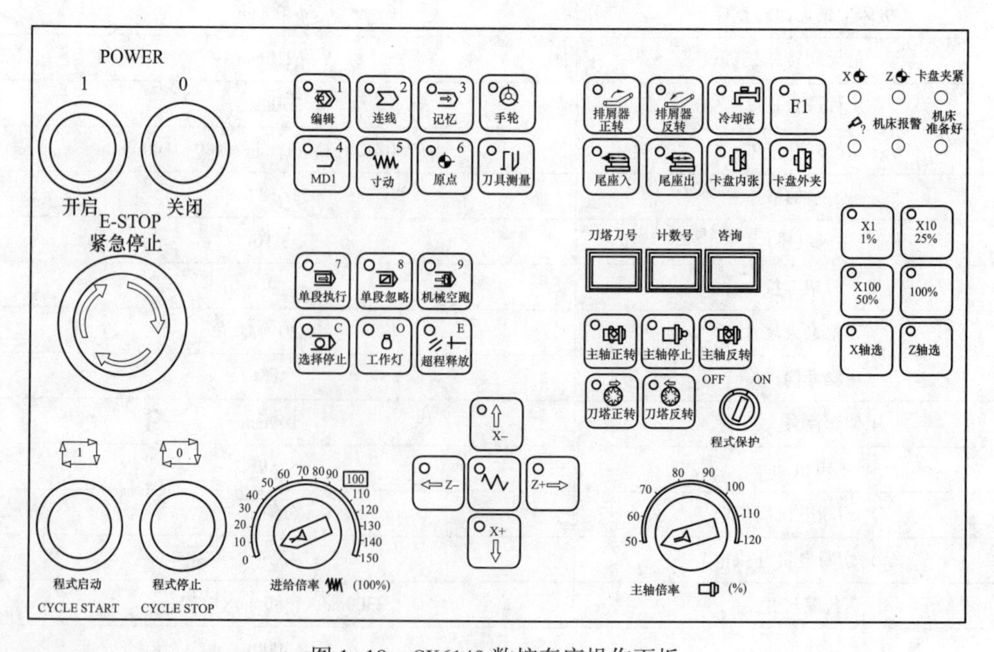

图 1-18 CK6140 数控车床操作面板

数控车床系统面板按键名称见表1-4，数控车床操作面板按键功能见表1-5。

表1-4　数控系统面板主要按键名称

图　标	按键名称	图　标	按键名称
◄□□□□□□►	软键	P O	地址和数字键
POS	位置显示键	PROG	程序键
OFS SET	偏置/设置键	⇧ SHIFT	切换键
CAN	取消键	INPUT	输入键
SYSTEM	系统参数键	? MASSAGE	信息键
CUS GRPH	图形显示键	ALTER	替换键
INSERT	插入键	DELETE	删除键
⇧ PAGE PAGE ⇩	翻页键	⇧ ⇦ ⇨ ⇩	移动光标键
HELP	帮助键	RESET	复位键
E EOB	段结束符键		

表1-5　数控车床操作面板按键功能

功能块名称	按　键	功　能　说　明
电源开关	1 开启	系统上电
	0 关闭	系统断电

续表 1-5

功能块名称	按键	功能说明
紧急停止	E-STOP 紧急停止	紧急停止：此按钮为红色，若遇到或将发生意外时按此钮，即切断电源停止伺服控制；欲恢复电源时，将此按钮轻轻向顺时针旋转后，再重新做原点复归
工作方式	1 编辑	编辑模式：对程序等进行编辑
	2 连线	连线模式：机台一方面可与个人计算机执行程序传输，另一方面可同时加工
	3 记忆	记忆模式：计算机执行内存存储程序
	4 MDI	MDI 模式：此键手动数据输入，用于加工中不记忆操作，通常仅限于输入一单节指令并执行
	5 寸动	寸动模式：使用机床各轴的手动移动，可选择轴向移动方向按键及移动速率，选择此按钮加以控制
	6 原点	原点模式：各轴向原点复归。（Z 轴执行原点复归前，须先执行 X 轴原点复归）
	手轮	手轮模式：当使用手轮操作方式时，选择此按键
主轴功能	主轴正转	主轴正转：（限手动模式）按下此键，主轴执行正转动作，指示灯亮起
	主轴反转	主轴反转：（限手动模式）按下此键，主轴执行反转动作，指示灯亮起
	主轴停止	主轴停止：按下此键，主轴立即停止转动
	50 60 70 80 90 100 110 120 主轴倍率 (%)	主轴旋转速调整：自动操作时，旋转此钮可调整主轴转速，切削螺纹时则无效，其调整范围在 50%～120%

功能块名称	按　键	功　能　说　明
自动操作	程序启动 CYCLE START	程序启动：此键是用来启动自动操作循环，在自动操作执行时，此灯会一直亮着，直到执行完毕
	程序停止 CYCLE STOP	程序停止：在自动操作中按此按钮，可立即停止刀具进给。当再按此按钮时又可继续执行未完成程序指令
	7 单段执行	单段执行：每按“程序启动”键一次，程序执行一个程序段
	8 单段忽略	单段忽略：程序执行到开头有“/”的程序段时，会不执行此段，而继续执行下一程序段
	9 机械空跑	机械空跑：程序中轴的切削速率由面板上的进给速率选择开关控制
	C 选择停止	选择停止：本按钮必须与程序中的 M01 指令配合，当执行该指令时，将此钮按下，当程序执行到 M01 时，机床将自动停止加工；若不按亮此钮，M01 指令将被忽略而不执行停止
	0 工作灯	工作灯：按下此键时，键指示灯立即亮起，同时工作照明灯会立即亮起。若再按一次，键指示灯则会消失，同时工作照明灯将会熄灭
	E 超程释放	超程释放：按住此键时，可以解除过行程状态或紧急停止状态
进给倍率	0 10 20 30 40 50 60 70 80 90 100 110 120 130 140 150 进给倍率 (100%)	进给倍率调整：在自动状态下，由 F 代码指定的进给速度可以用此开关调整，调整范围 0~150%；加工螺纹时不允许调整
快速移动倍率	X1 1%	当机床手动时，此钮可控制 *X/Z* 轴的移动速率，每按一次轴/位置移动 0.001mm； 当自动执行程序时，此钮可控制快速位移（G00）速率，降低至 1%； 当使用手摇操作时，每轮转动一格滑板移动 0.001mm
	X10 25%	当机床手动时，此钮可控制 *X/Z* 轴的移动速率，每按一次轴/位置移动 0.01mm； 当自动执行程序时，此钮可控制快速位移（G00）速率，降低至 25%； 当使用手摇操作时，每轮转动一格滑板移动 0.01mm

续表 1-5

功能块名称	按 键	功 能 说 明
快速移动倍率	X100 50%	当机床手动时，此钮可控制 X/Z 轴的移动速率，每按一次轴/位置移动 0.1mm； 当自动执行程序时，此钮可控制快速位移（G00）速率，降低至 50%； 当使用手摇操作时，每轮转动一格滑板移动 0.1mm
	100%	当机床手动时，此钮可控制 X/Z 轴的移动速率，每按一次轴/位置移动 1.0mm； 当机床自动执行时，选择 100% 其机床移动速度与程序设定 G00 速度一致； 当使用手摇操作时，每轮转动一格滑板移动 1.0mm
轴选择	X轴选	选择 X 坐标轴
	Z轴选	选择 Z 坐标轴
旋转手轮		沿“-”向旋转（逆时针）表示沿轴负方向进给，沿“+”向旋转（顺时针）表示沿轴正方向进给
轴/位置	⇧ X-	沿 X 轴负方向移动，刀具沿横向接近工件
	X+ ⇩	沿 X 轴正方向移动，刀具沿横向远离工件
	⇦ Z-	沿 Z 轴负方向移动，刀具沿纵向接近工件
	Z+ ⇨	沿 Z 轴正方向移动，刀具沿纵向远离工件
		快速移动：沿所选坐标轴快速移动
辅助功能	冷却液	手动冷却液：当按下此键时，键指示灯立即亮起，同时切削液会立即喷出；若再按一次，灯会熄灭，同时切削液会停止喷出
	OFF ON 程式保护	程序保护开关：此开关用来防止保存在内存的程序被错误修改。修改程序时需将程序保护开关转至“丨”的位置

续表 1-5

功能块名称	按　键	功 能 说 明
刀塔操作	刀塔正转	刀塔正转：按一下此键刀塔会执行正转一次，按住此键刀塔持续正转（限手动模式）
	刀塔反转	刀塔反转：按一下此键刀塔会执行反转一次，按住此键刀塔持续反转（限手动模式）
指示灯	机床准备好	系统接通电源，电源指示灯亮
	X	完成 X 向回参考点，X 回零指示灯亮
	Z	完成 Z 向回参考点，Z 回零指示灯亮
	机床报警	机械故障警示灯
	?	滑道油不足警示灯，用于润滑滚珠导螺杆及床台，灯亮时必须加油补充

1.2.3　数控车床的基本操作

1.2.3.1　机床启动和关闭

（1）开机操作顺序。

1）打开外部总电源开关。

2）切换机床电气箱总电源开关至“ON”状态，打开机床总电源。

3）按下操作面板上的 1 开启 开关，打开 NC 电源。

4）释放紧急停止 E-STOP 紧急停止 开关。

（2）关机操作顺序。

1）按下紧急停止 E-STOP 紧急停止 开关，使机床进入紧急停止状态。

2）按操作面板上（关闭）开关，关闭 NC 电源。

3）切换机床电气箱总电源开关至“OFF”状态，关闭机床总电源。

4）关闭外部总电源开关。

1.2.3.2 紧急停止开关

机床运行过程中，遇有危急情况时，应立即按紧急停止（E-STOP 紧急停止）按钮，机床将立即停止所有的控制。解除时，顺时针方向旋转此按钮，即可恢复待机状态。在解除紧急停止后，各轴必须返回机床参考点，方可继续操作。

1.2.4 手动返回机床参考点

采用增量式测量系统，机床工作前必须执行返回参考点操作。一旦机床出现断电、急停或超程报警信号，数控系统就失去了对参考点坐标的记忆，操作者在排除故障后也必须执行返回参考点操作。采用绝对式测量系统不需要回参考点。手动返回机床参考点操作步骤如下。

（1）按［原点］键。

（2）按［X+］键和［Z+］键，刀具快速返回参考点，回零指示灯亮，查看 CRT 显示器上机械坐标值是否为零。

应注意的是机床回参考点顺序是先 X 轴，后 Z 轴，防止刀架碰撞尾座。另外当滑板上的挡块距离参考点不足 30mm 时，要先按［寸动］键使滑板移向参考点的负方向，然后再返回机床参考点。

1.2.5 手动操作机床

1.2.5.1 滑板手动进给

快速移动功能，可以将各轴做长距离且快速移动。其操作步骤如下。

（1）选择［寸动］键。

（2）将快速移动速率百分比调至所需位置。

（3）根据轴/位置选择开关，使各轴沿需要的方向移动。

1.2.5.2 手动控制主轴转动

（1）主轴启动。

1）按［MDI］键；

2）按［PROG］键，CRT 显示器上出现 MDI 下的程序画面。

3）输入“M03”或“M04”，输入“S××”，如“M03 S500”，按 EOB、INSERT 键。

4）按循环启动 程式启动 CYCLE START 键，主轴按设定的转速转动。

（2）主轴停止。

1）在 MDI 画面中输入“M05”，按 EOB、INSERT 键。

2）按循环启动 程式启动 CYCLE START 键，主轴停止。

开机后首次主轴转动采用上面方法，后面操作可以在手动模式下直接按主轴正转、反转或停止。

1.2.5.3　手动操作刀架转位

（1）按 MDI 键。

（2）按 PROG 键。

（3）输入 T××，如“T01”，按 EOB、INSERT 键。

（4）按循环启动 程式启动 CYCLE START 键，1 号刀具转到工作位置。

1.2.6　数控车床的日常维护保养

数控车床在使用一段时间后，两个相互接触的零件之间会产生磨损，其工作性能逐渐受到影响，这时就应对数控车床的一些部件进行适当的调整、维护，使机床恢复到正常的技术状态。数控车床的日常保养与一级保养由操作人员完成，二级保养由操作人员与维修人员共同完成，其保养维护周期及每次保养维护的时间还应根据数控车床结构、粗精加工不同情况来确定，并结合实际加以调整。

1.2.6.1　数控车床操作人员每日进行相应的保养与维护

（1）检查润滑油箱的油量是否正常，开机检查润滑系统能否正常运转。

（2）检查液压油箱的油量是否正常，开机检查液压系统能否正常运转。

（3）检查冷却箱的切削液量是否正常，开机检查切削液系统能否正常运转。

（4）开机检查数控系统是否正常，各风扇排屑器运转是否正常，检查各压力表数值是否正常。

（5）清除数控车床内各导轨的铁屑及杂物，并检查其表面是否有划痕。

（6）开机后必须空转一定时间，检查其运转是否正常；完成一天加工任务后，关闭数控车床的数控系统，关闭电源，并清理工作地点。

1.2.6.2　数控车床的一级保养

（1）擦拭数控车床的外表，包括罩、盖和附件，保证外表无油污、无锈、蚀、无铁屑及杂物，保证内外清洁。

（2）清理回转刀架、尾座上的铁屑，检查其运转是否正常。

（3）检查车床上是否有螺钉松动，各油箱、管路、油标连接是否稳固，若有松动及时修补。

（4）检查主轴传动带、X 轴传动带、Z 轴传动带的磨损情况与张紧力。

（5）检查液压系统及该系统下液压泵、液压马达等部件运转是否正常，运转时是否有异响。

（6）检查冷却系统及该系统下的冷却泵、冷却马达等部件运转是否正常，运转时是否有异响。

（7）检查排屑系统及该系统下的排屑器、排屑电机等部件运转是否正常，运转时是否有异响。

（8）检查润滑系统运转是否正常，保证各油路畅通，通过检查并根据需要更换过滤器；检查各出油点是否正常出油并对机床（导轨、尾座）进行润滑。

（9）检查所有油箱及冷却水箱，根据需要进行补充，根据需要对机床各注油点注入润滑油脂，保证油箱及各油标、油窗明亮。夏天根据要求在切削液中加入防锈剂。

（10）检查数控系统启动后是否正常，数控车床各轴及回转刀架的运动是否正常。

（11）擦拭电器箱，保证内外清洁，检查各线路是否漏电，各触点接触是否良好，检查限位装置与接地是否安全可靠。

1.2.6.3 数控车床的二级保养

（1）常规检查。

机床运转 500 小时后，需进行常规检查和保养，一般应由操作工人为主、检修人员配合进行，检查时必须先切断电源。检查项目见表 1-6。

表 1-6 常规检查项目

序号	部 位	检 查 项 目
1	电器系统	急停按钮是否灵敏，可靠； 电动机是否运转正常，有无不正常的发热现象；电线，电缆有无破损； 行程开关、按钮功能是否正常，动作是否可靠
2	数控系统	数控系统启动、关闭是否正常可靠
3	冷却、润滑系统	切削液，润滑油是否符合要求； 油箱、切削液箱的液面是否达到规定要求； 各润滑点是否合理润滑； 切削液是否有明显的污染，润滑油质量是否合格； 刮屑板是否损坏
4	电机装置	三角皮带张紧力是否合适；表面是否有裂纹； 皮带轮运转是否正常
5	防护罩	检查防护罩是否已脏导致可视性下降

（2）定期检查。

机床运转一定时间后，由于相互接触的零件产生磨损，其工作性能逐渐受到影响，应定期检查调整机床。定期检查一般在检修人员的辅导下，由操作工人负责进行。检查和保养见表 1-7。

表 1-7　定期检查项目

序号	检查部位	检查与维修	间隔
1	电器装置	检查并紧固各接线螺钉；检查接地装置	6 个月
2	操作系统	检查数控系统启动、关闭是否正常可靠	每班检查
3	冷却系统	清理切屑盘	适时进行
		更换切屑液	2 个月（每天 8 小时工作制）
		清洗过滤网和水箱	6 个月
4	润滑系统	检查润滑泵、分油器；检查油路是否畅通 检查油质	1 年
5	安全保护	检查安全装置是否可靠	6 个月
6	三角带	外观检查，松紧程度检查；清理皮带轮	6 个月
7	防护罩	检查防护罩是否已脏而导致可视性下降； 用柔软的抹布蘸清洗剂轻轻擦洗污渍，然后用干净的抹布擦拭干净	1 个月

注：除特殊注明外，间隔时间是在两班制前提下确定的。

数控车床集电、机、液于一身，具有技术密集和知识密集特点，是一种自动化程度高、结构复杂且又昂贵的先进加工设备。为了充分发挥其效益，减少故障的发生，必须做好维护和保养工作，所以要求数控车床维护人员不仅要有机械、加工工艺以及液压气动方面的知识，也要具备电子计算机、自动控制、驱动及测量技术等知识，这样才能全面了解、掌控数控车床，及时做好维护工作。

1.2.6.4　机床工作环境

（1）环境温度：5~40℃范围内，且 24h 平均温度不超过 35℃。

（2）相对湿度：30%~95%范围内，且湿度变化的原则是不应引起冷凝。

（3）海拔高度：1000m 以下。

（4）大气：没有过多的灰尘、酸气、腐蚀气体和盐分。

（5）避免阳光直射机床，或热辐射机床而引起环境温度的变化。

（6）安装位置远离振动源。

（7）安装位置远离易燃、易爆物品。

1.2.6.5　机床精度

（1）被加工工件精度：IT6~IT7。

（2）被加工工件表面粗糙度：Ra1.6μm。

（3）定位精度（X/Z）：0.03/0.04mm。

（4）重复定位精度（X/Z）：0.012/0.016mm。

1.2.7 数控车床安全操作规程

1.2.7.1 学生守则

（1）上岗前必须穿戴好防护用品，加工时不准戴手套，女同学必须戴工作帽，不准将头发留在外边，不准穿高跟鞋，不准戴首饰。

（2）使用数控机床前做好设备的维护保养工作。

（3）毛坯、工量具应摆放在固定位置，图纸或指导书应放在便于使用处。

（4）工件必须卡牢，刀具要拧紧，防止松动甩出伤人。开机前，检查卡盘扳手是否拿离卡盘。

（5）加工时应关闭机床防护罩，禁止用手触摸正在转动的工件。

（6）装卸工件、测量加工表面及手动变挡调速时，必须先停车。

（7）加工过程中发现车床运转声音不正常或出现故障时，要立即停车检查并报告指导教师，以免出现危险。

（8）每日实习完毕要认真清扫机床，保证床面、导轨的清洁和润滑。

（9）整理好工具、量具和工件。

（10）遵守实习管理的各项规章制度，对违反纪律及规章制度的学生，指导老师要给予必要的批评教育，情节严重者，指导老师有权停止其实习。

1.2.7.2 安全操作规程

（1）学生必须在教师指导下进行数控机床操作。

（2）单人操作机床，禁止多人同时操作。

（3）手动回参考点时，机床各轴位置要距离参考点 30mm 以上。

（4）使用手轮或快速移动方式移动各轴时，一定要看清各轴“+”“-”方向后再移动，移动时先慢后快。

（5）学生遇到问题时，应立即向指导老师报告，禁止进行尝试性操作。

（6）程序运行前要检查光标在程序中的位置、机床各功能按钮的位置和导轨上是否有杂物、工具等。

（7）启动程序时，一定要一只手按开始按钮，另一只手放在急停按钮处，程序在运行当中手不能离开急停按钮，如有紧急情况立即按下急停按钮。

（8）机床在运行当中要关闭防护门，以防切屑、润滑油飞出。

（9）程序中有暂停指令，需要测量工件尺寸时，要待机床完全停止、主轴停转后进行测量；此时千万不要触及开始按钮，以免发生人身事故。

（10）注意手、身体和衣服不能靠近正在旋转的工件或机床部件。

（11）在高速切削时，不准用手直接清除切屑，用专门的钩子清除。

任务实施

（1）启动 CK6140 数控车床。

（2）分别在手动方式、MDI 方式下实现主轴正转。
（3）分别在手动方式、手摇模式下实现刀具进给。
（4）完成任意刀号的换刀操作。
（5）完成 CK6140 数控车床日常维护。

任务评价

任务评价见表 1-8。

表 1-8　任务评价表

项目	序号	评分标准		配分	得分
机床基本操作（90%）	1	机床启动、关闭	每一步操作错误扣除 6 分，扣完为止	12	
	2	手动方式下主轴正转	每一步操作错误扣除 2 分，扣完为止	14	
	3	MDI 方式下主轴正转	每一步操作错误扣除 4 分，扣完为止	12	
	4	手动方式下刀具进给	每一步操作错误扣除 4 分，扣完为止	12	
	5	手摇模式下刀具进给	每一步操作错误扣除 4 分，扣完为止	10	
	6	换刀操作	每一步操作错误扣除 5 分，扣完为止	10	
	7	日常维护	每日检查遗漏一项扣除 4 分，扣完为止	20	
职业能力（10%）	8	学习能力		2	
	9	表达沟通能力		2	
	10	团队合作		2	
	11	安全操作与文明生产		4	
合计					

同步思考与训练

（1）思考题：
1）CK6140 数控车床机械结构由哪些部分组成？
2）简述紧急停止按钮的作用。
3）简述数控车床关机的过程。
4）简述数控机床手动返回机床参考点的过程。
5）每日进行数控车床保养与维护的内容包括哪些？
（2）同步训练题：
1）在 MDI 模式下，实现主轴正转，转速 400r/min。
2）当前是 01 号刀具处于工作位置，要求将 4 号刀换到工作位置。

任务1.3 数控编程基本知识

任务描述

零件如图1-19所示，零件精加工程序见表1-9，任务要求：在数控机床上输入程序并模拟。

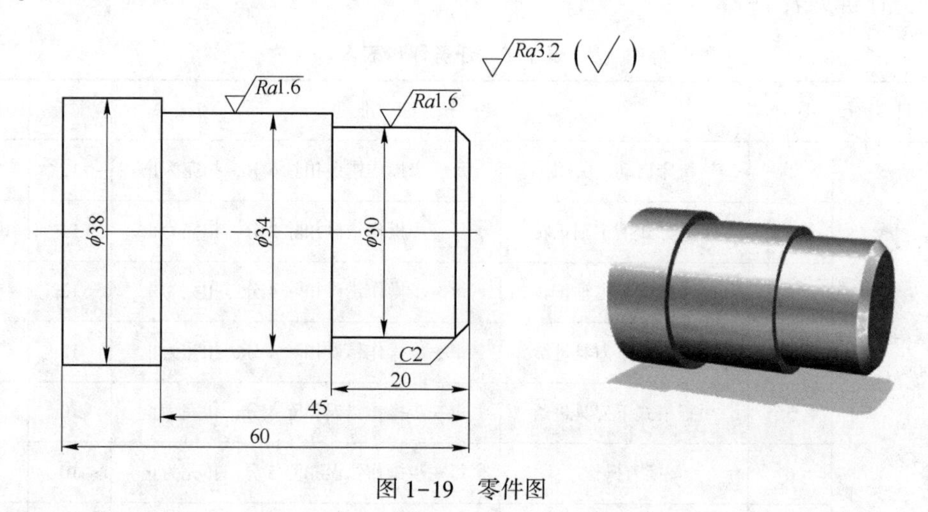

图1-19 零件图

表1-9 零件精加工程序

O131；	程序号
G40 G97 G99 M03 S1000；	主轴正转，转速为1000r/min
T0101；	换1号刀
M08；	打开切削液
G00 Z2.0；	快速定位
X26.0；	
G01 Z0 F0.1；	直线进给至工件端面
G01 X30.0 Z-2.0；	精车倒角
Z-20.0；	精车 φ30 外圆
X34.0；	精车端面
Z-45.0；	精车 φ34 外圆
X38.0；	精车端面

续表 1-9

Z-60.0;	精车 ϕ38 外圆
X40.0;	精车端面
G00 X100.0;	回换刀点（X100.0，Z100.0）
Z100.0;	
M30;	程序结束

任务目标

（1）知识目标：

1）了解数控编程内容与方法。

2）说明数控程序的结构与格式。

3）描述数控车削加工坐标系。

（2）技能目标：

1）能解释数控程序功能字的含义。

2）能利用数控机床系统面板，熟练完成数控程序的手工输入。

3）能利用数控机床操作面板，模拟数控程序。

（3）素质目标：

1）自觉遵守安全操作规范及 8S 工作要求。

2）培养认真负责的工作态度。

相关知识

1.3.1 数控车削编程基本知识

1.3.1.1 数控编程的内容及方法

（1）数控程序的编制一般包括以下几个方面。

1）分析零件图样，制定工艺方案。编程人员根据零件图，分析零件的材料、形状、尺寸、精度及毛坯形状和热处理要求等，明确加工的内容和要求，选择合适的数控机床，拟定零件加工方案，确定加工顺序、走刀路线、装夹方法、刀具及合理的切削用量等，结合所用数控机床的规格、性能、数控系统的功能等，充分发挥机床的功能。

2）数值计算。在确定了工艺方案后，就需要根据零件的几何尺寸、加工路线等，计算刀具中心运动轨迹，以获得刀位数据。数控系统一般均具有直线插补与圆弧插补功能，对于加工由圆弧和直线组成的较简单的零件，只需要计算出零件轮廓上相邻几何元素交点或切断的坐标值，得出各几何元素的起点、终点、圆弧的圆心坐标值等，就能满足编程要求。当零件的几何形状与控制系统的插补功能不一致时，就需要进行较复杂的数字计算，一般使用计算机辅助计算，否则难以完成。

3）编写零件加工程序。在完成上述工艺处理及数值计算工作后，编程人员根据使用数控系统规定的功能指令代码及程序段格式，编写零件加工程序。并填写有关的工艺文件，如数控加工工序卡片、数控刀具卡片、数控程序卡片等。

4）制备控制介质。将编写的加工程序内容记录在控制介质上，作为数控装置的输入信息，通过程序的手工输入或通信传输的方式输入到数控系统中。

5）程序校验和首件试切。在正式加工之前，必须对程序进行校验和首件试切。通常启动机床的空运行功能，来检查机床运动和运动轨迹的正确性，以校验程序。在具有 CRT 图形模拟显示功能的数控机床上，可通过显示走刀轨迹来模拟刀具对工件的切削过程，对程序进行检查。但这些方法只能检验出运动是否正确，不能检验出被加工零件的加工精度。因此，要进行零件的首件试切，当发现有加工误差时，要分析误差产生的原因，采取尺寸补偿措施，加以修正。数控编程的内容和步骤如图 1-20 所示。

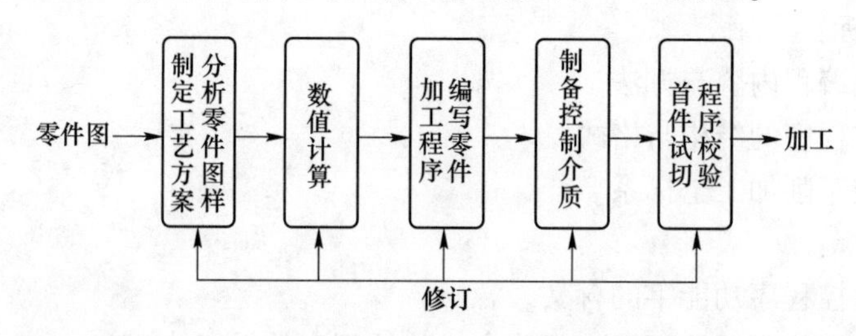

图 1-20　数控编程流程图

（2）数控编程的方法。

数控程序的编制方法一般有手工编程和自动编程两种。

1）手工编程。手工编程就是从分析零件图样、制定工艺方案、图形的数学处理、编写零件加工程序单、制备控制介质到程序的校验等主要由人工完成的编程过程。对于加工形状简单、计算量不大、程序段不多的零件采用手工编程即可实现，而且经济、及时。因此对于点位加工或由直线与圆弧组成的轮廓加工中，手工编程仍广泛应用。手工编程的缺点是耗费时间较长，容易出现错误，无法胜任复杂形状零件的编程工作。

2）自动编程。自动编程是指在编程过程中，除了分析零件图样和制定工艺方案由人工进行外，其余工作均由计算机辅助完成。采用计算机自动编程时，数学处理、编写程序、检验程序等工作是由计算机自动完成，由于计算机可自动绘制出刀具中心运动轨迹，编程人员可以及时检查程序是否正确，需要时可及时修改，以获得正确的程序，又由于计算机自动编程代替程序编制人员完成了烦琐的数值计算，可提高几十倍乃至上百倍编程效率，解决了手工编程难以解决的许多复杂零件的编程难题。因此，自动编程的特点就在于编程工作效率高，可解决复杂零件的编程难题。

1.3.1.2　数控程序的结构与格式

为了满足设计、制造、维修和普及的需要，在输入代码、坐标系统、加工指令、辅助功能及程序格式等方面，国际上形成了由国际标准化组织（ISO）和美国电子工程协会（EIA）分别制定的两种标准。

中国根据ISO标准制定了《数字控制机床用的七单位编码字符》（JB 3050—1982）、《数字控制机床坐标和运动方向的命名》（JB 3051—1999）、《数字控制机床穿孔带程序段格式中的准备功能G和辅助功能M代码》（JB 3208—1999）。由于数控机床生产厂家所用标准尚未完全统一，其所用的代码、指令及其含义也不完全相同，因此在数控编程时必须按所用数控机床编程手册中的规定进行。

（1）数控加工程序的结构。每一个程序都是由程序号、程序内容和程序结束三部分组成。程序内容则由若干程序段组成，程序段由若干字组成，每个字又由字母和数字组成。字组成程序段，程序段组成程序。下面是图1-21零件的精加工程序。

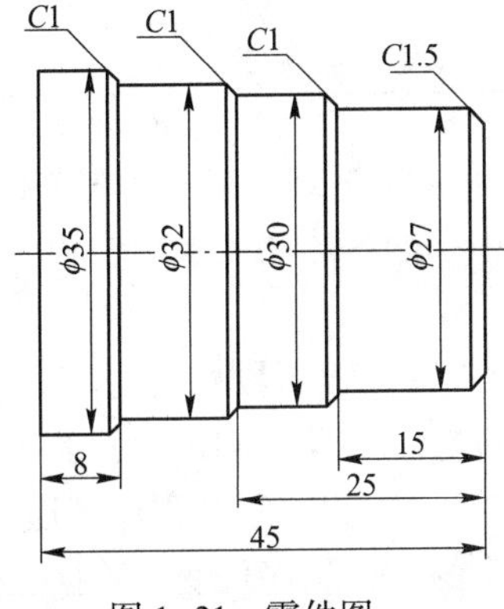

图1-21　零件图

1）程序号。程序号为程序的开始部分，为了区别存储器中的程序，每个程序都要有程序编号。在编号前采用程序编号地址符，不同的数控系统，程序地址符有所不同。在FANUC系统中，采用英文字母“O”作为程序编号地址，其他系统有采用“%”“P”等。

2）程序内容。程序内容是整个程序的核心，由许多程序段组成，每个程序段由一个或多个指令组成，表示机床要完成的全部动作。

3）程序结束。以程序结束指令M02或M30作为整个程序结束的符号，来结束整个程序。

```
O1121;                                                      程序号
G40 G97 G99 M03 S1000;   （主轴正转，转速1000r/min）          ┐
T0101;                   （调用01号刀具和01号刀补）            │
M08;                     （切削液开）                          │
G00 Z5.0;                （刀具快速点定位至精车起点Z5.0）       │
X30.0;                   （刀具快速点定位至精车起点X30.0）      │
G01 Z0 F0.1;             （直线插补至工件端面，进给量0.1mm/r）  │
X32.0 Z-1.0;             （直线插补切削倒角）                  ├ 程序内容
Z-20.0;                  （直线插补切削φ32外圆）               │
X34.0;                   （直线插补切削端面）                  │
X38.0 Z-22.0;            （直线插补切削倒角）                  │
Z-50.0;                  （直线插补切削φ38外圆）               │
G00 X100.0;              （快速退刀至换刀点）                  │
Z100.0;                  （快速退刀至换刀点）                  ┘
M30;                     （程序结束并返回起点）                程序结束
```

（2）程序段格式。零件的加工程序由程序段组成。程序段格式是指一个程序段中字、字符、数据的书写规则，通常有字—地址形式程序段格式、使用分隔符的程序段格式和固定程序段格式，最常用的为字—地址形式程序段格式。

字—地址形式程序段格式由程序段号、程序字和程序段结束符组成。字—地址形式程序段格式见表1-10。

表 1-10　字—地址可变程序段格式

1	2	3	4	5	6	7	8	9	10
N___	G___	X___ U___ P___	Y___ V___ Q___	Z___ W___ R___	I_J_K_ R___	F___	S___	T___	M___
程序段号	准备功能字	尺寸字				进给功能字	主轴功能字	刀具功能字	辅助功能字

程序段中包括的各种指令，并非在加工程序的每个程序段中都必须有，而是根据各程序段的具体功能来编入相应的指令。例如：

N10 G01 X30. 0 Z-25. 5 F0. 2;

1）程序段号。用以识别程序段的编号，位于程序段之首，由地址码 N 和后面的若干位数字组成。N10 表示该程序段的段号为 10。

2）程序字。程序字通常由地址符、数字和符号组成，字的功能类别由地址符决定，字的排列顺序可变，数据位数可多可少，不需要的字及上一程序段相同的程序字可以省略不写。

3）程序段结束。每一个程序段之后都有一个结束符，表示程序段结束。在国际标准化组织 ISO 标准代码中，结束符为“NL”或“LF”；在美国电子工业协会 EIA 标准代码中，结束符为“CR”；在 FANUC 数控系统中结束符用“;”，也有的数控系统用“ * ”或不设结束符，直接按回车键即可。

（3）程序功能字。

1）准备功能字 G。准备功能字 G 用于控制系统动作方式的指令。用地址 G 和两位数字表示，从 G00~G99 共 100 种。G 代码分为模态代码和非模态代码。模态代码表示该 G 代码在一个程序段中功能保持直到被取消或被同组的另一个 G 代码所代替。非模态代码只在有该代码的程序段中有效。

G 代码按其功能进行了分组，同一功能组的代码可互相代替，不允许写在同一程序段中。数控车床常用的 G 代码见表 1-11。

表 1-11　数控车床常用 G 代码

代码	组　别	功　能
G00	01	快速点定位
G01		直线插补
G02		顺时针圆弧插补
G03		逆时针圆弧插补
G04	00	暂停
G20	06	英寸输入
G21		毫米输入

续表 1-11

代码	组　别	功　能
G40	07	取消刀尖圆弧半径补偿
G41		刀尖圆弧半径左补偿
G42		刀尖圆弧半径右补偿
G50	00	1. 坐标系设定；2. 主轴最大速度限定
G65		调用宏指令
G70		精车复合循环
G71		粗车复合循环
G72		端面粗车复合循环
G73		固定形状粗车复合循环
G74		端面钻孔复合循环
G75		外圆切槽复合循环
G76		螺纹切削复合循环
G66	12	宏程序模态调用
G67		宏程序模态调用取消
G90	01	外圆切削循环
G92		螺纹切削循环
G94		端面切削循环
G96	02	恒线速度控制
G97		取消恒线速度控制
G98	05	每分钟进给量
G99		每转进给量

2）尺寸字。尺寸字用于确定加工时刀具运动的坐标位置。

所有坐标点的坐标值均从编程原点计量的坐标系，称为绝对坐标系。*X*、*Y*、*Z* 用于确定终点的直线坐标绝对尺寸。

坐标系中的坐标值是相对前一位置（或起点）来计算的，称为增量（相对）坐标。*U*、*V*、*W* 表示用于确定终点的直线坐标增量尺寸。

A、*B*、*C* 用于确定附加轴终点的角度坐标尺寸；*I*、*J*、*K* 用于确定圆弧的圆心坐标，R 用于确定圆弧半径。

3）进给功能字 F。进给功能字 F 由地址码 F 和后面数字构成，用于指定切削的进给速度，即刀具中心运动时的进给速度，单位为 mm/min 或 mm/r。

①每分钟进给模式 G98

格式：G98 F__；

F 后面的数字表示主轴每分钟进给量，单位为 mm/min，模态指令。

②每转进给模式 G99

格式：G99 F__；

F 后面的数字表示主轴每转进给量，单位为 mm/r，模态指令。

4）主轴转速功能字 S。主轴转速功能字 S 由地址码 S 和其后面的数字组成，用于指定主轴转速，单位为 r/min。

格式：S__；单位为 r/min。

【例 1-1】S600 表示主轴转速为 600r/min。

在具有恒线速功能的机床上，S 指令还有如下功能：

①主轴最高转速限制 G50

格式：G50 S__；

S 后面的数字表示主轴最高转速，单位为 r/min。

【例 1-2】G50 S2500 表示主轴最高转速为 2500r/min。

② 恒线速度控制 G96

格式：G96 S__；

S 后面的数字表示恒定的线速度，单位为 m/min。

【例 1-3】G96 S100 表示切削点的线速度控制在 100m/min。

该指令用于车削端面或直径变化较大的场合。此功能可保证当工件直径变化时，工件切削表面的线速度不变，从而保证切削速度不变，提高加工质量。

③ 恒线速度取消 G97

格式：G97 S__；

【例 1-4】G97 S1000 表示取消恒线速度控制，主轴转速为 1000r/min。

该指令可设定主轴转速并取消恒线速度控制，用于车削螺纹或工件直径变化较小的场合。

5）刀具功能字 T。刀具功能字 T 用于指定加工时所用刀具的编号。刀具功能的数字是指定的刀号，数字的位数由所用的系统决定，对于数控车床其后的数字还兼有指定刀具补偿作用。

格式：T××××；

前两位表示刀具号，后两位表示刀具补偿号。T××00 为取消刀具补偿。

【例 1-5】T0303 表示选用 3 号刀具及 3 号刀具补偿值。T0300 则取消 3 号刀位的刀补。

6）辅助功能字 M。辅助功能字 M 用于控制机床或系统的辅助装置的开关动作。由地址码 M 和后面的两位数字组成，从 M00~M99 共 100 种。各种机床的 M 代码规定有差异，必须根据说明书的规定进行编程，常用的 M 代码见表 1-12。

表 1-12　数控车床常用 M 功能

代　码	功　能	代　码	功　能
M00	程序暂停	M09	切削液关
M01	程序有条件暂停	M30	程序结束并返回起点
M02	程序结束	M41	低档
M03	主轴正转	M42	中档
M04	主轴反转	M43	高档
M05	主轴停止	M98	子程序调用
M08	切削液开	M99	子程序结束

1.3.2　数控车削加工坐标系

1.3.2.1　标准坐标系及运动方向

在数控编程时，为描述机床的运动、简化程序编制的方法及保证记录数据的互换性，数控机床的坐标系和运动方向实行标准化。国际标准化组织统一了标准坐标系，中华人民共和国机械工业部也颁布了 JB 3051—82《数字控制机床坐标系和运动方向的命名》的标准，对数控机床的坐标系和运动方向做出规定。

（1）机床相对运动的规定。为使编程人员不考虑机床上工件与刀具具体运动的情况，只需依据零件图样，确定机床的加工过程，规定：永远假定刀具相对于静止的工件而运动。

（2）坐标系的规定。数控加工时，机床的动作由数控系统发出的指令来控制，为确定机床的运动位移和方向，需要坐标系来实现，这个坐标系叫标准坐标系，也称为机床坐标系。

数控机床采用标准的右手笛卡尔直角坐标系，如图 1-22 所示。右手的大拇指、食指和中指保持相互垂直，大拇指的方向为 *X* 轴的正方向，食指为 *Y* 轴的正方向，中指的方向为 *Z* 轴的正方向。

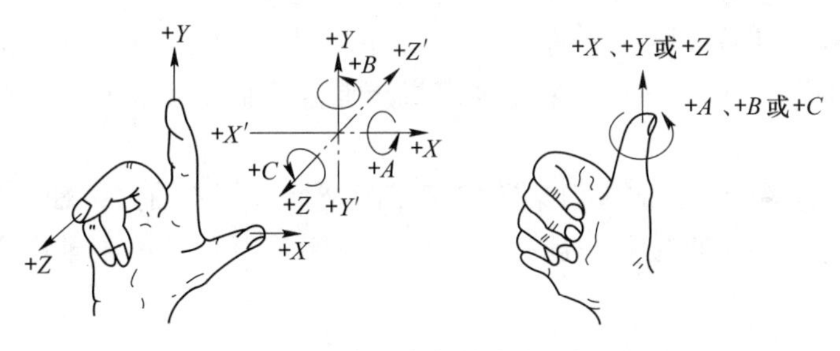

图 1-22　右手直角笛卡尔坐标系

围绕 *X*、*Y*、*Z* 坐标旋转的旋转坐标系分别用 *A*、*B*、*C* 表示，根据右手螺旋定则，大拇指的指向为 *X*、*Y*、*Z* 坐标中任意轴的正向，其余四指的旋转方向即为旋转坐标 *A*、*B*、*C* 的正向。

（3）运动方向的规定。JB 3051—82 中规定：机床某一部件运动的正方向是增大刀具与工件之间距离的方向。

1）*Z* 坐标的规定。*Z* 坐标的原点由传递切削力的主轴决定，即平行主轴轴线的坐标轴为 *Z* 坐标。若有多个主轴，则选垂直于工件装夹面的主轴为主要主轴，*Z* 坐标则平行于该主轴轴线。若机床无主轴，则规定垂直于工件装夹平面的方向为 *Z* 坐标。*Z* 坐标的正向为刀具离开工件的方向。

2）*X* 坐标的规定。*X* 坐标平行于工件的装夹面，一般是水平的。数控车床的 *X* 坐标在工件的径向上，且平行于横向拖板。刀具离开工件旋转中心的方向为 *X* 坐标的正方向。

3）*Y* 坐标的规定。*Y* 坐标轴垂直于 *X*、*Z* 坐标轴。*Y* 坐标轴的正方向根据 *X* 和 *Z* 坐标的正方向，按照右手笛卡尔直角坐标系来确定。

1.3.2.2 机床原点与机床参考点

（1）机床原点。机床原点是机床上的一个固定点，又称为机械原点，是数控机床进行加工运动的基准参考点，在机床装配、调试时就已经确定下来。数控车床的机床原点一般设在卡盘端面与主轴中心线的交点处，如图 1-23 所示。

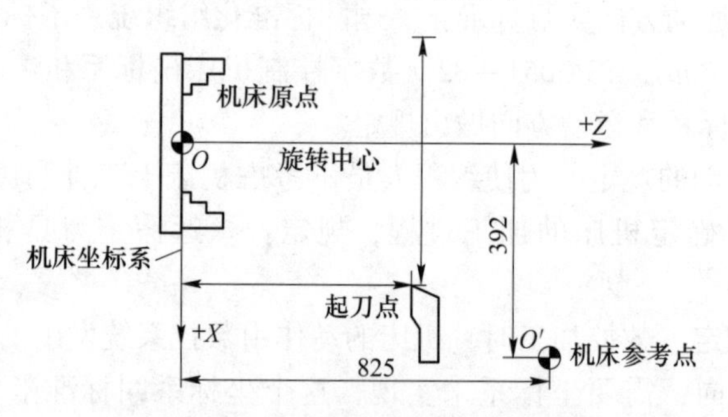

图 1-23 数控车床坐标系与参考点

（2）机床参考点。机床参考点是用于对机床运动进行检测和控制的固定位置点。机床参考点的位置是由机床制造厂家在每个进给轴上用限位开关精确调整好的，其坐标值已输入数控系统中。机床参考点与机床原点有着准确的位置关系，是一个已知数。机床参考点由机床行程限位开关和基准脉冲来确定，通常机床参考点位于行程的正极限点上，如图1-23 所示。数控机床开机时，一般需要先确定机床原点，而确定机床原点的运动就是回参考点的操作。

图 1-23 所示为平床身、前置刀架数控车床的坐标系。*Z* 轴为主轴，指向尾座的方向为正。*X* 轴的方向为工件的径向，且平行于横向滑座，刀具远离主轴中心的方向为 *X* 轴正向。

1.3.2.3 工件坐标系的建立

（1）编程坐标系。编程坐标系是供编程使用的坐标系，是编程人员根据零件图样及加工工艺等建立的坐标系。编程坐标系的原点需要根据零件图样及加工工艺要求选定，尽量选择在零件的设计基准或工艺基准上。编程坐标系的各轴方向应与所使用的数控机床相应的坐标轴方向一致，如图 1-24 所示。

（2）工件坐标系的建立。工件坐标系是指以确定的加工原点为基准所建立的坐标系。工件坐标系原点也称为编程原点，是指零件被装夹好后，相应的编程原点在机床坐标系中的位置。数控车床一般设在工件右端面的中点，如图 1-24 所示。

数控机床按照工件装夹后确定的加工原点位置和程序要求进行加工。编程人员在编制程序时，只要根据零件图样就可以选定编程原点、建立编程坐标系、计算坐标数值，而不必考虑工件毛坯装夹的实际位置。对加工人员来说，则应在装夹工件、调试程序时，将编程原点转换为加工原点，并确定加工原点的位置，在数控系统中设定加工坐标系后，就可以根据刀具当前位置，确定刀具起始点的坐标值。在加工时，工件各尺寸的坐标值都是相对于加工原点而言的，这样数控机床才能按照准确的加工坐标系位置开始加工。

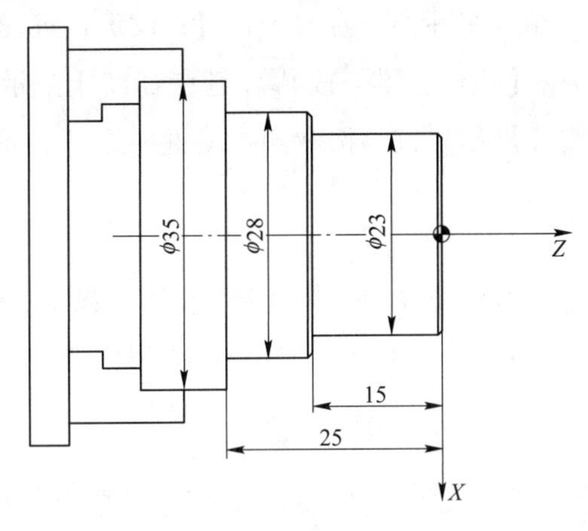

图 1-24　数控车床工件原点

1.3.3　数控程序输入与模拟

1.3.3.1　数控程序的输入

在数控机床上输入程序有两种方法，一种是手工输入程序，另一种是通过数控机床CF卡传输导入程序。

（1）手工输入程序。

按 [编辑] 键→按 [PROG] 键进入程序界面→输入程序名如“O111”→按 [INSERT] 键→按 [EOB] 键→按 [INSERT] 键→用鼠标或键盘输入 O111 程序的内容→输入结束后按 [RESET] 键回到程序起点。

输入、编辑程序常用功能如下。

1）换行：按 [EOB] 键→按 [INSERT] 键。

2）输入数据：按数字/字母键，如 M03 S500，数据被输入到输入区域；如果输入错误用 [CAN] 键删除输入区域内的数据。

3）移动光标：按 [⇧ PAGE] 键向上翻页，按 [PAGE ⇩] 键向下翻页；按 [⇧] 或 [⇩] 或 [⇦] 或 [⇨] 键向上、下、左、右移动光标。

4）删除、插入、替代：按 [DELECTE] 键删除光标所在位置的代码，按 [INSERT] 键输入区域的内容插入到光标所在代码后面，按 [ALTER] 键输入区域的内容替代光标所在位置的代码。

（2）CF卡导入程序。

CF卡导入程序步骤如下（FANUC Oi-TD 系统）。

1）确认输出设备已经准备好。

2）按 [编辑] 键→按 [PROG] 键→显示程序。

3）按【列表】键→按【操作】键→按右侧扩展键。

4）选择【设备】键→选择【M-卡】键→显示卡中内容。

5）按【F 读取】键→输入 M 卡程序序号→按【F 设定】键确认→输入机床中程序号→按【O 设定】键确认→按【执行】键→按 PROG 键在 CRT 上显示导入的程序。

6）程序传输完毕，按【操作】键→按右侧扩展键→按【设备】键→按【CNCMEM】键，回到原始状态。

（3）程序调用。

按 编辑 键→按 PROG 键进入程序界面→按【DIR】键，显示数控系统内所有的程序→输入程序名如“O111”→按【O 检索】键，调用 O111 程序。

1.3.3.2　数控程序的模拟

输入的程序必须进行检查，常用图形模拟检查程序是否正确。图形模拟操作步骤如下：

按 编辑 键→按 PROG 键→输入程序号，按 ⇩ 键显示程序→按 记忆 键→按 GRPH 键→按【图形】键→按 机械空跑 键→按循环启动键 程序启动 CYCLE START，观察程序的加工轨迹。

注意：图形模拟结束后，必须取消空运行和锁住功能，同时要进行全轴操作。

全轴操作步骤如下：

取消 机械空跑 →按 POS 键→按【绝对坐标】键→按【操作】键→按【W 预置】键→按【所有轴】键→CRT 面板坐标和实际坐标一致。

调用加工程序→按 记忆 键→按循环启动键 程序启动 CYCLE START，自动加工零件。

任务实施

（1）机床准备：

1）机床开机前例行检查。

2）系统启动。

3）手动返回机床参考点。

4）手动操作滑板进给、主轴转动、刀架转位，检查其功能运行是否正常。

（2）程序输入：

手工输入表 1-9 程序。

1）选中数控机床操作面板上的工作方式为 编辑，数控机床系统面板上的 PROG 键，进入程序输入状态。

2）输入程序号“O131”，点击 INSERT 键，输入程序号，点击 EOB 键，输入“；”。

3）依次输入程序内容：G40 G97 G99 M03 S1000；……直至程序结束。

4）点击 RESET 键，让光标回到程序号的位置，为程序模拟做准备。

（3）程序模拟：

1）程序模拟。

选中数控机床操作面板上的工作方式为 ，数控机床系统面板上的 键，【图形】键，按 和 键，按 键，观察程序的加工轨迹。

2）全轴操作。

任务评价

任务评价见表 1-13。

表 1-13　任务评价表

项目	序号	评分标准		配分	得分
编程基本知识（40%）	1	程序结构	正确划分程序结构	3	
	2	程序号	正确给程序命名（FANUC 系统）	3	
	3	程序段号	解释程序段号作用	2	
	4	准备功能 G	根据具体的程序段解释该功能	6	
	5	进给功能字 F	根据具体的程序段解释该功能	4	
	6	主轴转速 S	根据具体的程序段解释该功能	4	
	7	刀具功能 T	根据具体的程序段解释该功能	4	
	8	辅助功能 M	根据具体的程序段解释该功能	6	
	9	机床坐标系	解释机床坐标系、机床原点、参考点	6	
	10	编程坐标系	根据零件图确定编程坐标系	2	
程序输入与模拟（50%）	11	程序输入	在规定时间内，准确输入程序；每错一段扣 2 分	30	
	12	程序模拟	程序模拟操作正确；出现危险操作扣 10 分	15	
	13	全轴操作	操作有误，无分	5	
职业能力（10%）	14	学习能力		2	
	15	表达沟通能力		2	
	16	团队合作		2	
	17	安全操作与文明生产		4	
合计					

同步思考与训练

（1）思考题：

1）解释程序段“N10 G01 X26.0 Z-30.0 F0.1;”各地址含义。

2）什么是机床坐标系？什么是工件坐标系？

（2）同步训练题：

零件如图 1-25 和图 1-26 所示，加工程序见表 1-14，程序名分别为 O137、O138，在数控车床上分别输入加工程序并模拟。

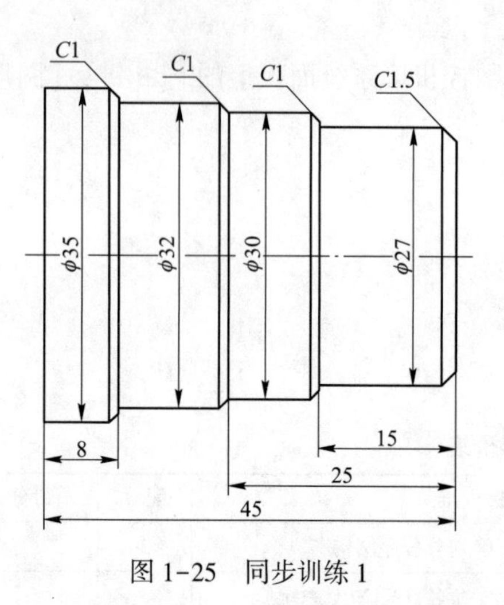

图 1-25　同步训练 1

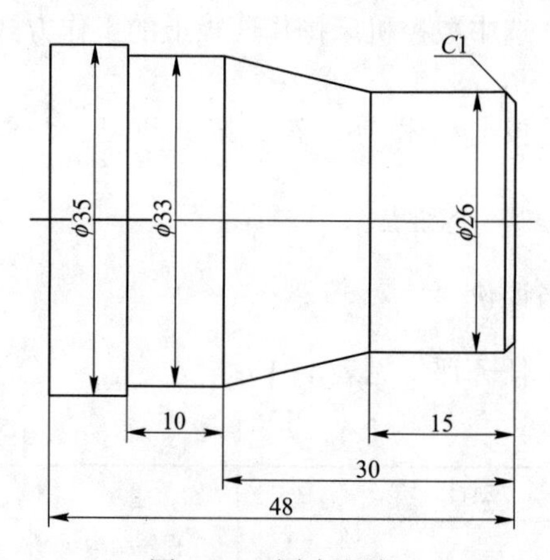

图 1-26　同步训练 2

表 1-14　同步训练加工程序

O137;	O138;
G40 G97 G99 M03 S1200;	G40 G97 G99 M03 S1200;
T0101;	T0101;
M08;	M08;
G00 Z5. 0;	G00 Z5. 0;
X24. 0;	X24. 0;
G01 Z0　F0. 1;	G01 Z0　F0. 1;
G01 X27. 0　Z-1. 5;	G01 X26. 0　Z-1. 0;
Z-15. 0;	Z-15. 0;
X30. 0 C1. . 0;	X33. 0 Z-30. 0;
Z-25. 0;	W-10. 0;
X32. 0 C1. 0;	X35. 0;
Z-37. 0;	Z-48. 0;
X35. 0 C1. 0;	G00 X100. 0;
Z-45. 0;	Z100. 0;
G00 X100. 0;	M30
Z100. 0;	
M30;	

项目 2　数控车削加工技术

数控车床能自动完成内外圆柱面、圆锥面、圆弧面、端面、螺纹等轮廓的切削加工，加工的尺寸精度可达到 IT5～IT6 级，表面粗糙度可达到 Ra1.6μm 以下，生产效率为普通机床的 3～5 倍，特别适于形状复杂的轴类或盘类零件加工，是目前应用最广泛的数控加工方法之一。

任务 2.1　台阶轴的数控加工

任务描述

图 2-1 所示零件，材料硬铝合金 LY15，毛坯为 ϕ40mm 棒料，单件生产，使用数控车床加工，编写加工程序，检测加工质量。

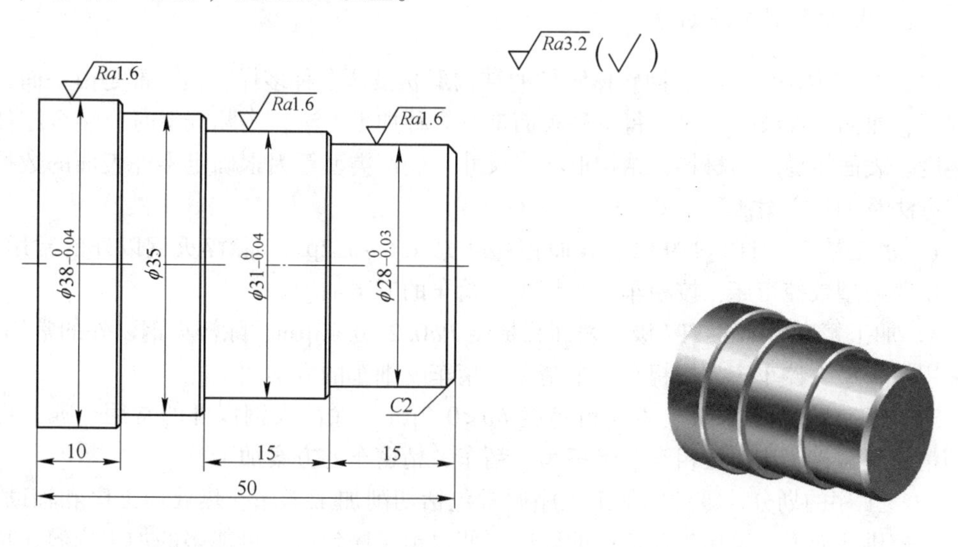

图 2-1　零件图

任务目标

（1）知识目标：

1）了解切削用量、车削刀具、数控加工工艺文件、台阶轴的车削方式。

2）解释 G20/G21，G00/G01 指令。

3）描述外圆车刀对刀过程。

4）认识游标卡尺的结构。

（2）技能目标：

1）能编写阶梯轴加工程序。

2）能完成外圆刀对刀操作。

3）能正确使用游标卡尺检测零件尺寸。

（3）素质目标：

1）正确执行安全技术操作规程，培养安全意识。

2）培养认真负责的工作态度和质量意识。

相关知识

2.1.1 工艺准备

数控车削加工工艺是结合数控车床的特点，综合运用工艺知识，在编程前对加工零件的工艺参数、刀具参数、切削用量进行工艺分析，解决数控车削加工过程中遇到的问题，实现数控加工的优质、高产与低耗。

2.1.1.1 数控车削工序划分

（1）加工方法的选择。回转体零件的结构形状虽然多种多样，但大都是由平面、内外圆柱面、曲面、螺纹等组成。每一种表面都有多种加工方法，实际选择时应结合零件的加工精度、表面粗糙度、材料、结构形状、尺寸、生产类型等因素确定零件表面的数控车削加工方法及加工方案。

1）加工精度为IT8~IT9级、表面粗糙度 $Ra1.6 \sim 3.2\mu m$、除淬火钢以外的常用金属，可采用普通型数控车床，按粗车、半精车、精车的方案加工。

2）加工精度为IT6~IT7级、表面粗糙度 $Ra0.2 \sim 0.63\mu m$、除淬火钢以外的常用金属，可采用精密型数控车床，按粗车、半精车、精车、细车的方案加工。

3）加工精度为IT5级、表面粗糙度 $Ra<0.2\mu m$、除淬火钢以外的常用金属，可采用高档精密型数控车床，按粗车、半精车、精车、精密车的方案加工。

（2）工序的划分。零件的加工工序通常包括切削加工工序、热处理工序和辅助工序。对于数控机床而言，采用工序集中的原则，使每道工序包含尽可能多的加工内容，可以减少工序数目、缩短工艺路线、提供生产效率。按工序集中原则划分工序方法如下。

1）按零件装夹定位方式划分。以一次安装完成的工艺过程为一道工序，适用于加工内容较少的零件，加工完后达到待检状态。

2）按所用刀具划分。以一把刀具加工的工艺过程为一道工序。

3）按粗、精加工划分。以粗加工中完成的工艺过程为一道工序，精加工中完成的工艺过程为一道工序。

4）按加工部位划分。以完成相同型面的工艺过程为一道工序，如外形、内腔、曲面或平面，每一部分的加工为一道工序。

（3）回转体零件非数控车削加工工序的安排。零件上有不适合数控车削加工的表面，如键槽、花键、渐开线齿形等，若有喷丸、滚压加工、抛光等特殊要求，需要安排相应的非数控车削加工工序。零件表面硬度及精度要求较高时，一般在数控车削加工之后安排热处理，再安排磨削加工。

（4）加工顺序的安排。数控车床上加工零件，按照先粗后精的原则安排零件的加工顺序，可逐步提高表面的加工精度，减少表面粗糙度。

1）粗加工阶段。主要任务是在较短的时间内，切除各表面上的大部分余量，目的是为了提高加工效率。

2）半精加工阶段。完成次要表面的加工，并为主要表面的精加工工作准备，尽量满足精加工余量均匀性要求。

3）精加工阶段。保证各主要表面达到图纸规定的要求，关键是保证加工质量。

2.1.1.2 数控车床常用装夹方法

数控车床常用三爪卡盘、四爪卡盘、一夹一顶、中心架等方式装夹工件。三爪卡盘夹紧力较小，装夹速度较快，适于装夹中小型圆柱形、正三边或正六边形工件。四爪卡盘夹紧力较大，装夹精度较高，不受卡爪磨损的影响，但夹持工件时需要找正，适于装夹形状不规则或大型的工件。采用三爪卡盘和机床尾座顶尖的一夹一顶方式，定位精度较高，装夹牢靠，适于装夹较长的轴类零件。中心架配合三爪卡盘或四爪卡盘来装夹工件，可以防止弯曲变形，适于装夹细长的轴类零件。

2.1.1.3 切削用量的选择

数控编程时，需要确定每道工序的切削用量，并以指令的形式写入程序中。选择切削用量的目的是在保证加工质量和刀具耐用度的前提下，使切削时间最短，生产率最高，成本最低。切削用量包括背吃刀量 a_p、进给量 F 和主轴转速 n（切削速度 v）。

（1）背吃刀量 a_p。背吃刀量主要根据机床、夹具、刀具、工件的刚度等因素决定。粗加工时，在条件允许的情况下，尽可能选择较大的背吃刀量，减少走刀次数，提高生产率；精加工时，通常选较小的背吃刀量，保证加工精度及表面粗糙度。

（2）进给量 F。进给量是指主轴每转一周，刀具沿着前进方向移动的距离。它根据零件的加工精度、表面粗糙度要求、刀具及工件的材料性质来选取。粗加工时，在保证刀具、机床、工件的刚度等前提下，为缩短切削时间，应选用尽可能大的进给量；精加工时，进给量主要受表面粗糙度的限制，当表面粗糙度要求较高时，应选较小的进给量。工件材料较软时，可选用较大的进给量，反之，应选较小的进给量。

（3）主轴转速 n。主轴转速是根据允许的切削速度来选择，在保证刀具的耐用度及切削负荷不超过机床额定功率的情况下选定切削速度。粗车时，背吃刀量和进给量均较大，故选较低的切削速度；精车时选较高的切削速度。切削速度与主轴转速的关系如下：

$$n = 1000v/\pi d$$

式中 n——主轴转速，r/min；

d——待加工表面直径，mm；

v——切削速度，m/min。

（4）切削用量选择的一般原则。切削用量的合理选择，对于实现优质、高产、低成本和安全操作有着非常重要的作用。切削用量选择的一般原则如下。

1）粗车时，一般以提高生产率为主，同时也考虑经济性和加工成本。首先选择大的背吃刀量，其次选择较大的进给量，增大进给量有利于断屑；最后根据已经选定的吃刀量和进给量，在工艺系统刚性、刀具寿命和机床功率许可的条件下选择一个合理的切削速度，减少刀具消耗，降低加工成本。

2）半精车或精车时，对加工精度和表面粗糙度要求较高，加工余量不大且均匀，应在保证加工质量的前提下，兼顾切削效率、经济性和加工成本，通常选择较小的背吃刀量和进给量，并选用切削性能高的刀具材料和合理的几何参数，以尽可能提高切削速度，保证零件加工精度和表面粗糙度。

切削用量的具体数值可参考切削用量手册，并结合机床说明书给定的允许切削用量范围确定，数控车削切削用量参考见表 2-1。

表 2-1　切削用量选择参考表

工件材料	加工方式	背吃刀量/mm	切削速度/m · min^{-1}	进给量/mm · r^{-1}	刀具材料
碳素钢 $\sigma_b>600MPa$	粗加工	5~7	60~80	0.2~0.4	YT 类
	粗加工	2~3	80~120	0.2~0.4	
	精加工	0.2~0.3	120~150	0.1~0.2	
	车螺纹		70~100	导程	
	钻中心孔		500~800r/min		W18Cr4V
	钻孔		~30	0.1~0.2	
	切断（宽度<5mm）		70~110	0.1~0.2	YT 类
合金钢 $\sigma_b>1470MPa$	粗加工	2~3	50~80	0.2~0.4	YT 类
	精加工	0.1~0.15	60~100	0.1~0.2	
	切断（宽度<5mm）		40~70	0.1~0.2	
铸铁 200HBS 以下	粗加工	2~3	50~70	0.2~0.4	YG 类
	精加工	0.3~0.15	70~110	0.1~0.2	
	切断（宽度<5mm）		50~70	0.1~0.2	
铝	粗加工	2~3	600~1000	0.2~0.4	YG 类
	精加工	0.2~0.3	800~1200	0.1~0.2	
	切断（宽度<5mm）		600~1000	0.1~0.2	
黄铜	粗加工	2~4	400~500	0.2~0.4	YG 类
	精加工	0.1~0.15	450~600	0.1~0.2	
	切断（宽度<5mm）		400~500	0.1~0.2	

2.1.1.4　车削加工刀具

（1）车削刀具材料。金属切削加工中常用的刀具材料有高速钢、硬质合金、陶瓷、立方氮化硼和金刚石等，目前数控加工中最常用的刀具是高速钢刀具和硬质合金刀具。

1）高速钢。高速钢是一种加入了较多的钨、铬、钒、钼等合金元素的高合金工具钢，有良好的综合性能。其强度是现有刀具材料中最高的，韧性也最好。高速钢的制造工艺简单，容易刃磨成锋利的切削刃，锻造、热处理变形小，目前在复杂的刀具如麻花钻、丝锥和成形刀具制造中，仍占有重要地位。

①普通高速钢。普通高速钢具有一定的硬度和耐磨性、有较高的强度和韧性，如W18Cr4V广泛用于制造各种复杂刀具。其切削速度一般不太高，切削普通钢料时为40~60m/min，不适合高速切削和硬材料的切削。

②高性能高速钢。高性能高速钢是在普通高速钢中再增加一些含碳量、含钒量及添加钴、铝等元素冶炼而成的，如W12Cr4V4Mo，它的耐用度为普通高速钢的1.5~3倍，但这类钢的综合性能不如普通高速钢。

2）硬质合金。硬质合金是由难熔金属碳化物（如TiC、WC、NbC等）和金属黏接剂（如Co、Ni等）经粉末冶金方法制成，按化学成分不同，可分为4类。

①钨钴类（WC+Co）。合金代号为YG，对应国际标准K类，适合切削短切削的黑色金属、有色金属和非金属材料。钴含量较高，韧性越好，适于粗加工；钴含量低的，适于精加工。

②钨钛钴类（WC+TiC+Co）。合金代号为YT，对应国际P类，此类合金由较高的硬度和耐热性，主要用于加工长切屑的钢件等塑性材料。合金中TiC含量高，耐磨性和耐热性提高，但强度降低。因此粗加工一般选择TiC含量少的牌号，精加工选择含量多的牌号。

③钨钛钽（铌）钴类（WC+TiC+TaC(NbC)+Co）。合金代号为YW，对应国际M类。此类硬质合金不但适用于加工冷硬铸铁、有色金属及合金的半精加工，也能用于高锰钢、淬火钢、合金钢及耐热合金钢的半精加工和精加工。

④碳化钛基类（WC+TiC+Ni+Mo）。合金代号为YN，对应于国际P01类。一般用于精加工和半精加工，对于大而长且加工精度较高的零件尤其适合，但不适于由冲击载荷的粗加工和低速加工。

（2）车削刀具类型。数控车削加工常用的刀具有外圆车刀、切槽刀、内孔车刀、外螺纹车刀、内螺纹车刀、中心钻、钻头等，如图2-2所示。

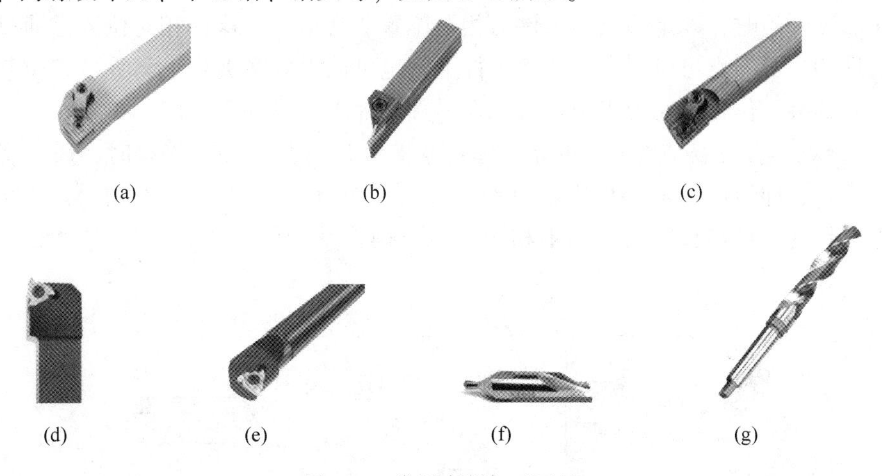

图2-2　常用车削加工刀具

（a）外圆车刀；（b）切槽刀；（c）内孔车刀；（d）外螺纹车刀；（e）内螺纹车刀；（f）中心钻；（g）麻花钻

外圆车刀（见图2-2(a)）用于切削圆柱面、圆锥面和端面；切槽刀（见图2-2(b)）用于切槽；内孔车刀（见图2-2(c)）用于加工内孔表面；外螺纹车刀（见图2-2(d)）用于切削外螺纹；内螺纹车刀（见图2-2(e)）用于切削内螺纹；中心钻（见图2-2(f)）用于钻中心孔；麻花钻（见图2-2(g)）用于孔的粗加工。

（3）机夹可转位车刀

数控车削加工时，常采用机夹可转位车刀。机夹可转位车刀是把经过研磨的可转位刀片用夹紧组件装夹在刀杆上，使用过程中，当一个切削刃磨钝后，松开夹紧机构，将刀片转位到另一切削刃，即可进行切削。当所有切削刃都磨损后再取下，换上新的同类型的刀片。

1）刀片夹紧方式。机夹可转位车刀的刀片紧固方式一般有上压式（代码为C）、上压与销孔夹紧式（代码M）、销孔夹紧式（代码P）和螺钉夹紧式（代码S）。

2）刀片外形。一般车削常用80°凸三边形（W型）、四方形（S型）和80°棱形（C型）刀片，仿形加工常用55°（D型）、35°（V型）和圆形（R型）刀片，如图2-3所示。在机床刚性、功率允许的条件下，粗加工或大余量加工时选用刀尖角较大的刀片，精加工或小余量加工时选用刀尖角较小的刀片。

图2-3　刀片外形

3）刀具形式。刀杆头部形式按主偏角和直头、弯头分有15~18种。加工直角台阶轴，选用主偏角大于或等于90°的刀杆。一般粗车可选用主偏角45°~90°的刀杆，精车可选用45°~75°的刀杆，在工件的中间切入、仿形车削选用45°~107.5°的刀杆。

4）切削方向。切削方向分为R(右手)、L（左手）、N（左右手）3种，选择时根据车床刀架前置还是后置、前刀面是向上还是向下、主轴的旋转方向及需要的进给方向来确定。

（4）对刀点、刀位点与换刀点。

1）对刀点。对刀点是指通过对刀确定刀具与工件相对位置的基准点。对于数控机床来说，在加工开始时，确定刀具与工件的相对位置非常重要，这一相对位置是通过确认对刀点来实现的。对刀点可设置在被加工零件上，也可以设置在夹具上与零件定位基准有一定尺寸联系的某一位置，对刀点往往选择在零件的加工原点处。

2）刀位点。刀位点是指刀具的定位基准点。在进行数控加工编程时，通常将整个刀具视为一个点，即刀位点，它是代表刀具位置的参照点。常用刀具的刀位点如图2-4所示。每把刀具的半径与长度尺寸均不相同，对刀操作就是使“刀位点”与“对刀点”重合的操作。

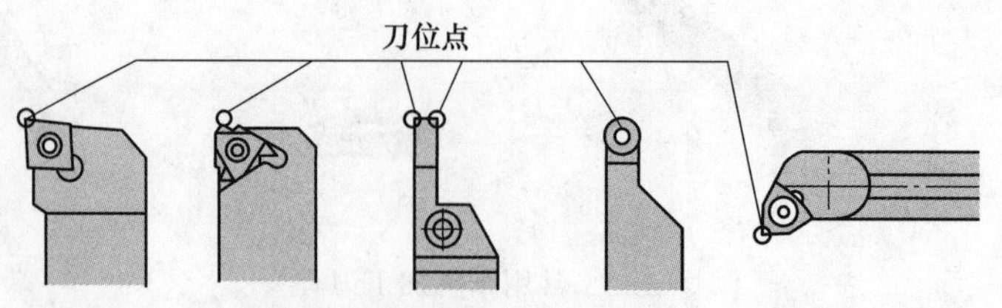

图2-4　常用刀具的刀位点

3）换刀点。换刀点是指刀架换刀时的位置。为防止换刀时碰伤零件及其他部件，换刀点通常设置在被加工零件或夹具的轮廓之外，并留有一定的安全量。

2.1.1.5 数控加工工艺文件

数控加工工艺文件是对数控加工的具体说明，目的是让操作者更明确加工程序的内容、装夹方式、各个加工部位所选用的刀具及其他技术问题。它既是数控加工、产品验收的依据，也是操作者遵守、执行的规程。数控加工工艺文件主要包括数控加工刀具卡、数控加工工序卡、数控加工程序单等。

（1）数控加工刀具卡。数控加工时，对刀具的要求十分严格，刀具卡片主要反映刀具编号、刀具名称及规格、刀片型号和材料、数量、加工表面等，它是组装刀具和调整刀具的依据。

（2）数控加工工序卡。数控加工工序卡对机床型号、程序编号进行简要说明，主要内容包括工步号、工步内容、各工步使用的刀具和切削用量等内容。

（3）数控加工程序单。数控加工程序单是数控加工的主要依据，它是编程人员按照机床特定的指令代码编制，记录数控加工工艺过程、工艺参数、位移数据等。

2.1.1.6 台阶轴的车削方式

当台阶轴相邻两圆柱体直径差较小时，可用车刀完成一次车削，加工路线为 $A \to B \to C \to D \to E$，如图 2-5（a）所示。

当台阶轴相邻两圆柱体直径差较大时采用分层切削，粗加工路线为 $A1 \to B1$、$A2 \to B2$、$A3 \to B3$，精加工路线为 $A \to B \to C \to D \to E$，如图 2-5（b）所示。

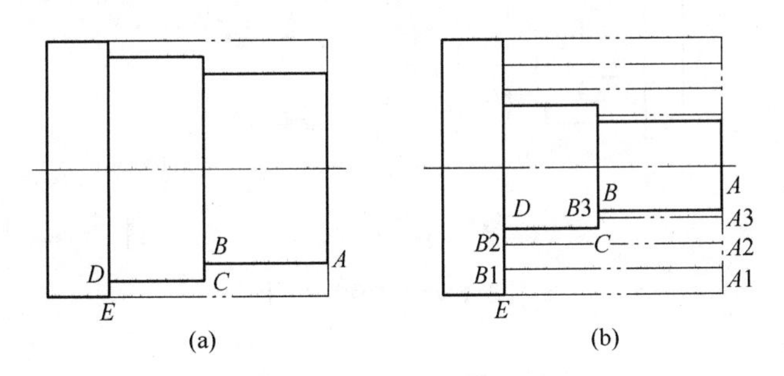

图 2-5　台阶轴车削方式

（a）车刀一次车削；（b）分层切削

2.1.2 编程指令

2.1.2.1 英制与公制指令 G20、G21

G20 指令表示坐标尺寸以英制输入，G21 指令表示坐标尺寸以公制输入。

在程序的开头指定 G20 或 G21，程序中间不能进行 G20 和 G21 的转换。系统通电后，CNC 保留上次关机时的 G20 或 G21。通常厂家设置为 G21 模式。

2.1.2.2　快速点定位指令 G00

（1）功能：刀具以点位控制方式从刀具所在位置快速移动到目标点。

（2）指令格式：

G00 X(U)__Z(W)__;

其中，X(U)、Z(W) 为目标点坐标值。

注意：

1）执行该指令时，刀具以生产厂家预先设定的速度从所在点，以点位控制方式移动到目标点，移动速度不能由 F 指令设定。

2）G00 为模态指令，只有遇到同组指令（G01、G02、G03）时才会被取替。

3）X、Z 后面跟的是绝对坐标值，U、W 后跟的是增量坐标值。

（3）直径与半径编程

X 坐标值用回转零件的直径值表示的编程方法称为直径编程。由于图纸上通常用直径表示零件的回转尺寸，采用直径编程比较方便。X 坐标值用回转零件的半径值表示的编程方法称为半径编程，符合直角坐标系的表示方法，较少采用。

【例 2-1】刀具从起点 *E* 快速移动到目标点 *A*1，起点 *D* 快速移动到目标点 *E*（见图 2-6）。

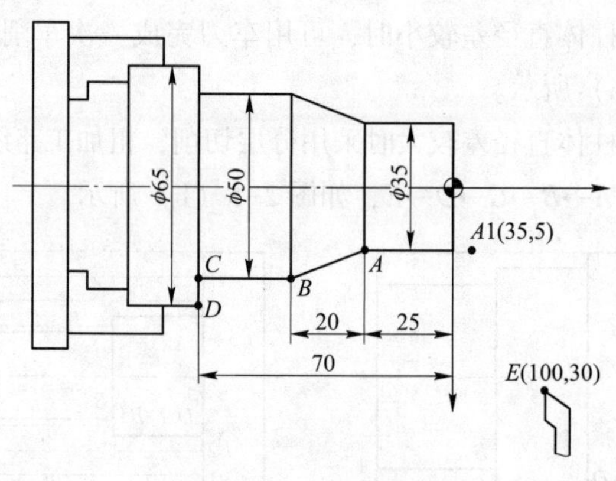

图 2-6　G00/G01 应用

E→*A*1　　　　*D*→*E*

绝对值编程：G00 X35.0 Z5.0；*E*→*A*1　　　　绝对值编程：G00 X100.0 Z30.0；*D*→*E*

其增量编程：G00 U-65.0 W-25.0；*E*→*A*1　　　　其增量编程：G00 U35.0 W100.0；*D*→*E*

2.1.2.3　直线插补指令 G01

（1）功能：该指令使刀具以给定的速度，从所在点出发，直线移动到目标点。

（2）指令格式：

G01 X(U)__Z(W)__F__;

其中，X(U)、Z(W) 为目标点坐标，F 为进给速度。

注意：

G01 指令的进给速度由 F 决定。如果在 G01 程序段之前没有 F 指令，当前 G01 程序段中也没有 F 指令，则机床不运动。

【例 2-2】精加工图 2-6 所示零件，选择右端面 O 点为编程原点，加工程序见表 2-2。

表 2-2　精加工程序

绝对值编程		增量值编程	
O0001；		O0002；	
N10	G40 G97 G99；	N10	G40 G97 G99；
N20	T0101；	N20	T0101；
N30	M03 S1000；	N30	M03 S1000；
N40	G00 X35.0 Z5.0；	N40	G00 U-65.0 W-25.0；
N50	G01 Z-25.0 F0.1；	N50	G01 W-30.0 F0.1；
N60	X50.0 Z-45.0；	N60	U15.0W-20.0；
N70	Z-70.0；	N70	W-25.0；
N80	X65.0；	N80	U15.0；
N90	G00X70.0；	N90	U35.0；
N100	Z30.0；	N100	W170.0；
N110	M30；	N110	M30；

【例 2-3】高台阶轴如图 2-7 所示，毛坯直径 40mm，背吃刀量 2.5mm，精加工余量 0.5mm，编写粗精车程序。高台阶轴粗精车程序见表 2-3。

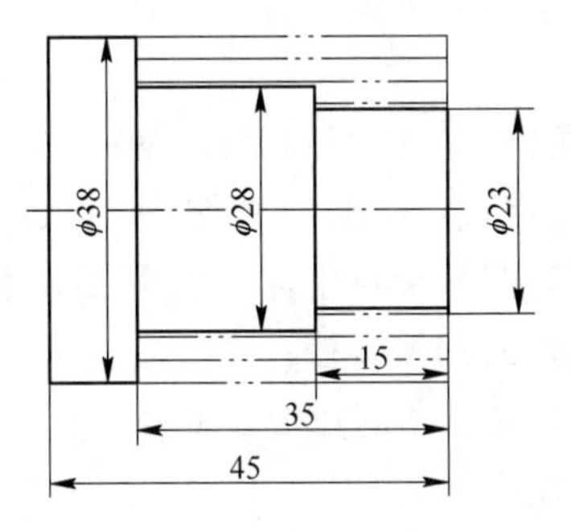

图 2-7　高台阶轴

表 2-3　高台阶轴粗精车程序

O121；	程序号
G40 G97 G99 M03 S600 F0.2；	G00 Z5.0；
T0101；	X24.0；
M08；	G01 Z-15.0；
G00 Z5.0；	X25.0；
X39.0；	G00 Z5.0；
G01 Z-45.0 F0.2；	X23.0；
X40.0；	M03 S1000 F0.1；
G00 Z5.0；	G01 Z-15.0；
X34.0；	X28.0；

续表 2-3

G01 Z-35.0；	Z-35.0；
X35.0；	X38.0；
G00 Z5.0；	Z-45.0；
X29.0；	G00 X100.0；
G01 Z-35.0；	Z100.0；
X30.0；	M30；

（3）G01 拓展功能。

1）功能：在相邻轨迹线之间自动插补倒直角或倒圆角，如图 2-8 所示。

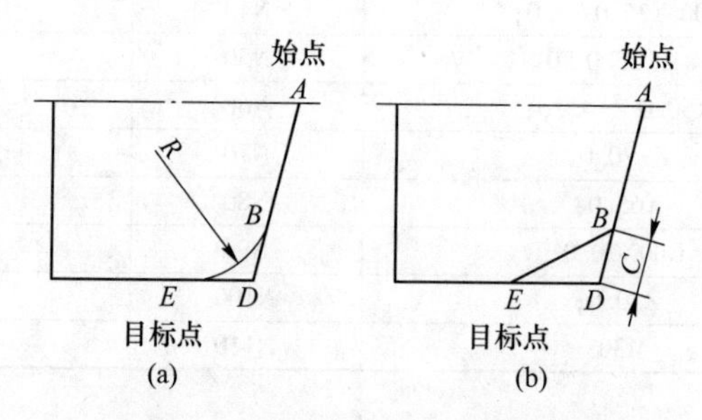

图 2-8　G01 拓展功能示意图

（a）G01 倒圆角；（b）G01 倒直角

2）指令格式：

倒圆角格式：G01 X(U)Z(W)_R_F_；

倒直角格式：G01 X(U)Z(W)_C_F_；

式中　X(U)，Z(W)——相邻直线的交点坐标（如图 2-8 中的 *D* 点）；

R——倒圆角的圆弧半径；

C——*D* 点相对倒角起点 *B* 的距离。

3）注意事项：R、C 为非模态代码。

【例 2-4】利用倒直角和倒圆角功能，编写图 2-9 所示零件的精加工程序。精加工程序见表 2-4。

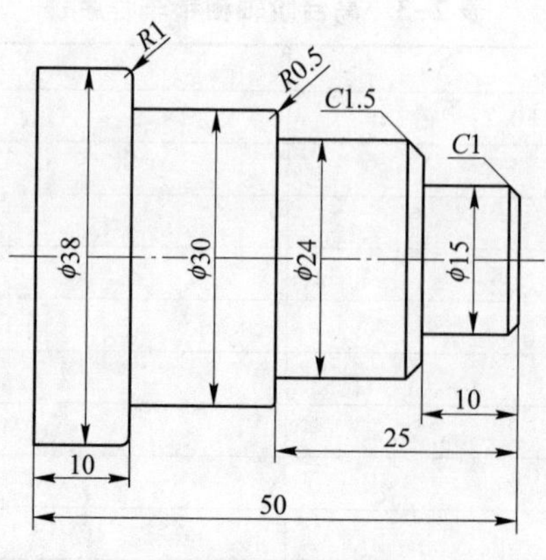

图 2-9　零件图

表 2-4　精加工程序

O0006;	程序号
G40 G97 G99 M03 S1000;	Z-25.0;
T0101;	X30.0 R0.5;
M08;	Z-40.0;
G00 Z5.0;	X38.0 R1.0;
X0;	Z-50.0;
G01 Z0 F0.1;	G00 X100.0;
X15.0 C1.0;	Z100.0;
Z-10.0;	M30;
X24.0 C1.5;	

2.1.3　加工操作

2.1.3.1　装夹、找正工件

采用三爪自定心卡盘夹住棒料外圆，进行外圆找正后，再夹紧工件。注意工件装卡一定要牢固。

找正方法一般为打表找正，常用的钟面式百分表如图 2-10 所示。百分表是一种指示式量仪，除用于找正外，还可以测量工件的尺寸、形状和位置误差。百分表的使用注意事项如下。

（1）使用前，应检查测量杆的灵活性。即轻轻推动测量杆时，测量杆在套筒内的移动要灵活，且每次放松后，指针能恢复到原来的刻度位置。

（2）使用百分表时，必须把它固定在可靠的夹持架上（如固定在万能表架或磁性表座上）。

（3）不要使测量头突然撞在零件上。

（4）不要使百分表受到剧烈的振动和撞击。

用百分表找正如图 2-11 所示，具体操作步骤如下。

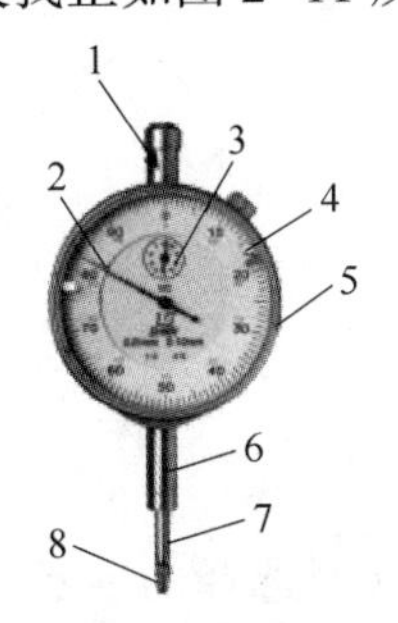

图 2-10　百分表的结构

1—手提测量杆；2—指针；3—转数指示盘；4—表盘；
5—表圈；6—套筒；7—测量杆；8—测量头

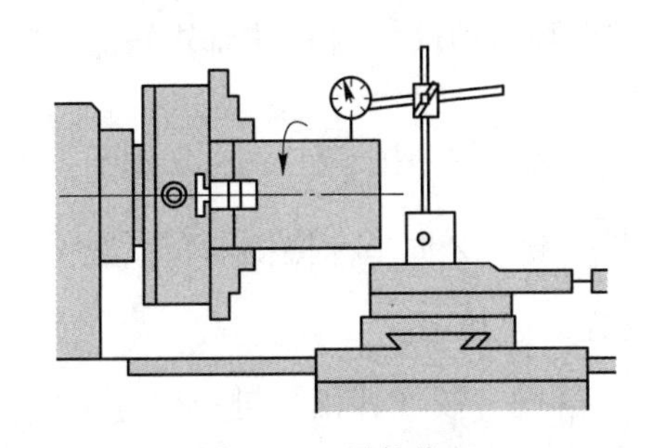

图 2-11　零件找正

（1）准备阶段：将钟面式百分表装入磁力表座孔内，锁紧，检查测头的伸缩性、测头与指针配合是否正常。

（2）测量阶段：百分表测头与工件的回转轴线垂直，用手转动三爪卡盘，根据百分表指针的摆动方向轻敲工件进行调整，使工件的回转轴线，即工件坐标系的 Z 轴与数控车床的主轴中心轴线重合。

工件装夹注意事项如下。

（1）装夹工件时应尽可能使基准统一，减少定位误差，提高加工精度。

（2）装夹已加工表面时，应在已加工表面包一层铜皮，以免夹伤工件表面。

（3）装夹部位应选在工件上强度、刚性好的表面。

2.1.3.2 安装刀具

机夹外圆车刀的安装步骤如下。

（1）将刀片装入刀体内，旋入螺钉，并拧紧。

（2）刀杆装上刀架前，先清洁装刀表面和车刀刀柄。

（3）车刀在刀架上伸出长度约为刀杆高度的 1.5 倍，伸出太长会影响刀杆的刚性。

（4）车刀刀尖应与工件中心等高。

（5）刀杆中心应与进给方向垂直。

（6）至少用两个螺钉压紧车刀，固定好刀杆。

2.1.3.3 外圆车刀对刀操作

手动或手摇轮均可完成对刀操作，外圆车刀对刀步骤如下。

（1）Z 向补偿。

1）移动外圆车刀切削端面。

①按 [寸动] 键→按 [X-] 或 [X+] 键→机床沿 X 向移动；同理使机床沿 Z 向移动接近毛坯。

②按 MDI 键 [MDI] →按 [PROG] 键→进入 MDI 界面→输入“M03 S600”→按 [EOB] 键→按 [INSERT] 键→按循环启动键 [程式启动 CYCLE START] →主轴正转。

③刀具接近工件→按 [X-] 键，切削工件端面。

④按 [X+] 键，Z 向坐标保持不变，沿 X 轴正向退刀。

2）设置 Z 向补正。

①按 [OFS SET] 键→按【补正】键→按【形状】键→移动光标至选择的刀具位置，如番号 G01，界面如图 2-12 所示。

②输入 Z0→按【测量】键。

3）验刀操作。完成 Z 向补正后，刀具沿 X 轴正向移动远离工件（Z 值不变），按 [MDI]

键→输入刀具号→按循环启动键，此时 CRT 屏幕上显示的 *Z* 坐标的绝对值为零。

（2）*X* 向补偿。

1）切削外圆直径。

①主轴正转。

②刀具接近工件→按键→机床沿 *Z* 轴负向移动，刀具切削工件外圆。

③按键，*X* 轴坐标保持不变，沿 *Z* 轴正向退刀。

2）测量切削直径。

主轴停止→测量试切削外圆，记下直径值。

3）设置 *X* 向补正。

①按键→按【补正】键→按【形状】键→移动光标至选择的刀具位置，如番号 G01，界面如图 2-12 所示。

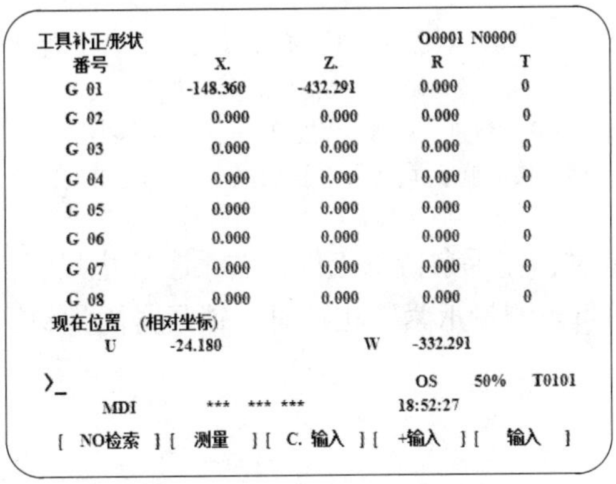

工具补正/形状　　　　O0001 N0000

番号	X.	Z.	R	T
G 01	-148.360	-432.291	0.000	0
G 02	0.000	0.000	0.000	0
G 03	0.000	0.000	0.000	0
G 04	0.000	0.000	0.000	0
G 05	0.000	0.000	0.000	0
G 06	0.000	0.000	0.000	0
G 07	0.000	0.000	0.000	0
G 08	0.000	0.000	0.000	0

现在位置　(相对坐标)

U　-24.180　　W　-332.291

>_　　OS　50%　T0101

MDI　***　***　***　18:52:27

[NO检索] [测量] [C. 输入] [+输入] [输入]

图 2-12　参数输入界面

② 输入 *X* 直径值（如 X33. 539）→按【测量】键。

4）验刀操作。

完成 *X* 向补正后，刀具沿 *Z* 轴正向移动远离工件（*X* 值不变），按键→输入刀具号→按循环启动键，此时 CRT 屏幕上显示的 *X* 坐标的绝对值为测量直径。

2. 1. 3. 4　自动加工

调用加工程序→按键→按循环启动键，自动加工零件。

自动加工前要进行全轴操作，并检查空运行和锁住按钮状态。

2.1.4 精度检验

通常使用游标卡尺检测零件的直径和长度，使用粗糙度比较样板检测表面粗糙度。

2.1.4.1 游标卡尺

（1）应用。游标卡尺可以测量长度、宽度、厚度、内径和外径、孔距、高度和深度等尺寸，是应用较广泛的量具。

（2）结构。

如图 2-13 所示。

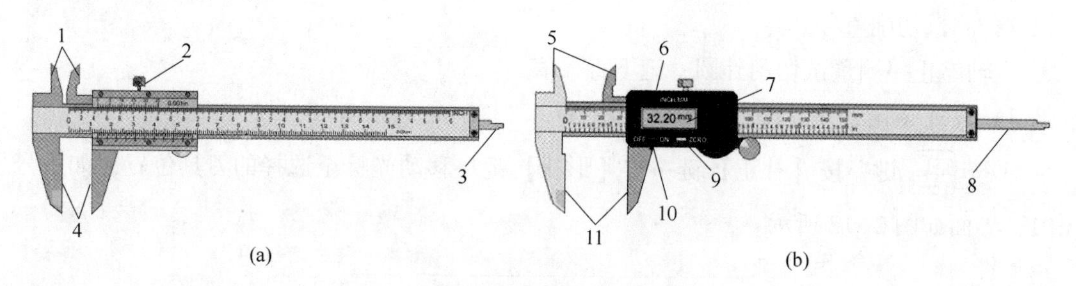

图 2-13 游标卡尺

（a）普通游标卡尺；（b）数显游标卡尺

1，5—内测量爪；2—紧固螺钉；3，8—深度尺；4，11—外测量爪；6—米/英制按钮；7—液晶显示屏；9—零点设定；10—电源开/关

（3）使用方法。测量时，左手拿待测工件，右手拿住主尺，大拇指移动游标尺，使待测工件位于测量爪之间，当与测量爪紧紧相贴时，锁紧紧固螺钉，即可读数。正确的测量方法如图 2-14 所示。

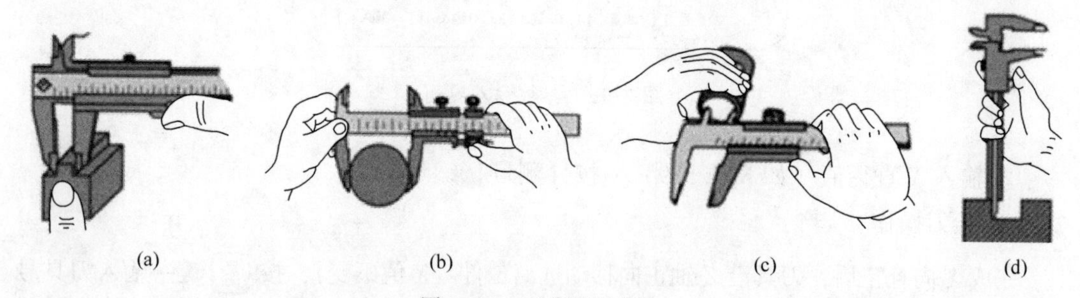

图 2-14 正确测量方法

（a）测宽度；（b）测外径；（c）测内径；（d）测深度

（4）读数。数显游标卡尺可以直接在液晶显示屏上读数。

普通游标卡尺按其测量精度可分为 0.10mm、0.05mm 和 0.02mm 三种。目前机械加工中常用精度为 0.02mm 的游标卡尺。游标卡尺是以游标零线为基线进行读数的，以图 2-15 为例，其读数方法分为三个步骤。

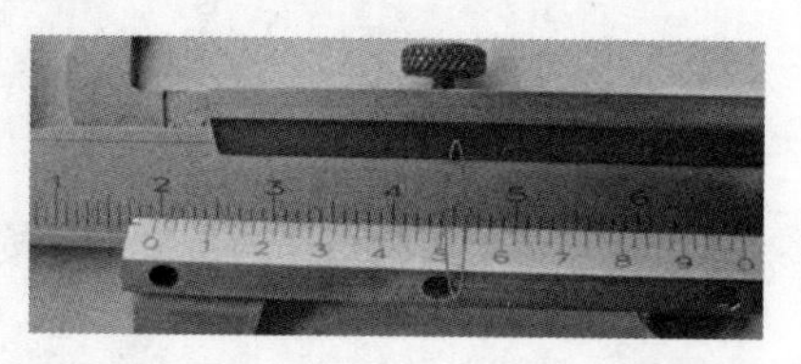

图 2-15 0.02mm 游标卡尺读数

1）读整数，即读出游标零线左面主尺上的毫米数

为整数值（19mm）；

2）读小数，即找出游标尺上与主尺上对齐的游标刻线，将对齐的游标刻线与游标尺零线间的格数乘以卡尺的精度为小数值（0.52mm）；

3）把整数值与小数值相加即为测量的实际尺寸（19.52mm）。

（5）注意事项：

使用游标卡尺时应注意以下事项。

1）测量前先将测量爪和被测工件表面擦拭干净，然后合拢两测量卡爪使之贴合，检查主尺、游标尺零线是否对齐。若未对齐，应在测量后根据原始误差修正读数或将游标卡尺校正到零位后再使用。

2）当测量爪与被测工件接触后，用力不宜过大，以免卡爪变形或磨损，降低测量的准确度。

3）测量零件尺寸时，卡尺两测量面的连线应垂直于被测量表面，不能歪斜。

4）不能用游标卡尺测量毛坯表面。

5）使用完毕后须把游标卡尺擦拭干净，放入盒内。

2.1.4.2 粗糙度检测

（1）粗糙度比较样板。粗糙度对比法是最早的检测机械加工工件表面粗糙度的方法，对比法就是用工件和粗糙度比较样板对比评定粗糙度是否合格，这种检测方法效率低、精准度差。粗糙度比较样板如图 2-16 所示，又称粗糙度比较板、粗糙度比较块或粗糙度对比样块等。

图 2-16 粗糙度比较样板

（2）粗糙度仪。对于表面质量比较高的零件可使用粗糙度仪进行检测。它具有测量精度高、测量范围宽、操作简便、便于携带、工作稳定等特点，可以广泛应用于各种金属与非金属的加工表面的检测，如图 2-17 所示。

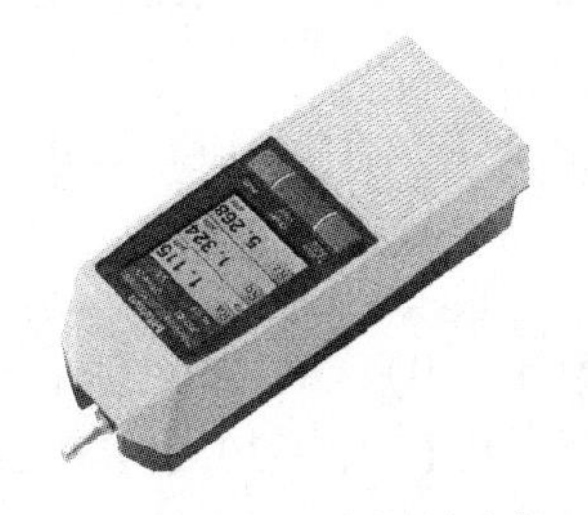

图 2-17 便携式粗糙度仪

任务实施

（1）图样分析：

图 2-1 所示零件材料为 LY15，总长度 50mm，加工表面有 $\phi28^{0}_{-0.03}$、$\phi31^{0}_{-0.04}$、$\phi35$、$\phi38^{0}_{-0.04}$外圆面，C2 倒角等，表面粗糙度分别为 $Ra1.6$ 和 $Ra3.2$。

（2）加工工艺方案制订：

1）加工方案：

①采用三爪自定心卡盘装卡毛坯，外伸卡盘长度约 60mm。

②加工零件外轮廓至尺寸要求。

2）刀具选用：

零件粗精加工用一把刀具，因此选用刀尖角 55°、主偏角 90°外圆车刀，数控加工刀具卡如表 2-5 所示。

表 2-5　数控加工刀具卡

零件名称		台阶轴零件		零件图号		2-1		
序号	刀具号	刀具名称	数量	加工表面		刀尖半径 R/mm	刀尖方位 T	备注
1	T01	90°外圆偏刀	1	粗精车外轮廓		0.4	3	
编制		审核	批准		日期		共 1 页	第 1 页

3）加工工序：

台阶轴零件加工路线如图 2-18 所示。

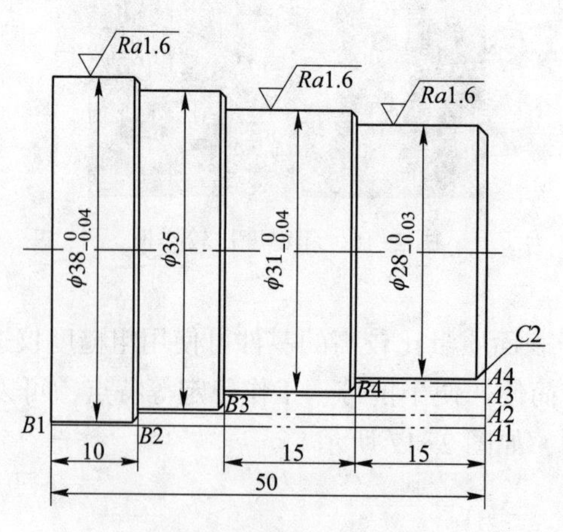

图 2-18　台阶轴加工线路图

坐标点：

A1(39，0)，A2(36，0)，A3(32，0)，A4(29，0)；

B1(39，-50)，B2(36，-40)，B3(32，-30)，B4(29，-15)

台阶轴零件数控加工工序见表 2-6。

表 2-6　数控加工工序卡

<table>
<tr><td rowspan="2">单位名称</td><td colspan="5" rowspan="2">天津工业职业学院</td><td colspan="2">零件名称</td><td colspan="2">零件图号</td></tr>
<tr><td colspan="2">台阶轴零件</td><td colspan="2">2-1</td></tr>
<tr><td colspan="2">程序号</td><td colspan="2">夹具名称</td><td colspan="2">使用设备</td><td colspan="2">数控系统</td><td colspan="2">场地</td></tr>
<tr><td colspan="2">O211</td><td colspan="2">三爪自定心卡盘</td><td colspan="2">CKA6140</td><td colspan="2">FANUCSERIES 0i</td><td colspan="2">数控实训中心</td></tr>
<tr><td>工序号</td><td colspan="4">工序内容</td><td>刀具号</td><td>主轴转速
$n/r\cdot min^{-1}$</td><td>进给量
$F/mm\cdot r^{-1}$</td><td>背吃刀量
a_p/mm</td><td>备注</td></tr>
<tr><td>1</td><td colspan="4">装卡零件并找正</td><td></td><td></td><td></td><td></td><td rowspan="3">手动</td></tr>
<tr><td>2</td><td colspan="4">对刀</td><td>T01</td><td></td><td></td><td></td></tr>
<tr><td>3</td><td colspan="4">粗车外轮廓，留余量 1mm</td><td>T01</td><td>600</td><td>0. 2</td><td>1. 5</td></tr>
<tr><td>4</td><td colspan="4">精车外轮廓</td><td>T01</td><td>1000</td><td>0. 1</td><td>0. 5</td><td>O211</td></tr>
<tr><td>编制</td><td></td><td>审核</td><td></td><td>批准</td><td></td><td>日期</td><td></td><td>共 1 页</td><td>第 1 页</td></tr>
</table>

（3）程序编制：

1）尺寸计算。

对于单件小批量生产，零件精加工尺寸一般取极限尺寸的平均值。

编程尺寸=基本尺寸+(上偏差+下偏差)÷2

$\phi28^{0}_{-0.03}$外圆的编程尺寸=28+(0-0. 03)÷2=27. 985

$\phi31^{0}_{-0.04}$外圆的编程尺寸=31+(0-0. 04)÷2=30. 98

$\phi38^{0}_{-0.04}$外圆的编程尺寸=38+(0-0. 04)÷2=37. 98

2）加工程序。

台阶轴零件精加工程序见表 2-7。

表 2-7　台阶轴零件加工程序

<table>
<tr><td>O121;</td><td>程序号</td></tr>
<tr><td>G40 G97 G99 M03 S600 F0. 2;</td><td>取消刀具半径补偿，取消主轴恒转速度，设每转进给量，主轴正转，转速为 600r/min，设进给量为 0. 2mm/r</td></tr>
<tr><td>T0101;</td><td>换 T01 外圆车刀，代入 01 号刀具补偿</td></tr>
<tr><td>M08;</td><td>打开切削液</td></tr>
<tr><td>G00 Z5. 0;</td><td rowspan="2">快速进刀至加工起点</td></tr>
<tr><td>X39. 0;</td></tr>
<tr><td>G01 Z-50. 0</td><td>切削至 B1 点</td></tr>
<tr><td>G01 X40. 0</td><td rowspan="2">退刀</td></tr>
<tr><td>G00 Z5. 0;</td></tr>
<tr><td>X36. 0;</td><td>快速进刀至加工起点</td></tr>
<tr><td>G01 Z-40. 0;</td><td>切削至 B2 点</td></tr>
<tr><td>G00 X40. 0;</td><td rowspan="2">退刀</td></tr>
<tr><td>Z5. 0;</td></tr>
<tr><td>G01 X32. 0;</td><td>快速进刀至加工起点</td></tr>
</table>

续表 2-7

Z-30.0;	切削至 *B*3 点
X33.0;	退刀
G00 Z5.0;	
X29.0;	快速进刀至加工起点
G01 Z-15.0;	切削至 *B*4 点
X30.0;	退刀
G00 Z5.0;	
X0;	快速进刀至加工起点
G01 Z0;	进刀至精加工起点
X27.985　C2;	精车轮廓
Z-15.0;	
X30.98　C1;	
Z-30.0;	
X35.0　C1;	
Z-40.0;	
X37.98　C1;	
Z-50.0;	
G00 X100;	快速退刀至换刀点
Z100.0;	
M30;	程序结束并返回起点

（4）工件加工：

系统启动→回参考点→装夹并找正工件→安装刀具→输入表 2-7 程序→模拟程序→对刀→自动加工。

（5）尺寸检测：

1）使用游标卡尺测量 $\phi28^{0}_{-0.03}$、$\phi31^{0}_{-0.04}$、$\phi35$、$\phi38^{0}_{-0.04}$外径尺寸；

2）采用游标卡尺测量 50、10、15 长度尺寸；

3）采用粗糙度比较板检测粗糙度。

任务评价

任务评价见表 2-8。

表 2-8　任务评价

项目	序号	评 分 标 准		配分	得分
加工工艺（10%）	1	加工步骤	符合数控车工工艺要求	3	
	2	加工部位尺寸	计算相关部位尺寸	3	
	3	刀具选择	刀具选择合理，符合加工要求	4	

续表 2-8

项目	序号	评分标准		配分	得分
程序编制（30%）	4	程序号	无程序号无分	2	
	5	切削用量	切削用量选择不合理无分	6	
	6	原点及坐标	标明程序原点及坐标，否则无分	2	
	7	程序内容	不符合程序逻辑及格式要求每段扣 2 分	20	
			程序内容与工艺不对应扣 5 分		
			出现危险指令扣 5 分		
加工操作（30%）	8	程序输入与检索	每错一段扣 2 分	10	
	9	加工操作 30 分钟	工件坐标系原点设定错误无分	20	
			误操作无分		
			超时无分		
尺寸检测（20%）	10	外圆	$\phi38^{0}_{-0.039}$超差 0.01 扣 3 分	3	
	11		$\phi35$ 降级无分	2	
	12		$\phi31^{0}_{-0.039}$超差 0.01 扣 3 分	3	
	13		$\phi28^{0}_{-0.033}$超差 0.01 扣 3 分	3	
	14	轴向长度	50、15、15、10	4	
	15	倒角	$C2$	1	
	16	表面粗糙度	$Ra1.6$ 降级无分	3	
	17		$Ra3.2$ 降级无分	1	
职业能力（10%）	18	学习能力		2	
	19	表达沟通能力		2	
	20	团队合作		2	
	21	安全操作与文明生产		4	
合计					

同步思考与训练

（1）思考题：

1）简述切削用量选择的一般原则。

2）数控车削加工常用的刀具有哪些？

3）简述如何装夹找正工件。

（2）同步训练题：

零件如图 2-19 和图 2-20 所示，编写程序并加工。

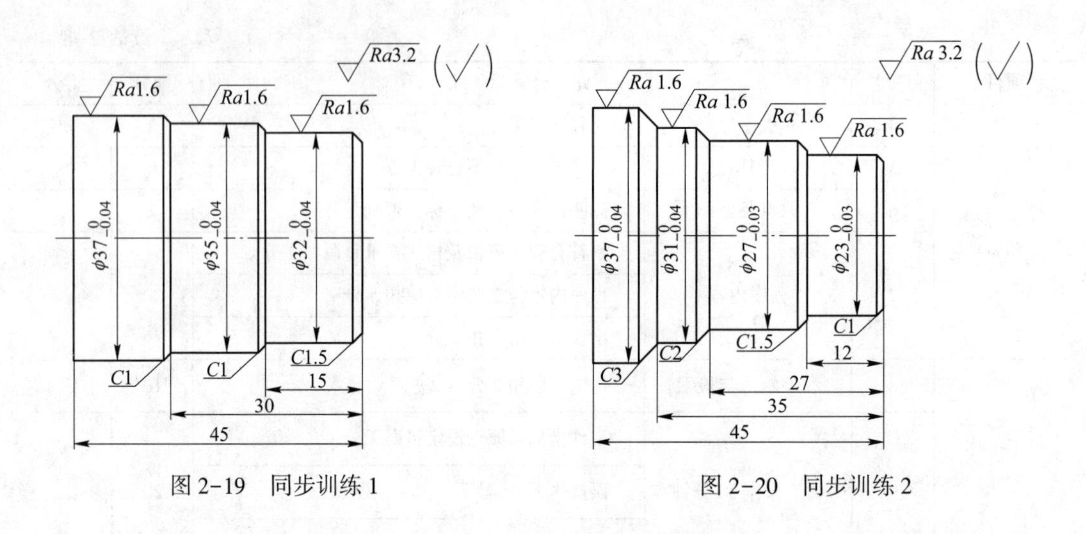

图 2-19　同步训练 1　　　　图 2-20　同步训练 2

任务 2.2　圆锥面的数控加工

任务描述

图 2-21 所示零件，材料硬铝合金 LY15，毛坯为 ϕ40mm 长棒料，单件生产，使用数控车床加工零件，编写加工程序，检测加工质量。

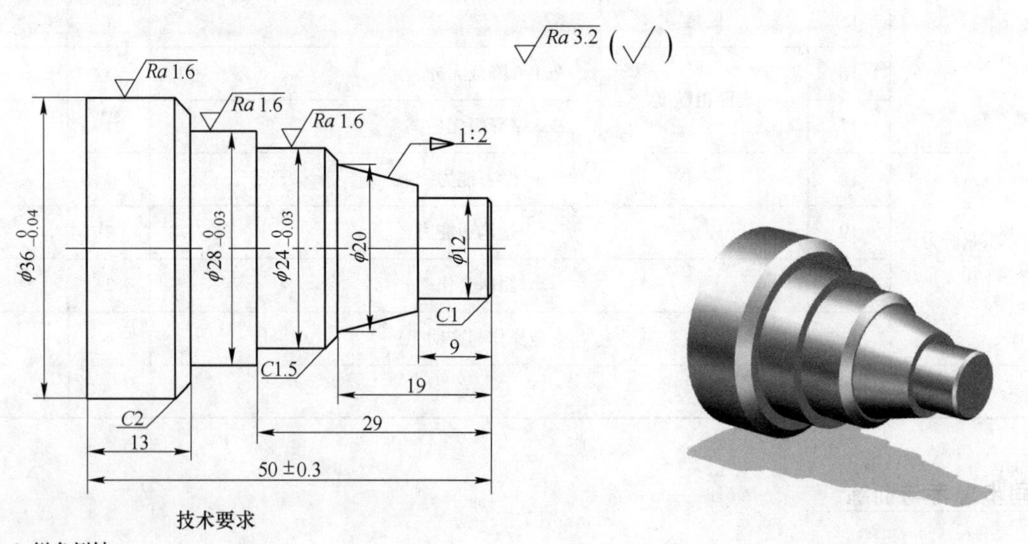

技术要求

1. 锐角倒钝。
2. 未注线性尺寸公差应符合 GB/T 1804—2000 的要求。

图 2-21　零件图

任务目标

（1）知识目标：

1）了解圆锥面的车削工艺、圆锥参数。

2）解释 G71/G70，G40/G41/G42 指令。

3）认识圆锥尺寸检测量具。

（2）技能目标：

1）能利用公式进行锥度尺寸计算。

2）能使用 G71/G70，G40/G41/G42 指令编写锥面轴加工程序。

3）能正确操作机床验证锥面轴加工程序。

4）能正确使用量具完成零件尺寸检测。

（3）素质目标：

1）培养学生安全作业以及自觉完成 8S 的工作要求。

2）培养质量意识和规范意识。

相关知识

2.2.1　工艺准备

圆锥面配合的同轴度较高，而且拆卸方便，当圆锥角较小（$\alpha<3°$）时能够传递很大的扭矩，因此在机械制造中应用广泛。

2.2.1.1　圆锥面的数控车削方式

图 2-22（a）所示圆锥的车削方式，走刀路线平行于圆锥面，切削量均匀。图 2-22（b）所示圆锥的车削方式，切削量不均匀，编程相对简单。

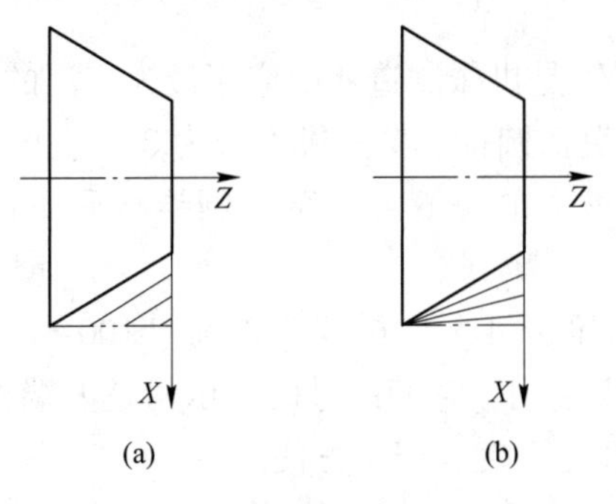

图 2-22　圆锥面的车削方式

（a）均匀切削；（b）不均匀切削

2.2.1.2　圆锥参数及尺寸计算

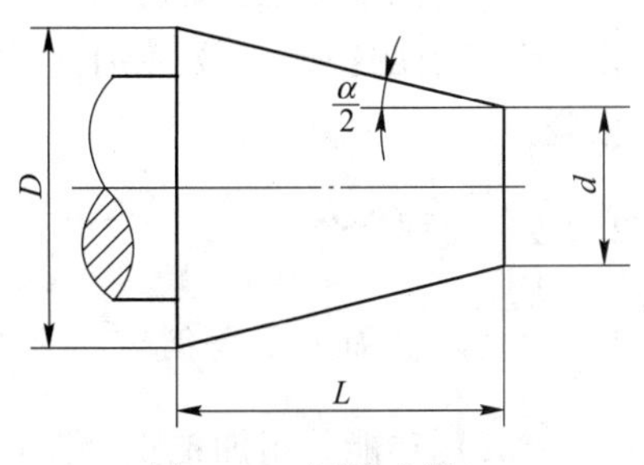

图 2-23　圆锥参数

常用的圆锥参数如图 2-23 所示：

（1）圆锥最大直径 D，简称大端直径。

（2）圆锥最小直径 d，简称小端直径。

（3）圆锥长度 L，最大圆锥直径与最小圆锥直径的轴向距离。

（4）锥度 C：锥度是圆锥最大直径和最小直径差值

与圆锥长度 L 的比值，即 $C=\dfrac{D-d}{L}$。

【例 2-5】如图 2-24 所示，圆锥面的锥度 $C=1:2.5$，求锥面大端直径。

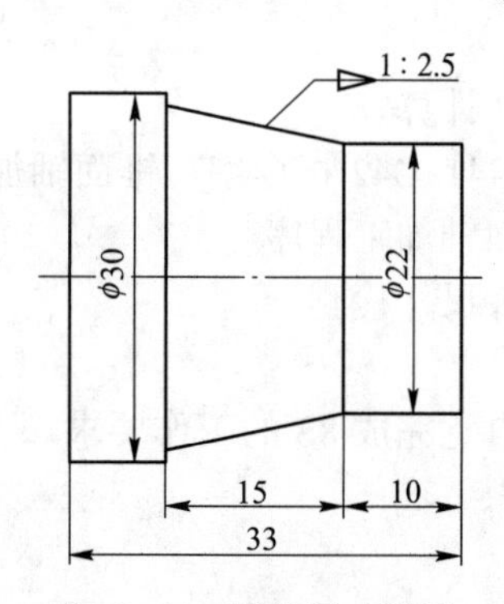

图 2-24　圆锥尺寸计算

图 2-24 中，圆锥最小直径 $d=22$，圆锥长度 $L=15$，锥度 $C=1:2.5=0.4$，根据公式 $C=\dfrac{D-d}{L}$，得出圆锥最大直径：

$$D=d+CL=22+0.4\times15=28$$

2.2.1.3　标准工具圆锥

为了方便制造和使用，降低生产成本，常用的工具、刀具上的圆锥实现了标准化。常用的标准工具的圆锥有两种。

（1）莫氏圆锥。莫氏圆锥是机械制造业中应用最为广泛的一种，如车床主轴锥孔、顶尖、钻头柄、铰刀柄等都是莫氏圆锥。莫式锥度分为 0 号、1 号、2 号、3 号、4 号、5 号、6 号七种，最小的是 0 号、最大的是 6 号。莫氏圆锥号码不同，圆锥尺寸和圆锥半角都不同。

（2）米制圆锥。米制圆锥分 4 号、6 号、80 号、100 号、120 号、140 号、160 号和 200 号八种，其中 140 号较少使用，它们的号码表示的是大端直径，锥度固定不变，即 $C=1:20$。米制圆锥的优点是锥度不变，记忆方便。

2.2.1.4　圆锥面加工刀具选择

用数控车床加工圆锥面时，使用的刀具一般与车削阶梯轴时的刀具相同，也采用偏刀。但车削倒锥时，要特别注意选用副偏角较大的刀具，使刀具副切削刃不能与锥面相碰。

2.2.2　编程指令

2.2.2.1　粗加工复合循环指令 G71

（1）功能。粗加工复合循环指令 G71 只需指定粗加工背吃刀量、精加工余量和精加工路线，数控系统便可自动给出粗加工路线和加工次数，完成内、外圆表面的粗加工。图

2-25 为 G71 指令循环路线，其中 A 为刀具循环起点，执行粗加工复合循环时，刀具从 A 点移动到 C 点，粗车循环结束后，刀具返回 A 点。

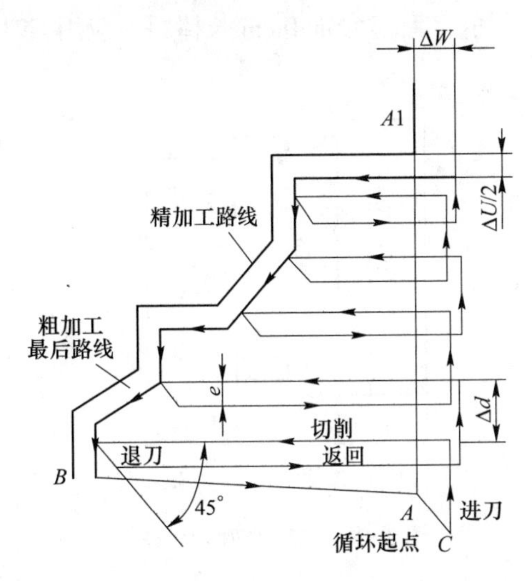

图 2-25　G71 指令循环路线

（2）指令格式。

G71 U(Δd) R(e)；

G71 P(ns)Q(nf)　U(Δu)　W(Δw)；

式中　Δd——每次的背吃刀量，用半径值指定，一般 45 钢件取 1~2mm，铝件取 1.5~3mm；

e——每次 X 向退刀量，用半径值指定，一般取 0.5~1mm；

ns——精加工轮廓程序段中的开始程序段号；

nf——精加工轮廓程序段中的结束程序段号；

Δu——X 方向精加工余量，一般取 0.5mm，加工内轮廓时为负值；

Δw——Z 方向精加工余量，一般取 0.05~0.1mm。

（3）注意事项。

1）使用 G71 粗加工时，包含在 ns~nf 程序段中的 F、S 指令对粗车循环无效。

2）顺序号为 ns~nf 的程序段中不能调用子程序。

3）零件轮廓必须符合 X 轴、Z 轴方向同时单调增大或单调减少。

4）精加工路线第一句必须用 G00 或 G01 沿 X 方向进刀。

（4）应用：棒料毛坯的粗加工。

2.2.2.2　精加工复合循环指令 G70

（1）功能：去除精加工余量。

（2）指令格式：

G70 P(ns)Q(nf)；

（3）注意事项：

1）在 ns~nf 之间的程序段中的 F、S 指令有效。

2）G70 切削后刀具回到 G71 的循环起点。

（4）应用：用于精加工，切除 G71 指令粗加工后留下的加工余量。

【例 2-6】如图 2-26 所示，毛坯为 ϕ40mm 长棒料，应用 G71/G70 指令编写零件粗精车程序。

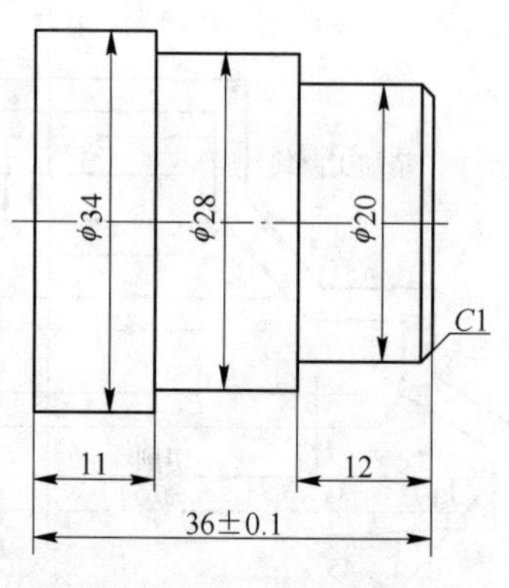

图 2-26　G71/G70 指令举例

加工程序见表 2-9。

表 2-9　零件程序及注释

O2001；	程序号
G40 G97 G99 M03 S600 F0.2；	取消刀具半径补偿，取消主轴恒转速度，设每转进给量，主轴正转，转速为 600r/min，设进给量为 0.2mm/r
T0101；	换 T01 外圆车刀，代入 01 号刀具补偿
M08；	打开切削液
G00 Z5.0；	快速进刀至粗车循环起点
X40.0；	
G71 U1.5 R0.5；	定义粗车循环，切削深度 1.5mm，退刀量 0.5mm；
G71 P10 Q20 U0.5 W0.05；	精车路线由 N10～N20 指定，X 方向精车余量 0.5mm，Z 方向精车余量 0.05mm
N10 G00 X0；	精车轮廓
G01 Z0；	
X18.0；	
X20.0 Z-1.0；	
Z-12.0；	
X28.0	
Z-25.0；	
X34.0	
N20 Z-40.0；	
G00 X100.0；	快速退刀至换刀点
Z100.0；	

续表 2-9

M05;	主轴停止
M00;	程序暂停
M03 S1000 F0.1;	主轴正转，转速 1000r/min
T0101;	代入 01 号刀具补偿
G00Z5.0	刀具快速点定位至粗加工复合循环起点
X40.0;	
G70 P10 Q20;	精加工复合循环
G00 X100.0;	快速退刀至换刀点
Z100.0;	
M30;	程序结束并返回起点

2.2.2.3 刀尖圆弧半径补偿指令 G41/G42/G40

（1）刀尖圆弧半径对加工精度的影响。理想状态下尖形车刀的刀位点是其尖点，如图 2-27 中的 *A* 点，该点即假想刀尖，对刀时也是以假想刀尖进行对刀。但实际加工中使用的车刀，为了提高刀具的使用寿命和降低工件的表面粗糙度，车刀的刀尖制成半径不大的圆弧，如图 2-27 中的 *BC* 圆弧。

如图 2-28 所示，车外圆、端面时，刀具实际切削刃的路线与工件轮廓一致，不产生误差。车削锥面时，工件轮廓为实线，实际车出形状为虚线，产生欠切误差 δ。若工件精度要求不高或留有精加工余量，可忽略此误差；否则应考虑刀尖圆弧半径对工件形状的影响。切削圆弧时，由于刀尖半径的存在而产生过切削或欠切削的现象，如图 2-29 所示。

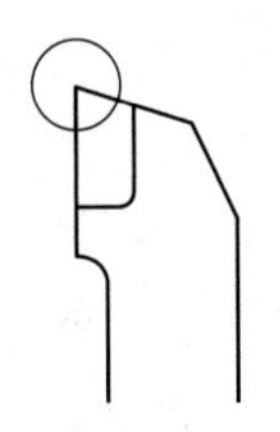

图 2-27　假想刀尖示意图

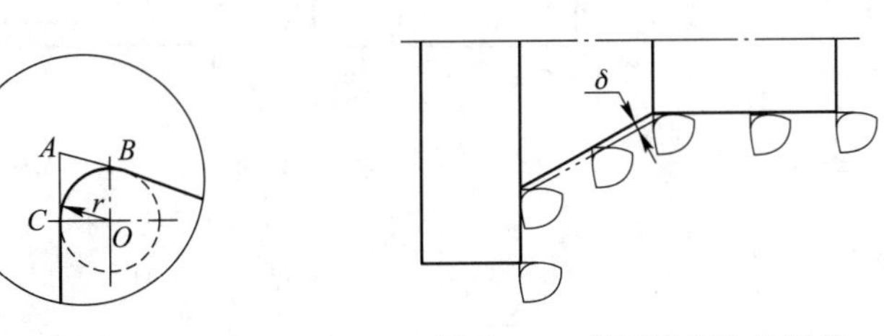

图 2-28　车圆锥产生的误差

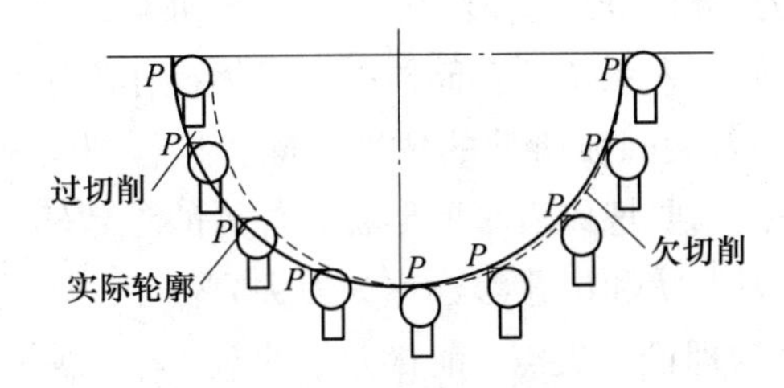

图 2-29　车圆弧产生的误差

（2）补偿方法。具有刀尖圆弧半径补偿功能的数控系统可以防止以上现象的出现，在编制零件加工程序时，先假想刀尖位置，按零件轮廓编程，在自动执行程序加工的过程中，数控系统根据补偿寄存器里设置的参数值或补偿指令进行刀尖圆弧半径补偿。

（3）刀尖圆弧半径补偿指令。G41 为刀尖圆弧半径左补偿指令，G42 为刀尖圆弧半径右补偿指令，G40 为取消刀尖圆弧半径补偿指令。指令格式如下：

$$\left.\begin{matrix}\text{G41}\\ \text{G42}\\ \text{G40}\end{matrix}\right\}\left.\begin{matrix}\text{G01}\\ \text{G00}\end{matrix}\right\}\ \text{X(U)\_Z(W)\_F\_;}$$

式中　X(U)、Z(W)——建立或取消刀尖圆弧半径补偿段的终点坐标；

F——指定 G01 的进给速度。

注意：

1）G41、G42、G40 指令与 G01、G00 指令可在同程序段出现，通过直线运动建立或取消刀补。

2）在 G41、G42、G40 所在程序段中，X 或 Z 至少有一个值变化，否则发生报警。

3）G41、G42 不能同时使用，即在程序中，前面程序段有了 G41 就不能继续使用 G42，必须先用 G40 指令解除 G41 刀补状态后，才可使用 G42 刀补指令。

4）在调用新的刀具前，必须用 G40 指令取消刀尖圆弧半径补偿，否则发生报警。

（4）G41 与 G42 选择。刀具与加工方向如图 2-30 所示时：顺着刀具运动方向看，工件在刀具的左边称左补偿，使用 G41 刀尖圆弧半径左补偿指令；工件在刀具的右边称右补偿，使用 G42 刀尖圆弧半径右补偿指令。

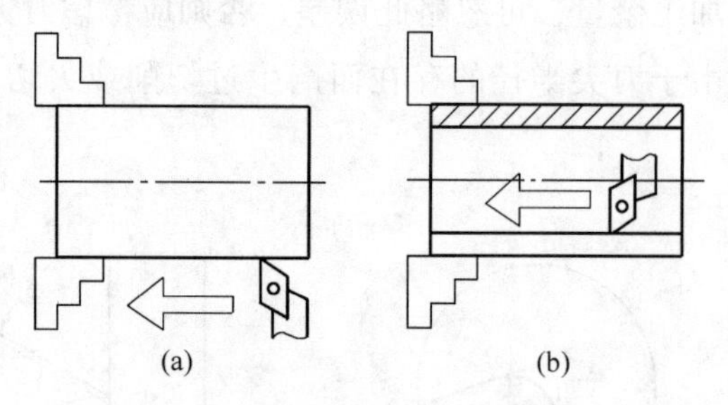

(a)　(b)

图 2-30　G42 与 G41 选择

（a）G42；（b）G41

（5）刀尖圆弧半径补偿参数及设置。

1）刀尖圆弧半径。刀尖圆弧半径补偿后，刀具自动偏离工件轮廓刀尖半径距离。因此，必须将刀尖圆弧半径值输入系统的存储器中，具体操作方法见加工操作部分。

2）刀尖方位的确定。刀尖圆弧所处的位置不同，执行刀具补偿时，刀具自动偏离工件轮廓的方向也不同。因此，要把代表车刀形状和位置的参数输入到存储器中，具体操作方法见加工操作部分。车刀形状和位置参数称为刀尖方位 T，如图 2-31 所示，共有 9 种，分别用参数 0~9 表示，P 为理论刀尖点。前置刀架的数控车床常用刀尖方位 T 为：外圆右偏刀 $T=3$，镗孔右偏刀 $T=2$。

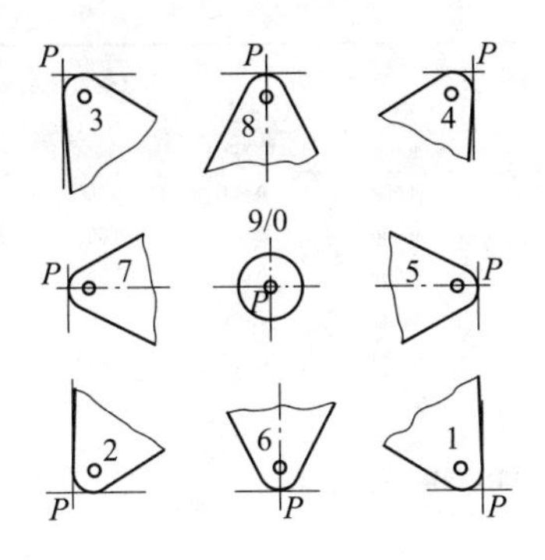

图 2-31　刀尖方位

【例 2-7】图 2-32 所示为零件编写刀尖圆弧半径补偿部分程序。

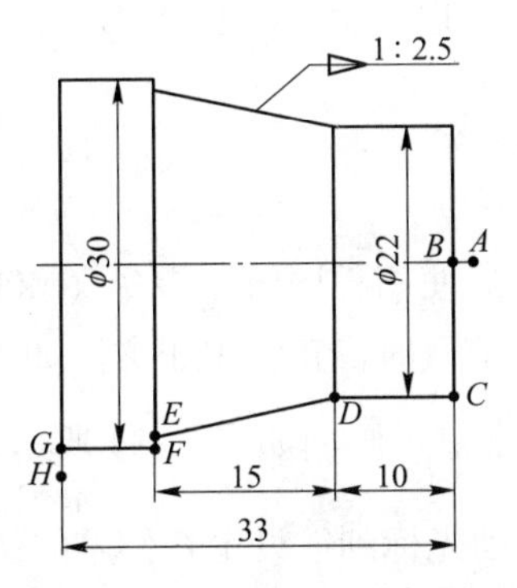

图 2-32　刀尖圆弧半径补偿例题

刀补过程分为三步：加工前建立刀补，图示 $A \to B$ 段；加工时执行刀补，图示 $B \to G$ 段；加工完成后取消刀补，图示 $G \to H$ 段，刀尖圆弧半径补偿部分程序见表 2-10。

表 2-10　刀尖圆弧半径补偿部分举例

G00 Z5. 0;	刀具快速点定位至 *A* 点
X0;	
G42 G01 X0 Z0;	直线插补至 *B* 点，建立刀尖圆弧半径右补偿（$A \to B$）
X22. 0;	直线插补至 *C* 点
……;	……
G01 Z-33. 0;	直线插补至 *G* 点
G40 X41. 0;	直线插补至 *H* 点并取消刀尖圆弧半径补偿（$G \to H$）

2. 2. 3　加工操作

2. 2. 3. 1　刀尖圆弧补偿参数输入

按 OFS SET 键→点击【补正】→点击【形状】→进入刀具补偿参数界面。

光标移至刀具 *R* 位置→输入刀尖圆弧半径（如 T0101 取值 0. 4)→点击【输入】；光标移至刀具 *T* 位置→输入刀具方位号（如 *T*=3)→点击【输入】，如图 2-33 所示，完成刀具补偿参数的输入。

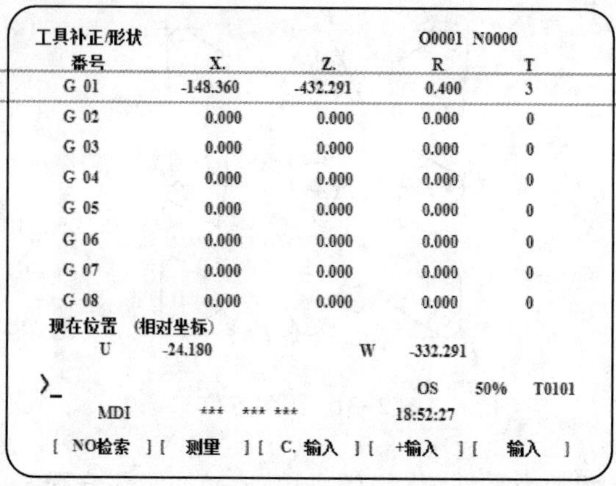

工具补正/形状				O0001 N0000
番号	X.	Z.	R	T
G 01	-148.360	-432.291	0.400	3
G 02	0.000	0.000	0.000	0
G 03	0.000	0.000	0.000	0
G 04	0.000	0.000	0.000	0
G 05	0.000	0.000	0.000	0
G 06	0.000	0.000	0.000	0
G 07	0.000	0.000	0.000	0
G 08	0.000	0.000	0.000	0

现在位置 (相对坐标)

U -24.180 W -332.291

>_ OS 50% T0101

MDI *** *** *** 18:52:27

[NO检索] [测量] [C. 输入] [+输入] [输入]

图 2-33 刀具补偿参数

2.2.3.2 尺寸精度控制方法

对于首件试切的工件，粗车后使用程序暂停指令（M00）停下机床，测量工件尺寸是否符合要求，如有偏差要在精车前及时修正，修正方法如下：

（1）按 [OFS/SET] 键，CTR 屏幕上显示画面如图 2-34 所示。

（2）用 [PAGE ⇧] 键或 [PAGE ⇩] 键移动光标到欲设定补偿号的位置。

（3）输入 X、Z 值。

如图 2-21 所示零件，用 90°外圆车刀（刀具号：T01），粗车后工件外圆直径大 0.02mm，端面长 0.03mm，按 [OFS/SET] 键，用 [PAGE ⇧] 键或 [PAGE ⇩] 键移动光标至 T01 对应的刀补号 W01，X 位置输入-0.02，点击【输入】；Z 位置输入-0.03，点击【输入】，完成刀具磨损补偿，如图 2-34 所示。若各向尺寸相应小，则输入“+”数据。通过上述修正后，再进行精加工，即可达到尺寸要求。

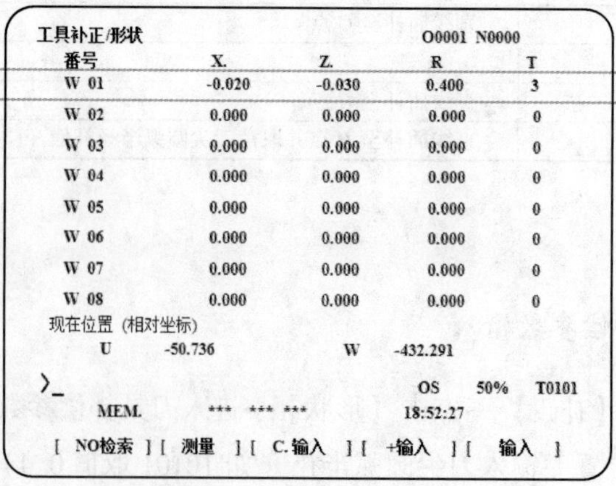

工具补正/形状				O0001 N0000
番号	X.	Z.	R	T
W 01	-0.020	-0.030	0.400	3
W 02	0.000	0.000	0.000	0
W 03	0.000	0.000	0.000	0
W 04	0.000	0.000	0.000	0
W 05	0.000	0.000	0.000	0
W 06	0.000	0.000	0.000	0
W 07	0.000	0.000	0.000	0
W 08	0.000	0.000	0.000	0

现在位置 (相对坐标)

U -50.736 W -432.291

>_ OS 50% T0101

MEM. *** *** *** 18:52:27

[NO检索] [测量] [C. 输入] [+输入] [输入]

图 2-34 刀具磨损补偿参数输入

进行批量生产时，因为刀具产生磨损测量工件尺寸偏大，当尺寸变化时，用上述方法可补偿刀具的磨损量。

2.2.4 精度检验

锥度的检验常用锥度量规、游标万能角度尺、正弦规检测锥度。

2.2.4.1 锥度量规

（1）结构。锥度量规是检验工件内、外锥度的量具。它分为锥度塞规和锥度环规两种，如图 2-35 所示。锥度塞规主要用于检验产品的大径、锥度和接触率，属于专用综合检具。锥度塞规可分为尺寸塞规和涂色塞规两种。锥度塞规规格：3~300mm。

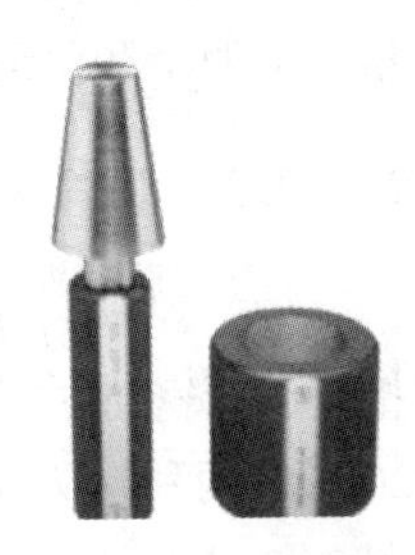

图 2-35 锥度量规

（2）使用方法。使用锥度环规检验锥度贴合率的步骤如下。

1）将锥度环规及工件的锥面擦拭干净，注意观察锥面不能有纤维；

2）在工件锥度部位的轴向方向均匀地涂抹 3 条极薄的红丹或蓝油（厚度在 0.01mm 内）；

3）套紧环规，旋转 90°；

4）取出环规，观察工件锥面接触后的痕迹来判断内圆锥的好坏。

接触面积越多，锥度越好，反之则不好，一般用标准量规检验锥度接触要在 75% 以上。

2.2.4.2 游标万能角度尺

（1）结构。游标万能角度尺是用来测量工件内、外角度的量具，其结构如图 2-36 所示。万能角度尺由主尺、90°角尺、游标、制动头、基尺、直尺、卡块等组成。分度值有 5′和 2′两种。

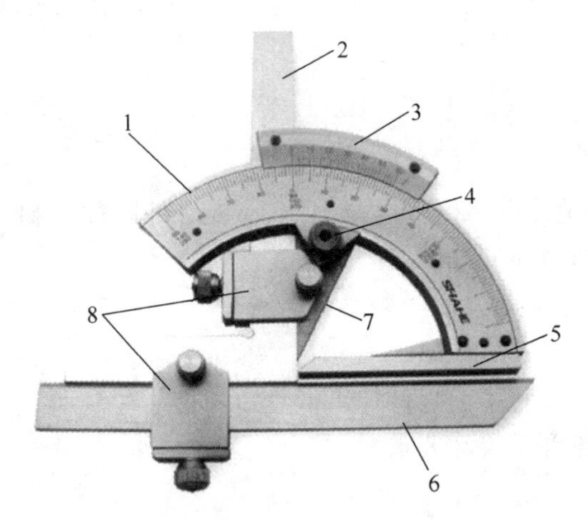

图 2-36 万能角度尺

1—主尺；2—角尺；3—游标；4—制动头；5—基尺；6—直尺；7—扇形板；8—卡块

（2）使用方法。测量前应先校准零位。万能角度尺的零位，是当角尺与直尺均装上，而角尺的底边及基尺与直尺无间隙接触，此时主尺与游标的“0”线对准。调整好零位后，通过改变基尺、角尺、直尺的相互位置可测试0°~320°范围内的任意角。其读数方法与游标卡尺基本相同。先读出游标零线前的角度，再从游标上读出角度“分”的数值，两者相加就是被测零件的角度数值。

用万能角度尺测量零件角度时，应使基尺与零件角度的母线方向一致，且零件应与量角尺的两个测量面的全长上接触良好，以免产生测量误差。

2.2.4.3　正弦规

（1）结构。正弦规是利用三角法测量角度的一种精密量具。正弦规由主体和两个直径相同的精密圆柱组成，两个精密圆柱的中心距要求很精确，中心距的尺寸有 100mm 和 200mm 两种。为便于被检工件在主体平面上定位和定向，主体上装有后挡板和侧挡板。如图 2-37 所示。

（2）测量原理。在正弦规工作面上放置圆锥角为 α 的被测圆锥工件，正弦规的一个圆柱与平板接触，另一个圆柱用量块垫高。如图 2-38 所示，图中 L 为正弦规中心距，H 为量块组高度尺寸，可以按被测零件的圆锥角度，根据公式 $\sin\alpha = H/L$ 算得。然后用百分表（或测微仪）检验工件圆锥面上母线两端的高度，若两端高度相等，说明工件的圆锥角度正确，若高度不等，说明工件的圆锥角度有误差。正弦规一般用于测量小于 45°的角度，在测量小于 30°的角度时，精确度可达 3″~5″。

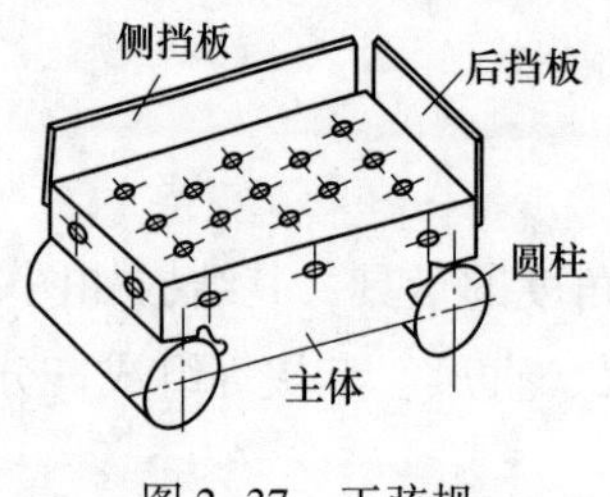

图 2-37　正弦规

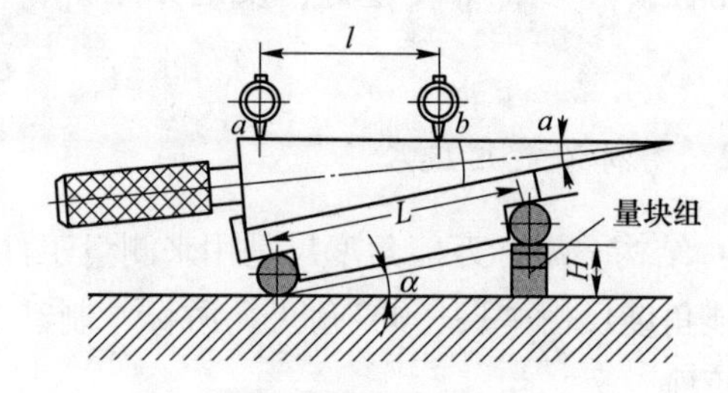

图 2-38　测量原理

任务实施

（1）图样分析：

零件加工表面有 $\phi36^{0}_{-0.04}$、$\phi28^{0}_{-0.03}$、$\phi24^{0}_{-0.03}$、$\phi12$ 外圆柱面和 1∶2 圆锥面及倒角等，表面粗糙度分别为 $Ra1.6$ 和 $Ra3.2$。

（2）加工工艺方案制订：

1）加工方案：

①采用三爪自定心卡盘装卡，零件伸出卡盘 60mm 左右 。

②粗精加工零件外轮廓至尺寸要求。

2）刀具选用：

T01　90°硬质合金外圆偏刀（刀尖角 55°）

T02　切槽刀

圆锥面零件数控加工刀具卡见表2-11。

表2-11　数控加工刀具卡片

零件名称		圆锥面零件			零件图号	2-21		
序号	刀具号	刀具名称		数量	加工表面	刀尖半径 R/mm	刀尖方位 T	备注
1	T01	90°外圆偏刀		1	粗精车外轮廓	0.4	3	
编制		审核		批准	日期		共1页	第1页

3）加工工序：

圆锥面零件数控加工工序卡见表2-12。

表2-12　数控加工工序卡片

单位名称	天津工业职业学院				零件名称		零件图号	
					圆锥面零件		2-21	
程序号		夹具名称		使用设备	数控系统		场地	
O221		三爪自定心卡盘		CKA6140	FANUC Oi		数控实训中心	
工序号	工序内容			刀具号	主轴转速 n/r·min^{-1}	进给量 F/mm·r^{-1}	背吃刀量 a_p/mm	备注
1	装夹零件并找正							手动
2	对外圆偏刀			T01				
3	对切槽刀			T02				
4	粗车外轮廓，留余量1mm			T01	600	0.2	1.5	O121
5	精车外轮廓			T01	1000	0.1	0.5	
编制		审核		批准	日期		共1页	第1页

（3）程序编制：

1）锥度尺寸计算。

圆锥最大直径 $D=20$，圆锥长度 $L=18-8=10$，锥度 $C=1/2=0.5$，根据公式：$C=\frac{D-d}{L}$，可以计算得出：圆锥最小直径 $d=D-CL=20-10\times0.5=15$。

2）加工程序。

加工程序如表2-13所示。

表2-13　圆锥面加工程序

O221;	程序号
G40 G97 G99 M03 S600 F0.2;	取消刀具半径补偿，取消主轴恒转速度，设每转进给量，主轴正转，转速为600r/min，设进给量为0.2mm/r
T0101;	换T01外圆车刀，代入01号刀具补偿
M08;	打开切削液

G42 G00 Z5.0;	快速进刀至粗车循环起点
X40.0;	
G71 U1.5 R0.5;	定义粗车循环，切削深度 1.5mm，退刀量 0.5mm； 精车路线由 N10～N20 指定，*X* 方向精车余量 0.5mm，*Z* 方向精车余量 0.05mm
G71 P10 Q20 U0.5 W0.05;	
N10 G00 X0;	精车轮廓
G01 Z0;	
X10.0;	
X12.0　Z-1.0;	
Z-8.0;	
X15.0;	
X20.0 Z-18.0;	
X22.0;	
X23.985 W-1.0;	
Z-27.0;	
X27.985;	
Z-35.0;	
X33.0;	
X33.0 Z-40.0;	
X35.0;	
X35.98 W-1.5;	
N20 Z-49.0;	
G40 G00 X100.0;	快速退刀至换刀点
Z100.0;	
M05;	主轴停止
M00;	程序暂停
M03 S1000 F0.1;	主轴正转，转速 1000r/min
T0101;	代入 01 号刀具补偿
G42 G00 Z5.0;	刀具快速点定位至粗加工复合循环起点
X40.0;	
G70 P10 Q20;	精加工复合循环
G40 G00 X100.0;	快速退刀至换刀点
Z100.0;	
M30;	程序结束并返回起点

（4）工件加工：

装夹并找正工件→装刀（T01、T02）→对刀→输入表 2-13 程序→模拟程序→自动加工。

(5) 尺寸检测：

1) 使用游标卡尺测量 $\phi36^{0}_{-0.04}$、$\phi28^{0}_{-0.03}$、$\phi24^{0}_{-0.03}$、$\phi12$ 外径尺寸。

2) 采用游标卡尺测量 8，18，27，10，45±0.1 长度尺寸。

3) 采用万能角度尺测量锥度。

4) 采用粗糙度比较板检测粗糙度。

任务评价

任务评价如表 2-14 所示。

表 2-14　任务评价

项目	序号	评分标准		配分	得分
加工工艺（10%）	1	加工步骤	符合数控车工工艺要求	3	
	2	加工部位尺寸	计算相关部位尺寸	3	
	3	刀具选择	刀具选择合理，符合加工要求	4	
程序编制（30%）	4	程序号	无程序号无分	2	
	5	程序段号	无程序段号无分	2	
	6	切削用量	切削用量选择不合理无分	4	
	7	原点及坐标	标明程序原点及坐标，否则无分	2	
	8	程序内容	不符合程序逻辑及格式要求每段扣 2 分	20	
			程序内容与工艺不对应扣 5 分		
			出现危险指令扣 5 分		
加工操作（20%）	9	程序输入与检索	每错一段扣 2 分	10	
	10	加工操作 30min	工件坐标系原点设定错误无分	10	
			误操作无分		
			超时无分		
尺寸检测（30%）	11	外圆	$\phi36^{0}_{-0.04}$ 超差 0.01 扣 3 分	3	
	12		$\phi28^{0}_{-0.03}$ 超差 0.01 扣 3 分	3	
	13		$\phi24^{0}_{-0.03}$ 超差 0.01 扣 3 分	3	
	14		$\phi12$	3	
	15	锥度	超差无分	6	
	16	轴向长度	45±0.1	3	
	17		8　18　27　10	4	
	18	倒角	$C1$　$C1.5$	3	
	19	表面粗糙度	$Ra1.6$ 降级无分	1	
	20		$Ra3.2$ 降级无分	1	
职业能力（10%）	21	学习能力		2	
	22	表达沟通能力		2	
	23	团队合作		2	
	24	安全操作与文明生产		4	
合计					

同步思考与训练

（1）思考题：

1）简述 G71 和 G70 指令的功能。

2）简述 G71 和 G70 指令格式及每个参数的含义。

3）简述刀尖圆弧半径对加工精度的影响。

4）简述刀尖圆弧半径补偿指令 G41、G42、G40 的使用方法。

5）简述圆锥面测量常用的量具。

（2）同步训练题：

图 2-39 和图 2-40 所示零件，材料硬铝合金，毛坯为 ϕ40mm 长棒料，编写零件加工加工程序，完成加工与检测。

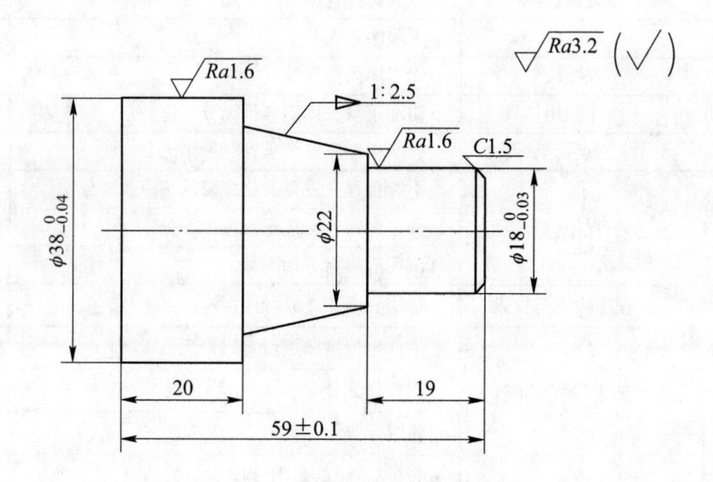

图 2-39　同步训练 1

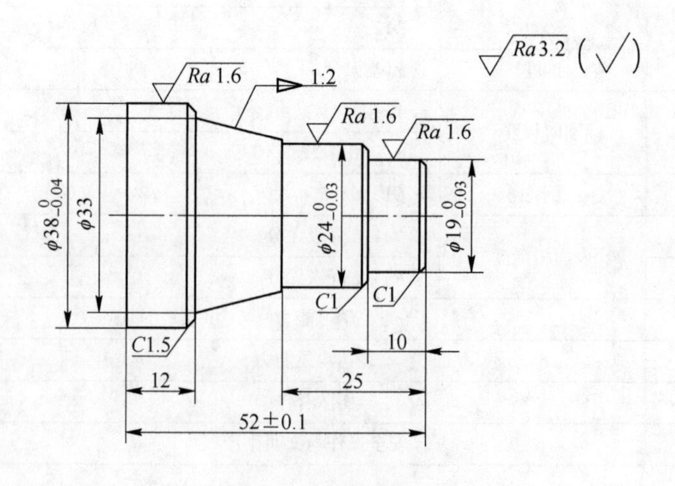

图 2-40　同步训练 2

任务 2.3　切槽与切断

任务描述

如图 2-41 所示的零件，材料硬铝合金，毛坯为 ϕ40mm 长棒料，单件生产，使用数控车床加工，编写加工程序，检测加工质量。

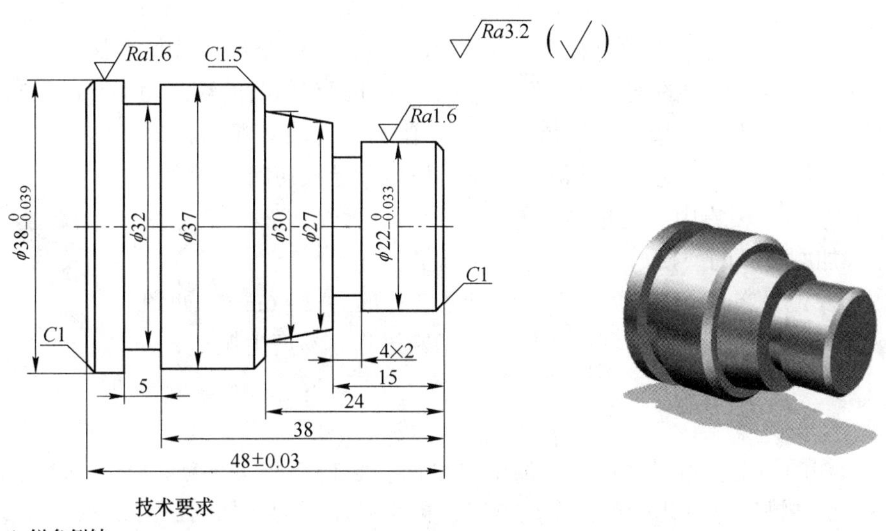

1. 锐角倒钝。
2. 未注线性尺寸公差应符合 GB/T 1804—2000 的要求。

图 2-41　零件图

任务目标

（1）知识目标：

1）认识零件中槽的类型和标注。

2）描述切槽的加工路径。

3）认识指令 G04。

4）认识外径千分尺的结构。

（2）技能目标：

1）能编写切槽和切断程序。

2）能完成切槽刀的对刀。

3）能使用外径千分尺测量。

（3）素质目标：

1）正确执行安全技术操作规程，培养安全意识。

2）培养认真负责的工作态度和质量意识。

相关知识

2.3.1 工艺准备

切槽与切断是数控加工的重要内容，轴类零件外螺纹一般都带有退刀槽、砂轮越程槽等，套类零件内螺纹也常带有内沟槽。

2.3.1.1 切槽加工的特点

（1）切削力大。由于切槽过程中切屑与刀具、工件的摩擦，被切金属的塑性变形较大，所以在切削用量相同的条件下，切削力比一般外圆车削时的切削力大20%~25%。

（2）切削变形大。切槽时，切槽刀的主切削刃和左右切削刃同时参与切削，切屑排出时，受到槽两侧的摩擦、挤压，导致切削变形大。

（3）切削热比较集中。切槽时，塑性变形大，摩擦剧烈，产生大量切削热，加剧刀具的磨损。

（4）刀具刚性差。切槽刀主切削刃较窄，一般为2~6mm，如图2-42所示，刀头狭长，所以刀具的刚性差，切断过程容易产生振动。

图2-42 切槽与切断刀具

2.3.1.2 槽的加工

（1）窄槽的加工。当刀头宽等于沟槽宽度时，该槽被称为窄槽。一般采用G01直进切削至槽底，在槽底停留几秒钟，光整槽底，再用G01退刀，如图2-43所示。

（2）宽槽的加工。当刀头宽度小于沟槽宽度时，该槽被称为宽槽。加工宽槽时常采用排刀的方式进行粗切（如图2-44所示①→②→③），然后用切槽刀沿槽的一侧切至槽底（图中④所示）；精加工槽底至槽的另一侧（图中⑤所示）；再沿侧面退出（图中6所示）。

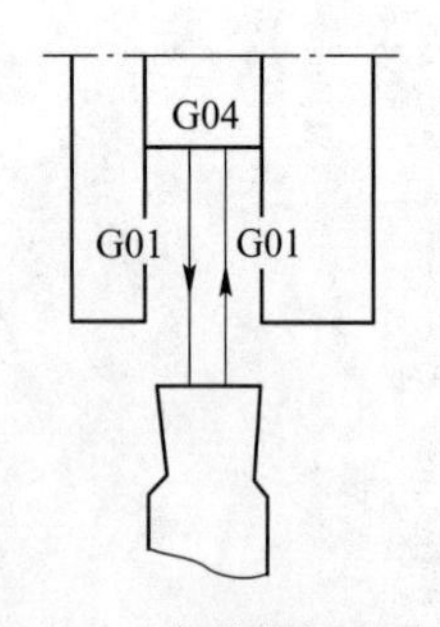

图2-43 窄槽的加工路线

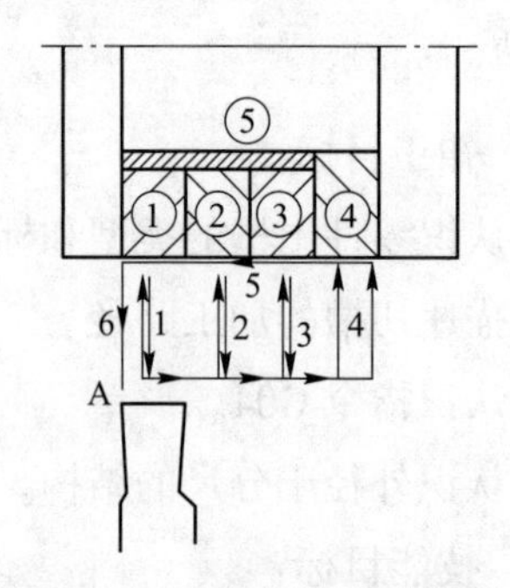

图2-44 宽槽的加工路线

2.3.1.3 槽加工切削用量选择

（1）背吃刀量 a_p。当横向切削时，切槽刀的背吃刀量等于刀的主切削刃宽度。

（2）进给量 f。如刀具刚性、强度及散热条件较差，应适当减小进给量。进给量太大时，刀具容易折断；进给量太小时，刀具与工件产生强烈摩擦会引起振动。用高速钢切槽刀车钢料时，f= 0.05 ~ 0.1mm/r；车铸铁时，f= 0.1 ~ 0.2mm/r。用硬质合金刀加工钢料时，f=0.1 ~ 0.2mm/r；加工铸铁时，f= 0.15 ~ 0.25mm/r。

（3）切削速度 v_c。用高速钢车刀切削钢料时，v_c = 30 ~ 40m/min；加工铸铁时，v_c = 15 ~ 25m/min。用硬质合金切削钢料时，v_c = 80 ~ 120m/min；加工铸铁时，v_c = 60 ~ 100m/min。

2.3.2 编程指令

2.3.2.1 暂停指令 G04

（1）功能。刀具相对于零件做短时间的无进给光整加工，主要用于槽的加工，以降低表面粗糙度，保证工件圆柱度。

（2）指令格式：

G04 P____；　　暂停时间 ms

G04 X____；　　暂停时间 s

G04 U____；　　暂停时间 s

式中　P，X，U ——暂停时间。

注意：

（1）X、U 后面可用小数点的数字，如 G04 X5.0 表示前面的程序执行后，要暂停 5s，才执行下面的程序段。

（2）P 后面不允许用小数点的数字，如 G04 P1000 表示暂停 1s。

2.3.2.2 窄槽与宽槽编程

【例 2-8】如图 2-45 所示的零件，毛坯为铝合金，工件的外圆与倒角已加工至图纸尺寸，编写窄槽部分的程序。参考程序如表 2-15 所示。

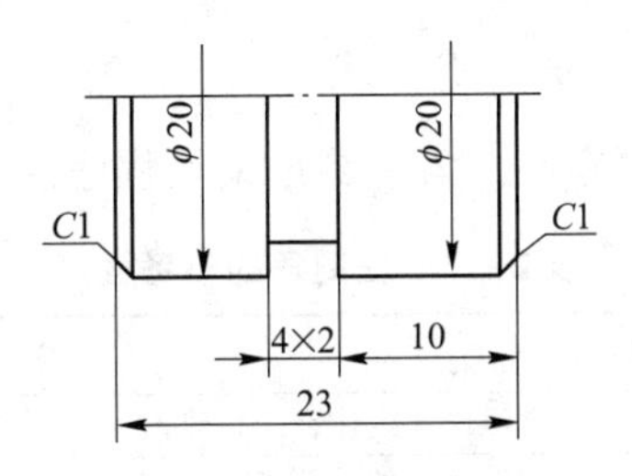

图 2-45　窄槽加工举例

表 2-15　窄槽加工部分程序（切槽刀刀宽 4mm）

程 序 段	说　明
T0202；	换 T02 切槽刀，代入 02 号刀具补偿
G00 X21.0；	快速移动到起始位置
Z-14.0；	
G01 X16.0；	进刀
G04 X2.0；	槽底暂停 2s，精车槽底
G01 X21.0；	退刀

【例 2-9】如图 2-46 所示的零件，毛坯材料为铝合金，工件外圆与倒角已加工至图纸尺寸，编写宽槽部分的程序。参考程序如表 2-16 所示。

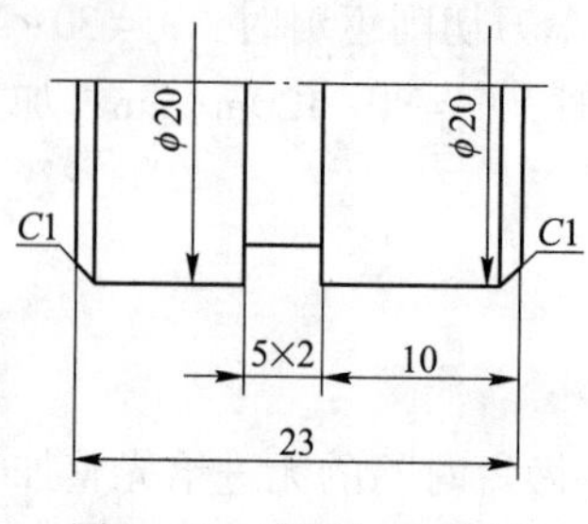

图 2-46　宽槽加工举例

表 2-16　窄槽加工部分程序（切槽刀刀宽 4mm）

程 序 段	说　明
T0202;	换 T02 切槽刀，代入 02 号刀具补偿
G00 X21.0;	快速移动到起始位置
Z-15.0;	
G01 X17.0;	进刀，离槽底差 0.5mm
X21.0;	退刀
W1.0;	向右进 1mm
X16.0;	切到槽底
W-1.0;	向左进刀 1mm，精车槽底
X21.0;	退刀

2.3.2.3　*左倒角编程*

【例 2-10】如图 2-46 所示零件，为左倒角编写程序。参考程序如表 2-17 所示。

表 2-17　左倒角的参考程序

程 序 段	说　明
G00 Z-27.0;	切槽刀快速定位在工件末端
X22.0;	定位于接近外圆一点
G01 X18.0 F0.08;	切外圆至 X18.0
X22.0;	退刀，*X* 方向单边退 2mm
W2.0;	切槽刀右移 2mm
X18.0 W-2.0;	用切槽刀右侧部分切左倒角
X-1.0;	切断工件

2.3.3　加工操作

2.3.3.1　安装切槽刀的注意事项

（1）切槽刀刃与工件中心线等高，安装方法同外圆偏刀。

（2）切槽刀安装时应注意刀刃平行工件轴线，不能歪斜，否则使工件侧壁不平直，严重歪斜造成切断刀折断。

2.3.3.2 切槽刀对刀方法

（1）*X* 向补偿。

1）主轴正转。

2）选择手轮，按键，刀具接近工件外圆面，如图 2-47（a）所示。

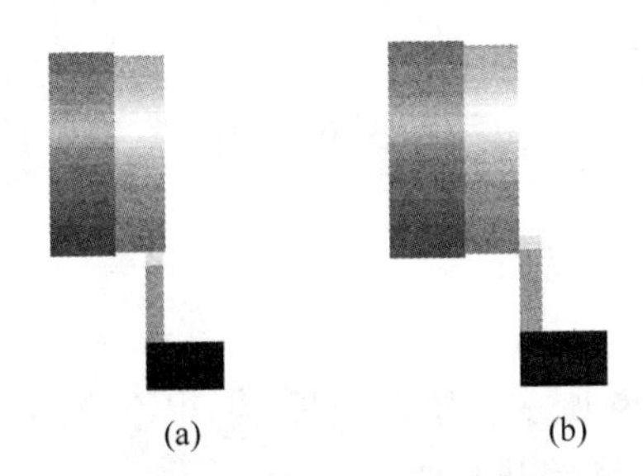

图 2-47 切槽刀对刀过程

（a）*X* 向对刀；（b）*Z* 向对刀

3）点击按键→按【补正】键→按【形状】键→移动光标至选择的刀具位置，如番号 G02，界面如图 2-48（a）所示。

工具补正/形状　　O0001 N0000

番号	X.	Z.	R	T
G 01	-148.360	-432.291	0.400	3
G 02	-144.235	0.000	0.000	0
G 03	0.000	0.000	0.000	0
G 04	0.000	0.000	0.000	0
G 05	0.000	0.000	0.000	0
G 06	0.000	0.000	0.000	0
G 07	0.000	0.000	0.000	0
G 08	0.000	0.000	0.000	0

现在位置 （相对坐标）

U -24.180　　W -332.291

>_　　OS 50% T0101

MDI *** *** ***　　18:52:27

[NO检索] [测量] [C. 输入] [+输入] [输入]

(a)

工具补正/形状　　O0001 N0000

番号	X.	Z.	R	T
G 01	-148.360	-432.291	0.400	3
G 02	-144.235	-414.340	0.000	0
G 03	0.000	0.000	0.000	0
G 04	0.000	0.000	0.000	0
G 05	0.000	0.000	0.000	0
G 06	0.000	0.000	0.000	0
G 07	0.000	0.000	0.000	0
G 08	0.000	0.000	0.000	0

现在位置 （相对坐标）

U -24.180　　W -332.291

>_　　OS 50% T0101

MDI *** *** ***　　18:52:27

[NO检索] [测量] [C. 输入] [+输入] [输入]

(b)

图 2-48 参数输入界面

（a）*X* 向补偿输入；（b）*Z* 向补偿输入

4）输入 *X* 测量值，也就是之前外圆刀车削之后，测量的外圆值，如 *X*39. 875→按【测量】键，完成 *X* 向补偿。

（2）*Z* 向补偿。

1）主轴正转。

2）选择手轮，按键，刀具接近工件端面，如图 2-47（b）所示。

3）点击按 OFS/SET 键→按【补正】键→按【形状】键→移动光标至选择的刀具位置，如番号 G02，界面如图 2-48（b）所示。

4）输入 Z0→按【测量】键，完成 *Z* 向补偿。

注意：操作中，当切槽刀接近外圆或端面时，将手轮进给倍率逐渐降至最低，慢速摇动手轮进刀，直至接触表面有碎屑出现为止。

2.3.4 精度检验

2.3.4.1 外径千分尺应用

外径千分尺是一种比游标卡尺更为精密的量具。常用外径千分尺测量零件的外径、凸肩厚度、板厚和壁厚等。

2.3.4.2 外径千分尺结构

外径千分尺结构如图 2-49 所示，主要由尺架、固定测砧、测微螺杆、固定刻度套筒、微分筒、棘轮旋柄、锁紧螺钉和绝热板组成。

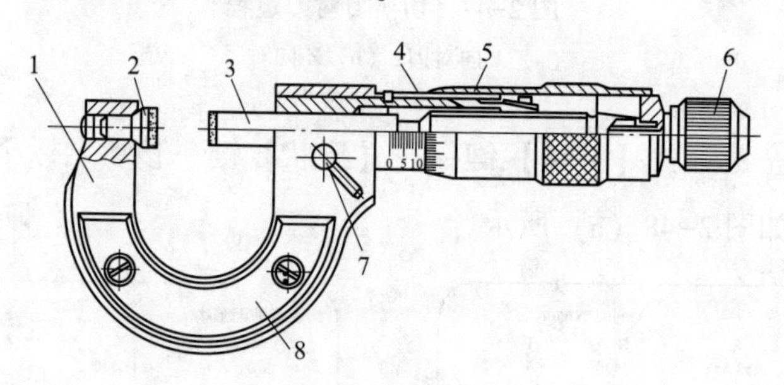

图 2-49 外径千分尺

1—尺架；2—固定测砧；3—测微螺杆；4—固定刻度套筒；5—微分筒；6—棘轮旋柄；7—锁紧螺钉；8—绝热板

2.3.4.3 使用方法

外径千分尺的使用方法如下。

（1）千分尺使用时轻拿轻放，被测物体需擦拭干净。

（2）松开千分尺锁紧装置，校准零位，转动旋钮，使测砧与测微螺杆之间的距离略大于被测物体。

（3）一只手拿住千分尺的尺架，将待测物置于测砧与测微螺杆的端面之间，另一只手转动旋钮，当螺杆要接近物体时，改旋测力装置直至听到“喀喀”声后再轻轻转动 0.5~1 圈。

（4）旋紧锁紧装置，以防止移动千分尺时螺杆转动，即可读数。

2.3.4.4 读数

以图 2-50 为例，千分尺读数方法分为如下三个步骤。

（1）以微分筒的端面为基准线，读出固定套筒上的数值为 8.5mm。

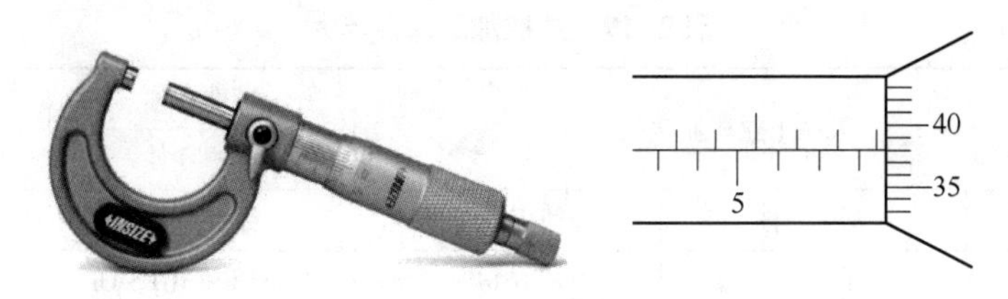

图 2-50　千分尺读数

（2）以固定套筒上的中心线作为读数基准线，读出微分筒与固定套筒的基准线对齐的刻线数，将其乘以 0.01，即 0.380mm。

（3）两部分的数值相加即为测量的实际尺寸 8.880mm。

2.3.4.5　注意事项

使用千分尺测量时应注意以下事项。

（1）不能测量旋转的工件，这样会造成严重损坏。

（2）测量前被测工件表面应擦拭干净。

（3）测量时，应该握住弓架，旋转微分筒的力量要适当。

（4）储存千分尺前，要使测微螺杆离开测砧，用布擦净千分尺外表面，并抹上黄油。

任务实施

（1）图样分析：

图 2-41 所示零件，加工表面有 $\phi22^{0}_{-0.033}$、$\phi37$、$\phi38^{0}_{-0.039}$外圆面，4×2 和宽度为 5mm 的两个槽，此外在零件左端有 1×45°左倒角等，表面粗糙度为 $Ra1.6$ 和 $Ra3.2$。

（2）加工工艺方案制订：

1）加工方案：

①采用三爪自定心卡盘装卡，零件伸出卡盘 65mm 左右 。

②粗精车加工零件外轮廓至尺寸要求。

③加工 4×2 的窄槽和宽度为 5mm 的宽槽。

④加工左倒角，切断工件。

2）刀具选用：零件数控加工刀具卡片如表 2-18 所示。

表 2-18　数控加工刀具卡片

零件名称		多槽零件		零件图号		2-41		
序号	刀具号	刀具名称		数量	加工表面	刀尖半径 R/mm	刀尖方位 T	备注
1	T01	90°外圆偏刀		1	粗精车外轮廓	0.4	3	
2	T02	4mm 切槽刀		1	切槽和切断			
编制		审核		批准	日期		共 1 页	第 1 页

3）加工工序：零件数控加工工序卡片如表 2-19 所示。

表 2-19　数控加工工序卡片

单位名称	天津工业职业学院			零件名称	零件图号	
				多槽零件	2-41	
程序号	夹具名称	使用设备		数控系统	场地	
O231	三爪自定心卡盘	CK6140		FANUCSERIES Oi	数控实训中心	
工序号	工序内容	刀具号	主轴转速 n/r·min^{-1}	进给量 F/mm·r^{-1}	背吃刀量 a_p/mm	备注
1	装夹零件并找正					手动
2	对外圆偏刀	T01				
3	对切槽刀	T02				
4	粗车外轮廓，留余量 0.5mm	T01	600	0.2	1.5	O231
5	精车外轮廓	T01	1000	0.1	0.5	
6	切窄槽和宽槽	T02	400	0.05	4	
6	左倒角和切断	T02	400	0.05	4	
编制	审核	批准		日期	共 1 页	第 1 页

（3）程序编制：

零件数控加工参考程序见表 2-20。

表 2-20　数控加工程序

程序	说明
O231;	程序号
G40 G97 G99 M03 S600 F0.2;	取消刀尖圆弧半径补偿，取消主轴恒转速度，设每转进给量，主轴正转，转速为 600r/min，设进给量为 0.2mm/r
T0101;	换 T01 外圆车刀，代入 01 号刀具补偿
M08;	打开切削液
G00 Z5.0;	快速进刀至粗车循环起点
X42.0;	
G71 U1.5 R0.5;	设置外圆粗车循环
G71 P10 Q20 U0.5 W0.05;	
N10 G00 X0;	精加工轮廓
G01 Z0;	
X19.984;	
X21.984 Z-1.0;	
Z-15.0;	
X27.0;	
X30.0 Z-24.0;	
X34.0;	
X37.0 W-1.5;	
Z-43.0;	

续表 2-20

X37.981；	精加工轮廓
Z-53.0；	
N20 X40.0；	
G00 X100.0；	返回换刀点
Z100.0；	
M05；	主轴停转
M00；	程序暂停
M03 S1000 F0.1；	主轴正转，转速为 1000r/min，设进给量为 0.1mm/r
T0101；	代入 01 号刀具补偿
G42 G00 Z5.0；	快速进给到精车循环起点，采用刀尖圆弧右补偿
X42.0；	
G70 P10 Q20；	精车循环
G40 G00 X100.0；	取消刀尖圆弧右补偿，返回换刀点
Z100.0；	
M05；	主轴停转
M00；	程序暂停
M03 S400 F0.05；	主轴正转，转速为 400r/min，设进给量为 0.05mm/r
T0202；	换 T02 切槽刀，代入 02 号刀具补偿
G00 Z5.0；	快速进给到起刀点
X28.0；	
Z-15.0；	快速进给到窄槽位置
G01 X17.984；	切削窄槽 4×2
G04 X2.0；	
G01 X27.0；	
G00 X39.0；	快速退刀
Z-43.0；	快速进给到宽槽位置
G01 X32.5；	切削宽槽 $\phi32$
X38.0；	
W1.0；	
X32.0；	
W-1.0；	
X38.0；	
G00 X100.0；	返回换刀点
Z100.0；	
M05；	主轴停转
M00；	程序暂停

续表 2-20

M03 S400 F0.05;	主轴正转，转速为400r/min，设进给量为0.05mm/r
T0202;	代入02号刀具补偿
G00 Z5.0;	快速进给到起刀点
X39.0;	
Z-52.0;	快速进给到切断位置
G01 X35.981;	切削左倒角
X37.981;	
W1.0;	
X35.981 W-1.0;	
X-1.0;	切断
G00 X100.0;	返回换刀点
Z100.0;	
M30;	程序结束并返回起点

（4）工件加工：

1）三爪卡盘装夹工件，外伸60mm，找正工件。

2）分别将外圆车刀、切槽刀安装在刀架的1号、2号位置。

3）分别进行外圆车刀、切槽刀的对刀操作。

4）输入表2-20程序。

5）模拟程序。

6）自动加工。

（5）尺寸检测。

1）使用外径千分尺测量 $\phi22^{0}_{-0.033}$、$\phi38^{0}_{-0.039}$ 外径尺寸。

2）使用游标卡尺测量4×2槽和 $\phi32$ 槽的尺寸。

3）使用游标卡尺测量总长。

4）使用粗糙度比较板检测粗糙度。

任务评价

任务评价如表2-21所示。

表 2-21　任务评价

项目	序号	评分标准		配分	得分
加工工艺（10%）	1	加工步骤	符合数控车工工艺要求	3	
	2	加工部位尺寸	计算相关部位尺寸	3	
	3	刀具选择	刀具选择合理，符合加工要求	4	

续表 2-21

项目	序号	评分标准		配分	得分
程序编制（30%）	4	程序号	无程序号无分	2	
	5	程序段号	无程序段号无分	2	
	6	切削用量	切削用量选择不合理无分	4	
	7	程序内容	不符合程序逻辑及格式要求每段扣 2 分	20	
			程序内容与工艺不对应扣 5 分		
			出现危险指令扣 5 分		
	8	原点及坐标	标明程序原点及坐标，否则无分	2	
加工操作（30%）	9	程序输入与检索	每错一段扣 2 分	10	
	10	加工操作（30min 以内）	工件坐标系原点设定错误无分	20	
			误操作无分		
			超时无分		
尺寸检测（20%）	11	外圆	$\phi22^{0}_{-0.033}$超差 0.01 扣 3 分	3	
			$\phi38^{0}_{-0.039}$超差 0.01 扣 3 分	3	
	12	锥度	超差无分	2	
	13	轴向长度	48±0.3	3	
	14	沟槽	4×2 槽	3	
	15	沟槽	$\phi32$ 槽	3	
	16	倒角	*C*1	1	
	17	表面粗糙度	*Ra*1.6 降级无分	1	
			*Ra*3.2 降级无分	1	
职业能力（10%）	18	学习能力		2	
	19	表达沟通能力		2	
	20	团队合作		2	
	21	安全操作与文明生产		4	
合计					

同步思考与训练

（1）思考题：

1）切槽加工有什么特点？

2）简述切削窄槽和宽槽的路径。

3）如何选择切槽加工时的切削用量？

4）安装切槽刀架时应注意哪些事项？

5）外径千分尺由哪些部分组成？

（2）应会训练：

图 2-51 和图 2-52 所示零件，材料硬铝合金，毛坯为 ϕ40mm 长棒料，单件生产，使用数控车床 CK6140 加工零件，图中未注倒角为 *C*0.5，其余表面粗糙度为 *Ra*3.2，要求编写加工程序，加工并检测加工尺寸。

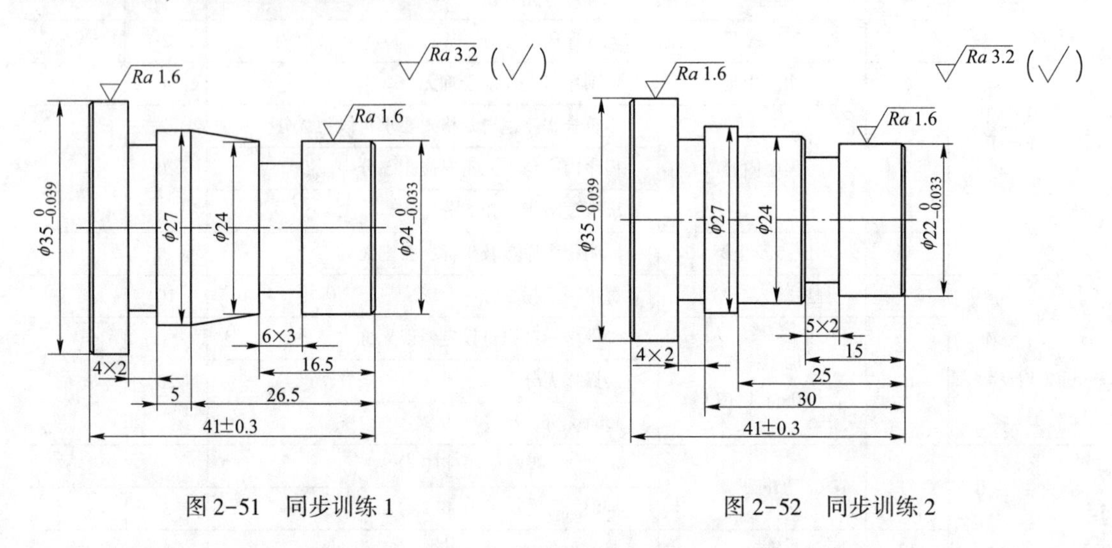

图 2-51　同步训练 1　　　　图 2-52　同步训练 2

任务 2.4　圆弧面的数控加工

任务描述

如图 2-53 所示的零件，材料硬铝合金 LY15，毛坯为 ϕ40mm 长棒料，单件生产，使用数控车床加工，编写加工程序，检测加工质量。

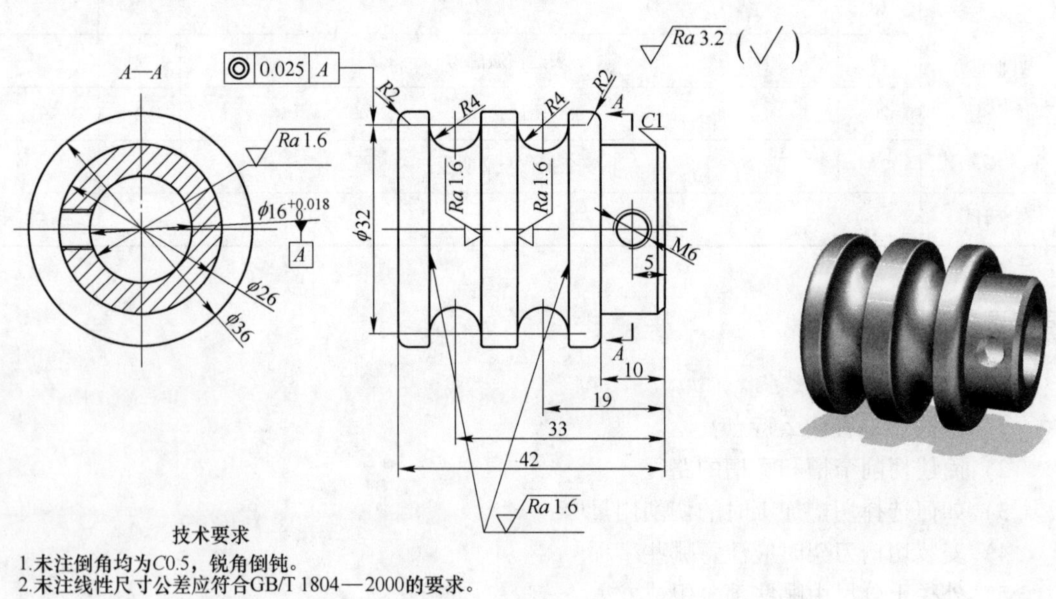

图 2-53　零件图

任务目标

（1）知识要求：

1）熟悉成型面加工工艺。

2）解释 G02/G03 和 G73 指令的含义。

3）认识圆弧加工刀具。

（2）技能目标：

1）能够识读典型轴类零件图。

2）能够使用 G02/G03 指令编写简单的圆弧面零件加工程序。

3）能够使用 G73 指令编写简单的成型面零件加工程序。

4）能够使用量具和样板检测零件质量。

（3）素质目标：

1）树立安全意识、质量意识和效率意识。

2）培养学生勇于开拓和创新的精神。

相关知识

2.4.1 工艺准备

2.4.1.1 圆弧的车削方式

（1）凸圆弧车削方式。凸圆弧车削方式有行切法、仿形法和组合法三种，如图 2-54 所示，其各自的车削特点见表 2-22。

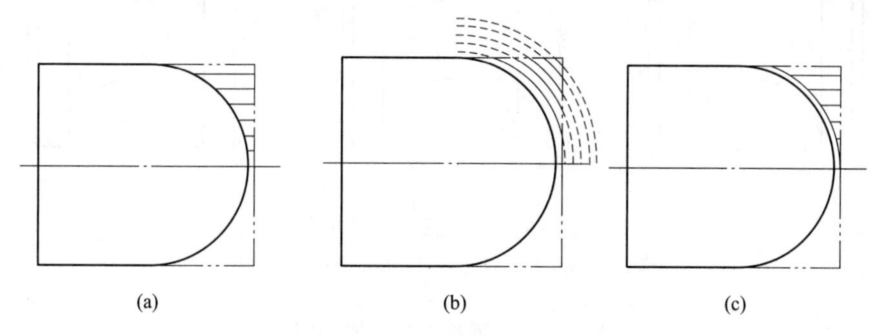

图 2-54 凸圆弧表面车削方式

（a）行切法；（b）仿形法；（c）组合法

表 2-22 凸圆弧车削方式比较

方式	特　点
行切法	走刀路线短，数值计算复杂，精车余量不均匀。加工路线不能超过圆弧面，否则会伤到圆弧的表面。
仿形法	数值计算简单，精车余量均匀，空行程多
组合法	走刀路线处于两者之间，精车余量均匀

(2) 凹圆弧车削方式。如图 2-55 所示的凹圆弧车削方式分别为同心圆式车削和阶梯式车削，其各自的特点见表 2-23。

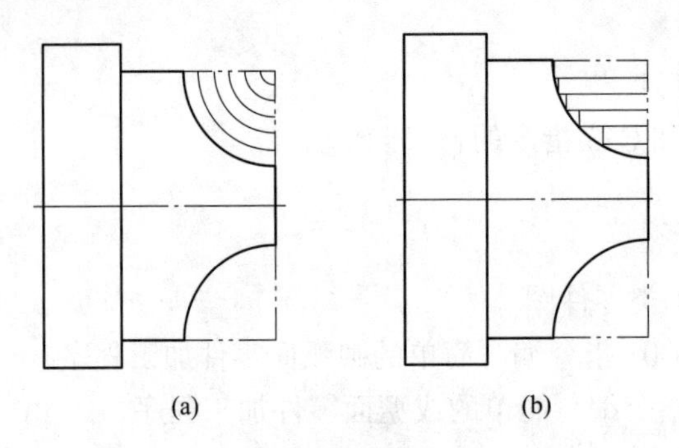

图 2-55　凹圆弧车削方式

(a) 同心圆弧形式；(b) 阶梯形式

表 2-23　凹圆弧车削方式比较

形式	特　点
同心圆弧形式	数值计算简单，编程方便，精车余量均匀
阶梯形式	走刀路线短，数值计算复杂，编程工作量增多

如图 2-56 所示的凹圆弧车削方式有同心圆弧形式、等弦长形式和阶梯形式三种，各自的特点见表 2-24。

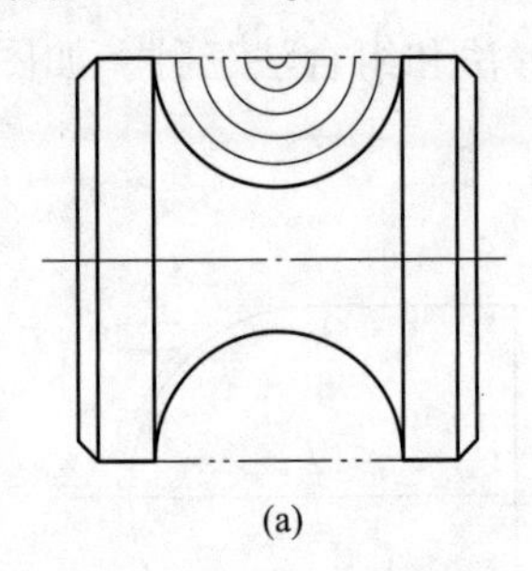

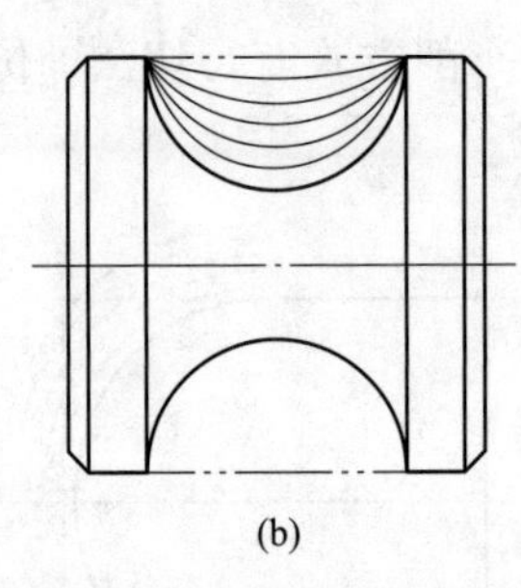

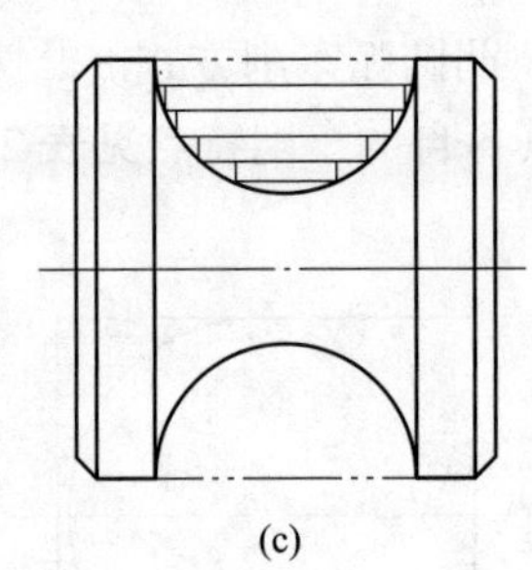

图 2-56　凹圆弧 2 车削方式

(a) 同心圆弧形式；(b) 等弦长形式；(c) 阶梯形式

表 2-24　凹圆弧 2 车削方式比较

形式	特　点
同心圆弧形式	走刀路线短，精车余量最均匀
等弦长形式	计算和编程最简单，走刀路线最长
阶梯形式	切削力分布合理，切削率最高

2.4.1.2　圆弧的车削刀具

常见的外圆圆弧类型有四种形式，如图 2-57 所示。常用的加工圆弧的刀具有三种类型，如图 2-58 所示。

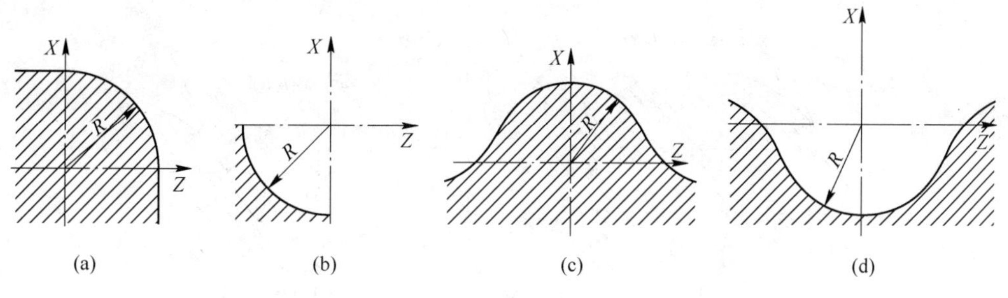

图 2-57　外圆圆弧类型

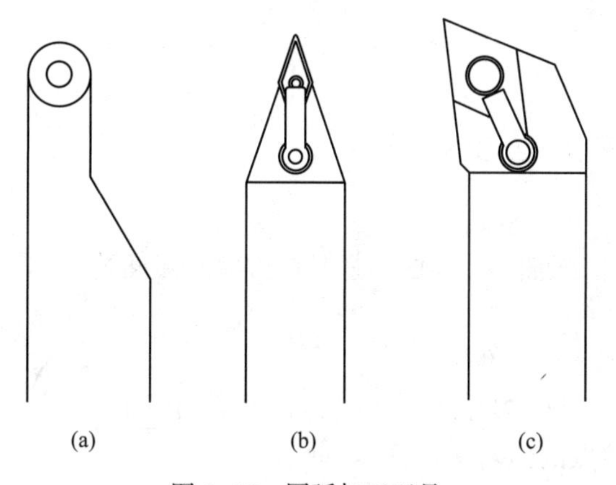

图 2-58　圆弧加工刀具

在图 2-57 中（a）、（b）两种类型的圆弧，在加工时可使用主偏角大于 90°的普通外圆车刀进行加工，刀具如图 2-58 中（c）所示。

在加工图 2-57（c）、（d）两种类型圆弧时，所选用加工刀具要考虑刀具的副偏角，因为副偏角太小会导致加工过切现象，加工过切如图 2-59 所示。

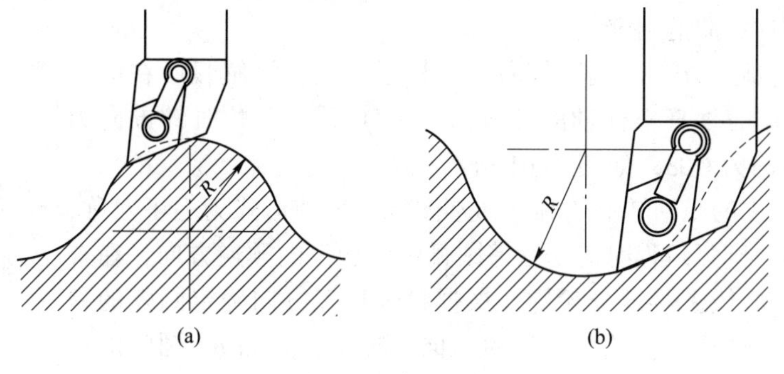

图 2-59　加工过切

在副偏角允许的情况下可选用图 2-58（a）圆弧车刀或图 2-58（b）尖刀两种刀具进行加工。

圆弧加工进行刀具选择时，可根据圆弧半径和圆弧深度计算出最小选用的副偏角，如图 2-60 所示，选用的刀具副偏角必须大于计算出的角度，才能保障加工时不会出现过切现象。

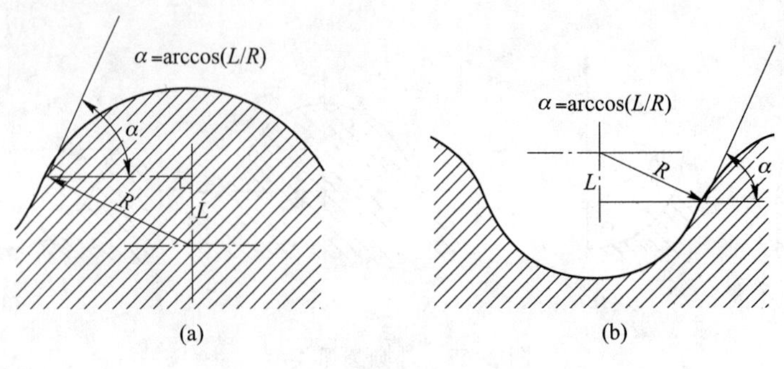

图 2-60　刀具副偏角计算示意图

2.4.2　编程指令

2.4.2.1　圆弧插补指令 G02/G03

（1）功能。

G02 为顺时针方向圆弧插补，G03 为逆时针方向圆弧插补。

（2）指令格式。

1）格式 1：用圆弧半径 R 编程。

G02/G03　X(U)_Z(W)_R_F_;

式中　X，Z——圆弧终点的绝对坐标；

U，W——圆弧终点相对于圆弧起点的增量坐标；

R——圆弧半径，圆心角为 0°~180°取正值，大于 180°取负值。

2）格式 2：用 I，K 指定圆心位置编程。

G02/G03　X(U)_Z(W)_I_K_F_;

式中　I，K——圆心相对于圆弧起点的增量值。

（3）注意事项。

1）圆弧顺逆方向的判定

对于圆弧的顺逆方向的判断按右手坐标系确定：沿圆弧所在的平面（*XOZ* 平面）的垂直坐标轴的负方向（-*Y*）看去，顺时针方向为 G02，逆时针方向为 G03，如图 2-61 所示。

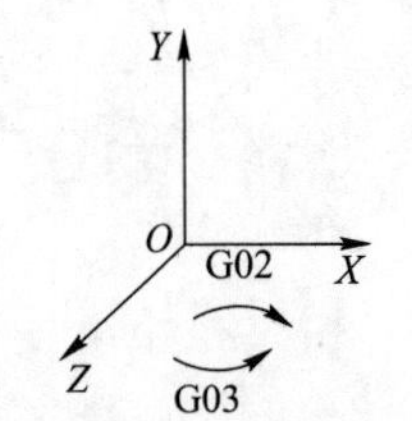

图 2-61　圆弧顺逆方向判定

由于车床加工的是回转形轮廓，通常图纸的表现形式为上下对称，如图 2-62 所示。圆弧编程时需要考虑圆弧轮廓加工的顺逆方向，在查看图样时要看中心线上半部分轮廓，根据轮廓加工方向确定圆弧的顺逆方向，如图 2-62 所示。轮廓一般由右向左加工，右侧 *R*2 圆弧的加工走向为逆时针，两个 *R*4 圆弧的加工走向为顺时针。

2）I、K 值

不论是用绝对尺寸编程还是用增量尺寸编程，I、K 都是圆心相对于圆弧起点的增量值，I 为 *X* 方向增量值，I=X 圆心坐标-*X* 起点坐标；K 为 *Z* 方向增量值，K=Z 圆心坐标-*Z* 起点坐标。当 I、K 为零时可以省略；I、K 和 R 同时指定的程序段，R 优先，I、K 无效。

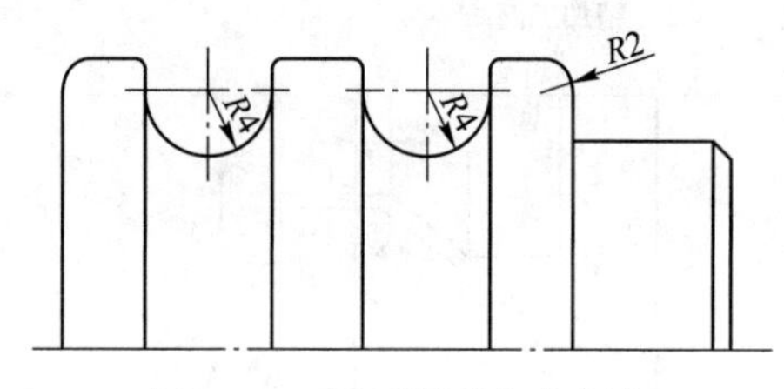

图 2-62　圆弧顺逆方向判定

【例 2-11】零件如图 2-63 所示，利用圆弧插补指令编写圆弧部分精加工程序段。加工程序见表 2-25。

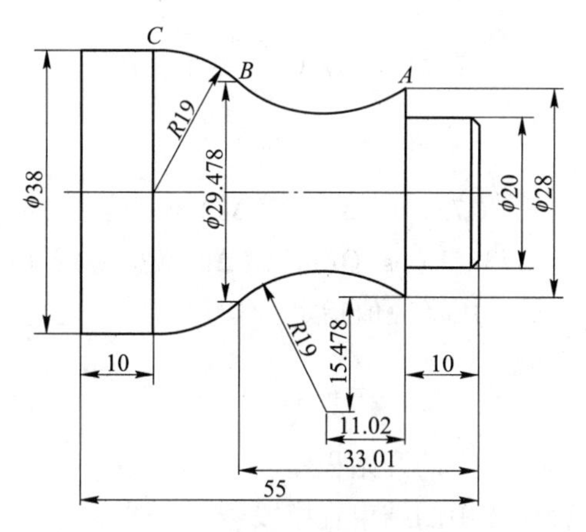

图 2-63　圆弧插补指令举例

表 2-25　圆弧加工程序

R 形式	I、K 形式
G01 X0 Z0 F0. 1;	G01 X0 Z0 F0. 1;
G01 X20. 0 C1. 0;	G01 X20. 0 C1. 0;
G01 Z-10. 0;	G01 Z-10. 0;
G01 X28. 0;	G01 X28. 0;
G02 X29. 478 Z-33. 01 R19. 0;	G02 X29. 478 Z-33. 01 I15. 478 K-11. 02;
G03 X38. 0 Z-45. 0 R19. 0;	G03 X38. 0 Z-45. 0 I-14. 739 K-11. 99;
G01 Z-55. 0;	G01 Z-55. 0;

2. 4. 2. 2　固定形状粗车复合循环指令 G73

（1）功能。固定形状粗车复合循环指令 G73 可以有效的车削固定的图形，如铸造成型、锻造成型或已经粗车成型的工件。该指令只须指定粗加工循环次数、精加工余量和精加工路线，系统自动算出粗加工的切削深度，给出粗加工路线，完成各表面的粗加工。G73 指令粗车循环路线如图 2-64 所示。

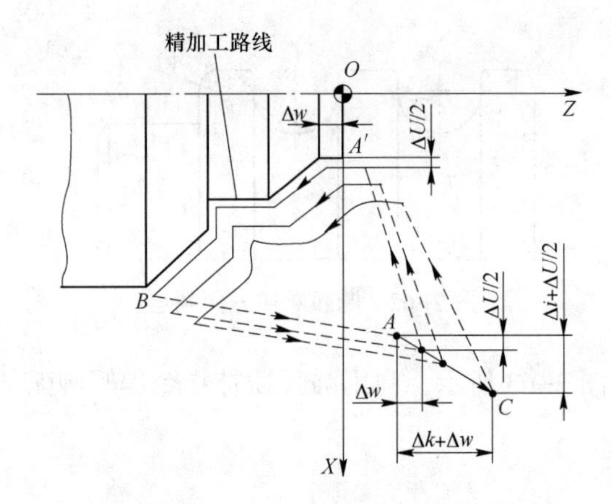

图 2-64　G73 指令循环路线

（2）指令格式：

G73　U(Δi) W(Δk)R(d)；

G73　P(ns)Q(nf)U(Δu)W(Δw)；

式中　Δi——X 方向总退刀量，用半径值指定；

Δk——Z 方向总退刀量；

d——循环次数；

ns——精加工轮廓程序段中的开始程序段号；

nf——精加工轮廓程序段中的结束程序段号；

Δu——X 方向上的精加工余量和方向，用直径值指定，一般取 0. 5mm；

Δw——Z 方向上的精加工余量，一般取 0. 05～0. 1mm。

（3）注意事项：

编写 G73 指令时应注意如下事项。

1）G73 和 G71 指令都是粗加工，不同之处是 G73 可以加工任意形状轮廓的零件。

2）G73 也可以加工未去除余量的棒料，但是空走刀较多。

3）ns、nf 程序段不必紧跟在 G73 程序段后编写，系统能自动搜索到 ns～nf 程序段并执行，完成 G73 指令后，会接着执行 G73 后面程序段。

【例 2-12】如图 2-63 所示的零件，毛坯为 ϕ40mm 棒料，用 G73 和 G70 指令编写零件粗、精加工程序。加工程序见表 2-26。

表 2-26　G73 编程示例

O153;	程序号
G40 G97 G99 M03 S600 F0. 2;	主轴正转，转速 600r/min，进给量 0. 2mm/r
T0101;	换 01 号外圆车刀，代入 01 号刀具补偿
M08;	切削液开
G00 Z5. 0;	刀具快速点定位至固定形状粗加工复合循环起点
X45. 0;	
G73 U20. 0 W0. 0 R15. 0;	定义 G73 粗车循环，X 方向总退刀量 20mm，Z 方向总退刀量 0mm，循环 15 次

续表 2-26

G73 P10 Q20 U0.5 W0.0;	精加工路线由 N10～N20 指定，*X* 向精加工余量 0.5mm，*Z* 向精加工余量 0mm
N10 G42 G01 X0 F0.1;	精车轮廓
Z0;	
G01 X20 C1.0;	
G01 Z-10.0;	
G01 X28.0;	
G03 X29.478 Z-33.01 R19.0;	
G02 X38.0 Z-45.0 R19.0;	
G01 Z-55;	
N20 G40 G01 X40;	
M05;	主轴停止
M00;	程序暂停
M03 S1000 T0101;	主轴正转，转速 1000r/min，代入 01 号刀具补偿
G42 G00 Z5.0;	刀具快速点定位至固定形状粗加工复合循环起点，建立刀尖圆弧半径右补偿
X45.0;	
G70 P10 Q20;	精加工复合循环
G40 G00 X100.0;	快速退刀至换刀点，取消刀尖圆弧半径补偿
Z100.0;	
M30;	程序结束并返回程序起点

2.4.3 加工操作

加工圆弧常用的刀具为外圆偏刀、尖刀和圆弧车刀，尖刀对刀需要用对刀工具才能保证刀具 Z 值位置，对此常用对刀仪来完成，如图 2-65 所示。

圆弧刀的理想刀尖点为圆弧顶点，如图 2-66 所示，圆弧刀对刀的过程与切槽刀基本

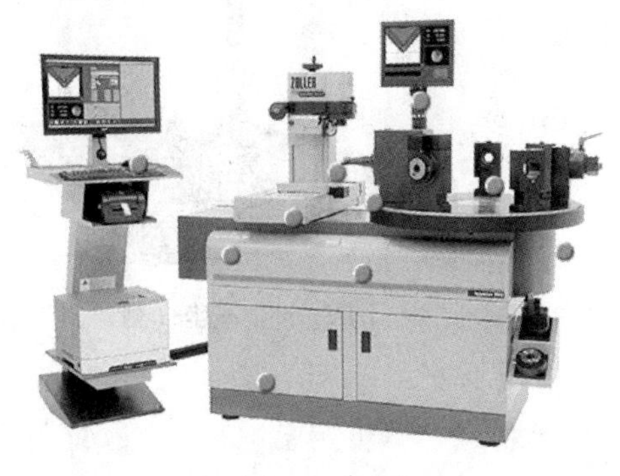

图 2-65　对刀仪

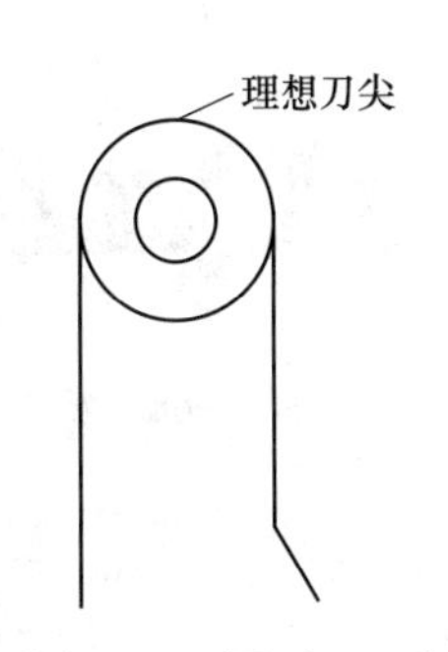

图 2-66　圆弧刀刀尖

相同，不同的是在确定 Z 值时，切槽刀输入 Z0 再点击【测量】，圆弧刀则输入刀具的圆弧半径 Z_R 再点击【测量】，确定 Z 值位置，如图 2-67 所示。

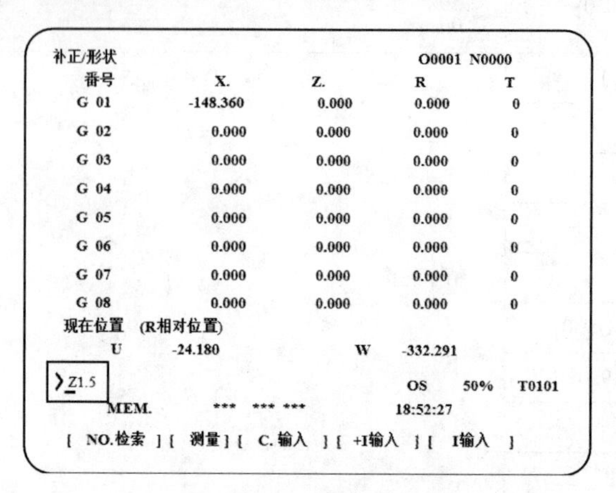

图 2-67　圆弧刀对刀

使用圆弧刀加工圆弧时，刀具的圆弧半径补偿 R，由圆弧刀的圆弧半径决定，如圆弧刀的圆弧半径 $r=2\text{mm}$，则数控系统内刀具补偿 $R=2.000$，刀尖方位 T 取 8。

2.4.4　精度检验

对于一般精度要求的圆弧轮廓，可使用半径规通过光隙法进行检测，如图 2-68 所示，测量时必须使半径规的测量面与工件的圆弧完全紧密地接触，当测量面与工件的圆弧中间没有间隙时，工件的圆弧度数为半径规上所示的数字。由于是目测，故准确度不是很高，只能做定性测量。

高精度的轮廓可使用三坐标测量机进行检测，三坐标测量机可用测通扫描轮廓表面检测出轮廓度，如图 2-69 所示。

图 2-68　半径规

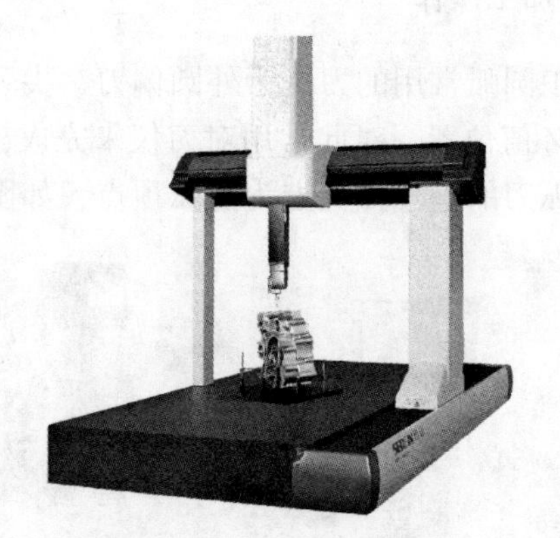

图 2-69　三坐标测量

任务实施

（1）图样分析：

如图 2-53 所示的零件为圆形带轮，有两个 $R4$ 圆弧槽，每个圆弧槽与外圆相交处有 $R0.5$ 的圆弧过渡，外圆与端面有 $R2$ 的圆弧过渡，中间为安装电机轴的 $\phi16$ 通孔，在 $\phi26$ 外圆上有一个 M6 螺纹孔，表面粗糙度分别为 $Ra1.6$ 和 $Ra3.2$。本任务主要对零件外圆和圆弧槽进行加工。

（2）加工工艺方案制定：

1）加工方案：

①使用三爪自定心卡盘装卡，零件伸出卡盘 60mm 左右。

②加工零件外轮廓至尺寸要求。

③切断工件。

2）刀具选用：

T01　90°硬质合金外圆偏刀（刀尖角 55°）

T02　R1.5 圆弧刀

T03　3mm 切刀

带轮零件数控加工刀具卡如表 2-27 所示。

表 2-27　数控加工刀具卡

<table>
<tr><td colspan="2">零件名称</td><td colspan="3">带　轮</td><td colspan="2">零件图号</td><td colspan="3">2-53</td></tr>
<tr><td>序号</td><td>刀具号</td><td colspan="2">刀具名称</td><td>数量</td><td colspan="2">加工表面</td><td>刀尖半径
R/mm</td><td>刀尖方位
T</td><td>备注</td></tr>
<tr><td>1</td><td>T01</td><td colspan="2">90°外圆偏刀</td><td>1</td><td colspan="2">粗精车带轮主体</td><td>0.4</td><td>3</td><td></td></tr>
<tr><td>2</td><td>T02</td><td colspan="2">$R1.5$mm 圆弧刀</td><td>1</td><td colspan="2">加工 $R4$ 圆弧槽</td><td>1.5</td><td>8</td><td></td></tr>
<tr><td>3</td><td>T03</td><td colspan="2">3mm 切断刀</td><td>1</td><td colspan="2">切断</td><td></td><td></td><td></td></tr>
<tr><td>编制</td><td></td><td>审核</td><td></td><td>批准</td><td></td><td>日期</td><td></td><td>共 1 页</td><td>第 1 页</td></tr>
</table>

3）加工工序：

带轮数控加工工序如表 2-28 所示。

表 2-28　数控加工工序卡

<table>
<tr><td rowspan="2">单位名称</td><td colspan="4" rowspan="2">天津工业职业学院</td><td colspan="2">零件名称</td><td colspan="2">零件图号</td></tr>
<tr><td colspan="2">带轮</td><td colspan="2">2-53</td></tr>
<tr><td colspan="2">程序号</td><td>夹具名称</td><td colspan="2">使用设备</td><td colspan="2">数控系统</td><td colspan="2">场地</td></tr>
<tr><td colspan="2">O2410</td><td>三爪自定心卡盘</td><td colspan="2">CKA6140</td><td colspan="2">FANUCSERIES 0i</td><td colspan="2">数控实训中心</td></tr>
<tr><td>工序号</td><td colspan="3">工序内容</td><td>刀具号</td><td>主轴转速
n/r · min^{-1}</td><td>进给量
F/mm · r^{-1}</td><td>背吃刀量
a_p/mm</td><td>备注</td></tr>
<tr><td>1</td><td colspan="3">装夹零件外伸 55mm，并找正</td><td></td><td></td><td></td><td></td><td rowspan="4">手动</td></tr>
<tr><td>2</td><td colspan="3">对外圆偏刀</td><td>T01</td><td></td><td></td><td></td></tr>
<tr><td>3</td><td colspan="3">对 $R4$ 圆弧刀</td><td>T02</td><td></td><td></td><td></td></tr>
<tr><td>4</td><td colspan="3">对切槽刀</td><td>T03</td><td></td><td></td><td></td></tr>
</table>

续表 2-28

工序号	工序内容	刀具号	主轴转速 $n/\mathrm{r\cdot min^{-1}}$	进给量 $F/\mathrm{mm\cdot r^{-1}}$	背吃刀量 a_p/mm	备注
5	粗加工带轮主体留 0.5mm 余量	T01	800	0.2	1.5	
6	精加工带轮主体保证尺寸	T01	1000	0.1	0.25	
7	粗加工 *R*4 圆弧槽留 0.2mm 余量	T02	800	0.1	1	*R*1.5mm
8	精加工 *R*4 圆弧槽至尺寸	T02	1200	0.08	0.1	
9	切断工件长度留 0.5 余量	T03	600	0.1	3	3mm
编制	审核　　　批准		日期		共 1 页	第 1 页

（3）编制加工程序：

编制加工程序见表 2-29。

表 2-29　简单成型面零件加工程序

O2410;	程序号
G40 G97 G99 M03 S600 F0.2;	主轴正转，转速 600r/min，进给量 0.2mm/r
T0101;	换 T01 号外圆车刀，代入 01 号刀具补偿
M08;	切削液开
G00 G42.0 Z5.0;	刀具快速点定位至粗加工复合循环起点
X40.0	
G71 U1.5 R0.5;	定义粗车循环，切削深度 1.5mm，退刀量 0.5mm
G71 P10 Q20 U0.5 W0.05;	精车路线由 N10～N20 指定，*X* 方向精车余量 0.5mm，*Z* 方向精车余量 0.05mm
N10 G00 X0;	精车轮廓
G01 Z0 F0.1;	
X26.0 C1.0;	
Z-10.0;	
X32.0;	
G03 X36.0 Z-12.0 R2.0;	
G01 Z-46.0;	
N20 G01 X40.0;	
G40 G00 X100.0;	快速退刀至换刀点
Z100.0;	
M05;	主轴停止
M09	切削液关
M01;	程序选择性暂停
M03 S1000 F0.1;	主轴正转，转速 1000r/min

续表 2-29

T0101;	代入 01 号刀具补偿
M08;	切削液开
G42 G00 Z5.0;	刀具快速点定位至粗加工复合循环起点，建立刀尖圆弧半径右补偿
X40.0;	
G70 P10 Q20;	精加工复合循环
G40 G00 X100.0;	快速退刀至换刀点，取消刀尖圆弧半径补偿
Z100.0;	
M05;	主轴停止
M09	切削液关
M01;	程序选择性暂停
M03 S800 F0.1;	主轴正转，转速 400r/min
T0202;	换 T02 号 *R*1.5 圆弧车刀，代入 02 号刀具补偿
M08;	切削液开
G00 Z5.0;	刀具快速点定位至粗加工复合循环起点
X45.0;	
Z-10.0;	
G73 U6 R6;	定义粗车循环，*X* 总退刀量 6mm，走刀次数 6 次
G73 P30 Q40 U0.2 W0.0;	精车路线由 N30～N40 指定，*X* 方向精车余量 0.5mm，*Z* 方向精车余量 0.05mm
N30 G42 G01 X36.0;	精车轮廓
G01 Z-14.5;	
G03 X35.0 Z-15.0 R0.5;	
G01 X32.0;	
G02 Z-23.0 R4.0;	
G01 X35.0;	
G03 X36.0 Z-23.5 R0.5;	
G01 Z-28.5;	
G03 X35.0 Z-29.0 R0.5;	
G01 X32.0;	
G02 Z-37.0 R4.0;	
G01 X35.0;	
G03 X36.0 Z-37.5 R0.5;	
G01 Z-42.0;	
N40 G01 G40 X40.0;	

续表 2-29

G00 X100.0;	快速退刀至换刀点
Z100.0;	
M05;	主轴停止
M09	切削液关
M01;	程序选择性暂停
M03 S1000 F0.08;	主轴正转，转速 1000r/min
T0202;	代入 02 号刀具补偿
M08;	切削液开
G42 G00 Z5.0;	刀具快速点定位至粗加工复合循环起点
X40.0;	
Z-15.0;	
G70 P30 Q40;	精加工复合循环
G40 G00 X100.0;	快速退刀至换刀点，取消刀尖圆弧半径补偿
Z100.0;	
M05;	主轴停止
M09	切削液关
M01;	程序选择性暂停
M03 S600 F0.08	主轴正转，转速 6000r/min
T0303;	换 T03 号切槽刀，代入 03 号刀具补偿
G00 X45.0;	刀具快速点定位至切断处
G00 Z-45.5;	
X41.0;	
G01 X0 F0.05;	切断工件
G00 X100.0;	快速退刀至换刀点
Z100.0;	
M30;	程序结束并返回起点

（4）工件加工：

1）按照刀具表选择刀具，并安装刀具。

2）进行对刀工作，注意输入刀尖半径及刀尖方位号。

3）输入程序并进行图形模拟，保证程序的正确性。

4）对零件进行加工，加工效果如图 2-70 所示。

（5）尺寸检测：

1）采用游标卡尺测量 $\phi26$、$\phi36$、10、42、$R4$ 槽宽尺寸。

2）采用半径规检测 $R4$ 圆弧尺寸。

3）采用粗糙度比较板检测粗糙度。

图 2-70　加工效果图

任务评价

任务评价如表 2-30 所示。

表 2-30　任务评价

项目	序号	评分标准		配分	得分
加工工艺（10%）	1	加工步骤	符合数控车工工艺要求	3	
	2	加工部位尺寸	计算相关部位尺寸	3	
	3	刀具选择	刀具选择合理，符合加工要求	4	
程序编制（30%）	4	程序号	无程序号无分	2	
	5	程序段号	无程序段号无分	2	
	6	切削用量	切削用量选择不合理无分	4	
	7	原点及坐标	标明程序原点及坐标，否则无分	2	
	8	程序内容	不符合程序逻辑及格式要求每段扣 2 分	20	
			程序内容与工艺不对应扣 5 分		
			出现危险指令扣 5 分		
加工操作（20%）	9	程序输入与检索	每错一段扣 2 分	10	
	10	加工操作 30min	工件坐标系原点设定错误无分	10	
			误操作无分		
			超时无分		
尺寸检测（30%）	11	直径	$\phi26\pm0.2$ 超差不得分	4	
	12		$\phi36\pm0.3$ 超差不得分	4	
	13	长度	10 ± 0.2 超差不得分	4	
	14		42 ± 0.3 超差不得分	4	
	15	$R4$ 槽宽	8 ± 0.2 超差不得分	4	
	16		8 ± 0.2 超差不得分	4	
	17	轮廓度	$R4$ 两处	2	
	18		$R2$ 两处	2	
	19	粗糙度	$Ra1.6$ 四处，超差不得分	2	
职业能力（10%）	20	学习能力		2	
	21	表达沟通能力		2	
	22	团队合作		2	
	23	安全操作与文明生产		4	
合计					

同步思考与训练

（1）思考题：

1）简述按工序集中原则划分工序方法。

2）简述 G73 指令的功能及用法。

3）简述如何选择 G41、G42 指令。

4）简述如何判定 G02、G03 加工方向。

5）简述在加工时如何保证半径公差。

（2）同步训练题：

零件如图 2-71 和图 2-72 所示，毛坯 $\phi40$mm 长棒料，材料硬铝合金 LY15，编写加工程序，使用机床验证程序正确性并进行加工。

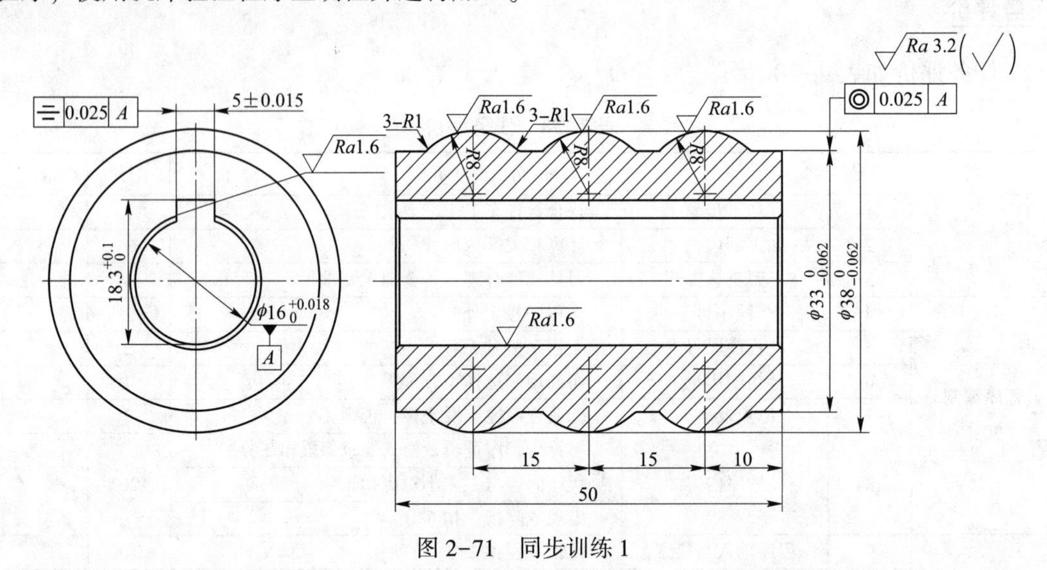

图 2-71　同步训练 1

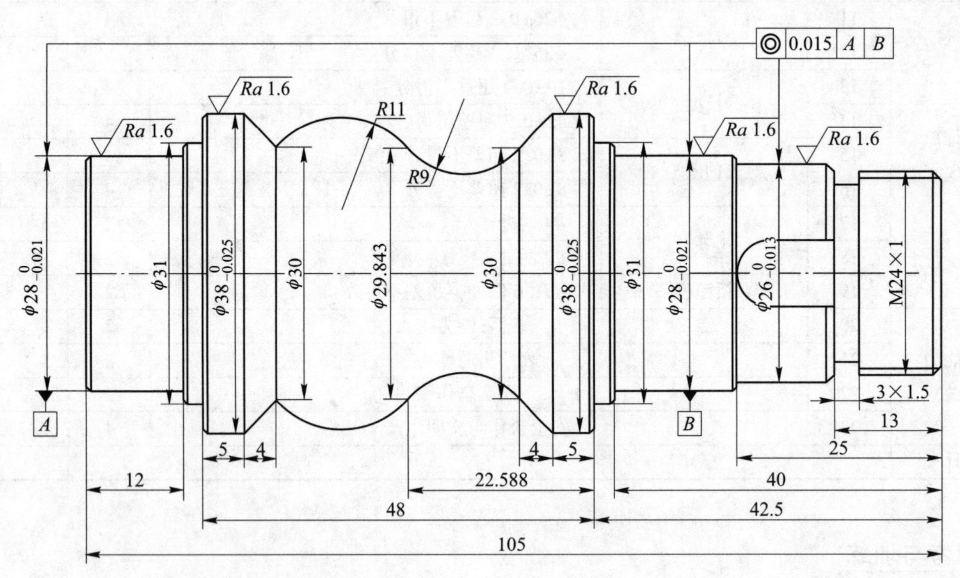

技术要求

1.未注倒角均为$C1$，锐角倒钝。

2.未注线性尺寸公差应符合GB/T 1804—2000的要求。

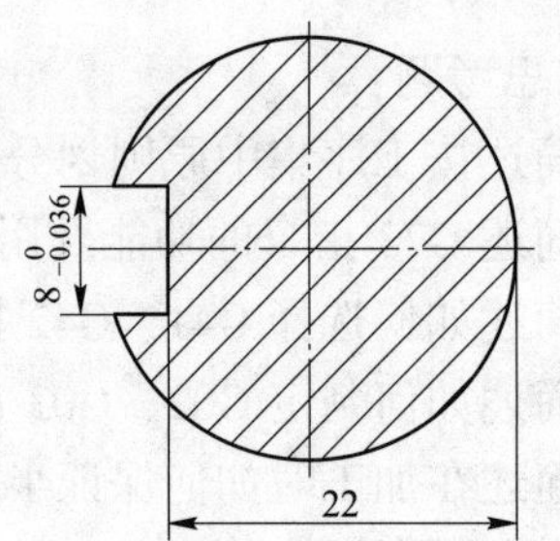

图 2-72　同步训练 2

任务 2.5 螺纹零件的数控加工

任务描述

如图 2-73 所示的零件，材料硬铝合金 LY15，毛坯为 ϕ40mm×75mm 棒料，单件生产，请按照图纸要求，与任务 2.6 零件配合，使用数控车床加工，编写加工程序，检测加工质量。

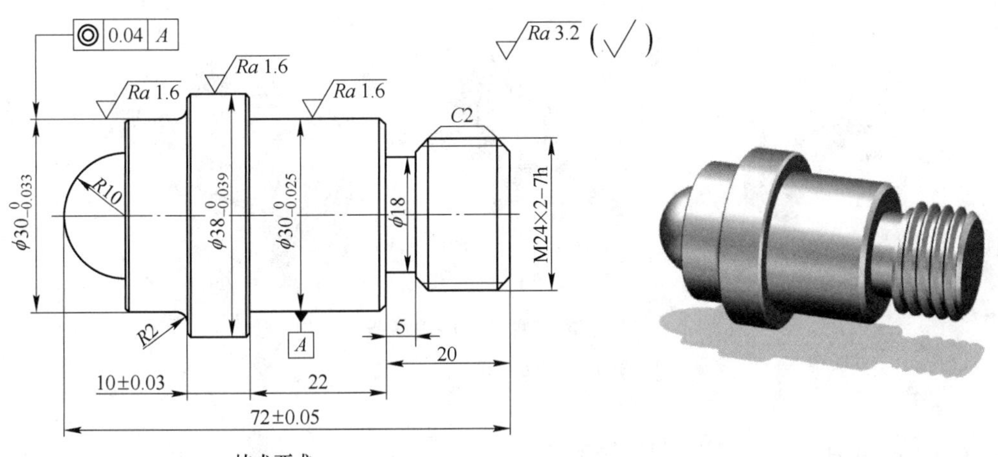

图 2-73 零件图

任务目标

(1) 知识目标：

1) 认识螺纹车削工艺及螺纹参数。

2) 解释 G92 和 G76 指令的含义。

3) 认识螺纹量具的种类和结构。

(2) 技能目标：

1) 能完成螺纹车刀的对刀。

2) 能使用 G92 和 G76 指令编写螺纹零件加工程序。

3) 能使用螺纹量规、螺纹千分尺的检测螺纹尺寸。

4) 能够完成零件调头加工对刀操作 。

(3) 素质目标：

1) 具有高度的责任心、爱岗敬业、团结合作精神。

2) 正确执行安全技术操作规程。

相关知识

2.5.1 工艺准备

2.5.1.1 螺纹应用

螺纹是一种使用广泛的可拆卸固定连接，具有结构简单、连接可靠、拆装方便等优点，主要应用在联接、紧固和传动等场合，如图 2-74 所示。

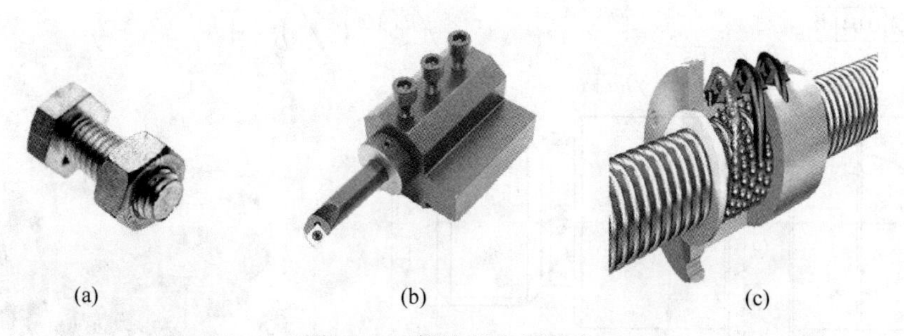

图 2-74 螺纹的应用场合

(a) 联接；(b) 紧固；(c) 传动

2.5.1.2 螺纹种类

螺纹有外螺纹和内螺纹两类。起连接作用的螺纹称为联接螺纹，起传动作用的螺纹称为传动螺纹。按螺纹的旋向可分为左旋螺纹与右旋螺纹，日常中使用最多的为右旋螺纹。螺纹的螺旋线有单线、双线及多线，连接螺纹的一般为单线。

常用螺纹的类型主要有普通三角螺纹、管螺纹、梯形螺纹、矩形螺纹、锯齿形螺纹。前两种主要用于联接，后三种主要用于传动，如图 2-75 所示。

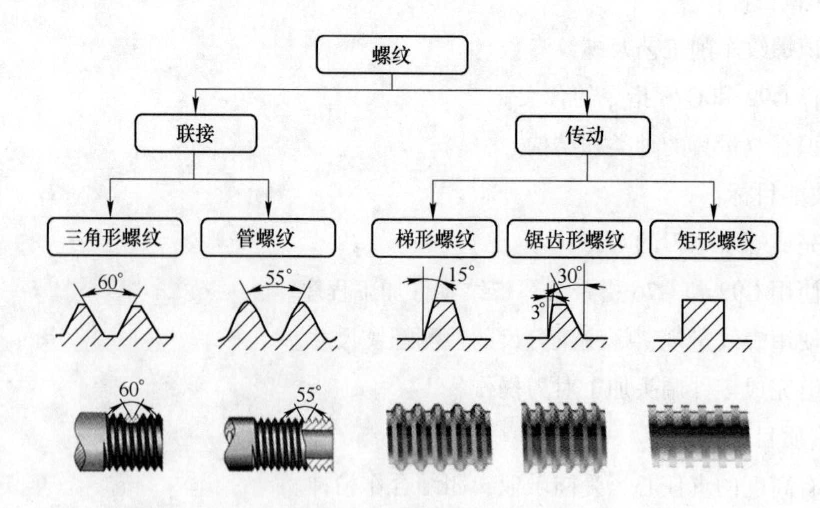

图 2-75 螺纹的种类

2.5.1.3　螺纹的基本要素

（1）螺纹大径（d、D）。外螺纹牙顶直径 d，内螺纹牙底直径 D，是螺纹的最大直径。

（2）小径（d_1、D_1）。外螺纹牙底直径 d_1，内螺纹牙顶直径 D_1，是螺纹的最小直径，如图 2-76 所示。

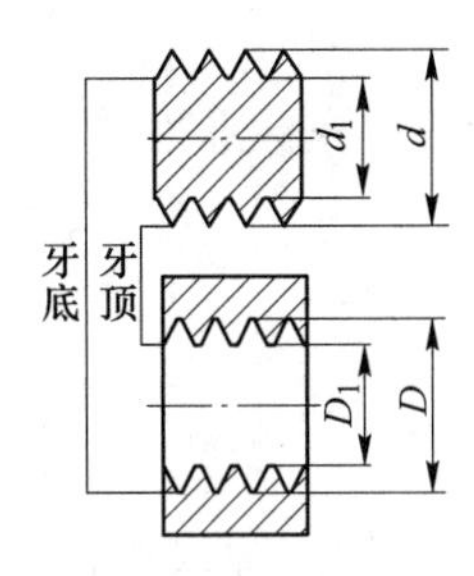

图 2-76　螺纹的大径、小径

（3）螺距（P）。螺纹上相邻两牙之间的轴向距离，如图 2-77 所示。

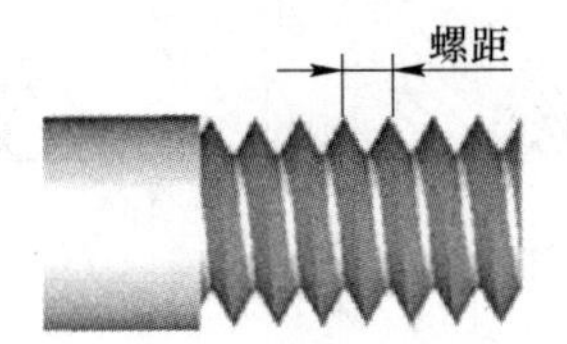

图 2-77　螺距

（4）导程（L）。沿同一条螺旋线形成的螺纹上相邻两牙之间的轴向距离，如图 2-78 所示。

导程=螺距×线数

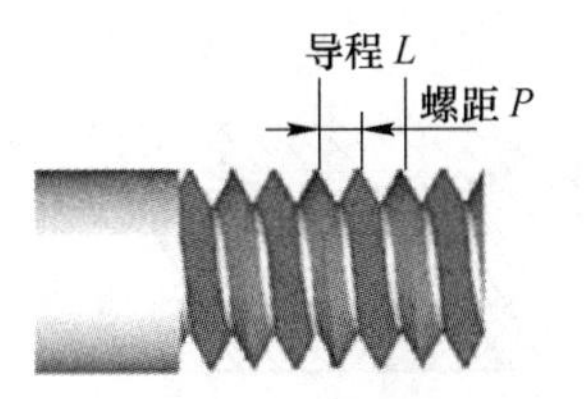

图 2-78　导程

（5）牙型角 α。在通过螺纹轴向的剖面上，螺纹的轮廓形状称螺纹牙型，相邻两牙侧之间的夹角即为牙型角，如图 2-79 所示。

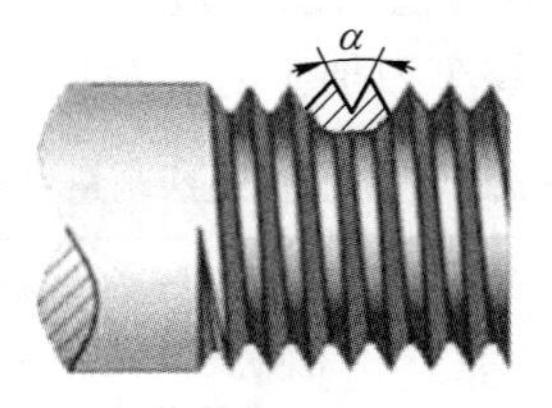

图 2-79　牙型角

2.5.1.4 外螺纹加工尺寸计算

（1）实际车削外圆柱面直径。车削三角形外螺纹时，受刀具挤压会使螺纹尺寸胀大，因此车螺纹前的外圆柱直径应比螺纹大径小。实际车削时外圆柱面的直径 d 实为：

$$d_{实} = d - 0.1P$$

式中，d 为外螺纹大径，P 为螺距。

（2）螺纹牙型高度。根据普通螺纹国家标准规定，三角形螺纹的牙型高度 $h_{牙}=0.65P$。

（3）螺纹小径。

$$d_1=d-2h_{牙}=d-1.3P$$

2.5.1.5 螺纹的车削方法

（1）螺纹加工进刀方式。数控车床加工螺纹的进刀方式通常有直进法和斜进法两种，如图 2-80 所示。当螺距 $P<3$mm 时，一般采用直进法；螺距 $P\geqslant 3$mm 时，一般采用斜进法。

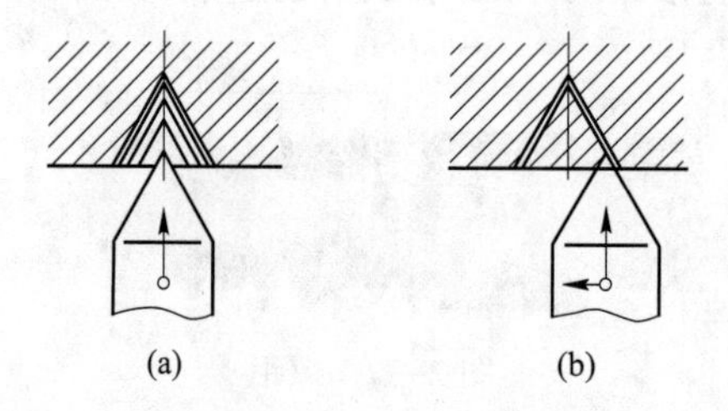

图 2-80 进刀方式

（a）直进法；（b）斜进法

螺纹加工中的走刀次数和背吃刀量大小直接影响螺纹的加工质量，车削时应遵循递减的背吃刀量分配方式，如图 2-81 所示。

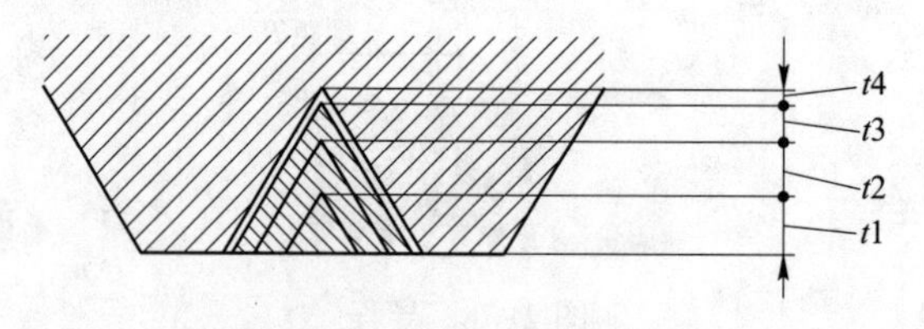

图 2-81 背吃刀量分配方式

（2）分层切削方法。螺纹加工时，切削力大，所以需要分层切削，常用的进给次数与背吃刀量如表 2-31 所示。加工时为防止切削力过大，可适当增加切削次数。但为了提高螺纹的表面粗糙度，用硬质合金螺纹车刀加工时，最后一刀的切削深度不能小于 0.1mm。

表 2-31 切削次数及切削量推荐值

螺距/mm	1.0	1.5	2.0	2.5
牙深/mm	0.65	0.975	1.3	1.625
总切深/mm	1.3	1.95	2.6	3.25

续表 2-31

每次背吃刀量/mm	1 次	0.7	0.8	0.9	1.0
	2 次	0.5	0.65	0.7	0.8
	3 次	0.1	0.4	0.6	0.6
	4 次		0.1	0.3	0.5
	5 次			0.1	0.25
	6 次				0.1

（3）螺纹加工升速进刀段和减速退刀段。由于车削螺纹起始时有一个加速过程，结束前有一个减速过程。因此在车螺纹时，两端必须设置足够的升速进刀段和减速退刀段，如图 2-82 所示 δ_1 为升速进刀段距离，δ_2 为减速退刀段距离。

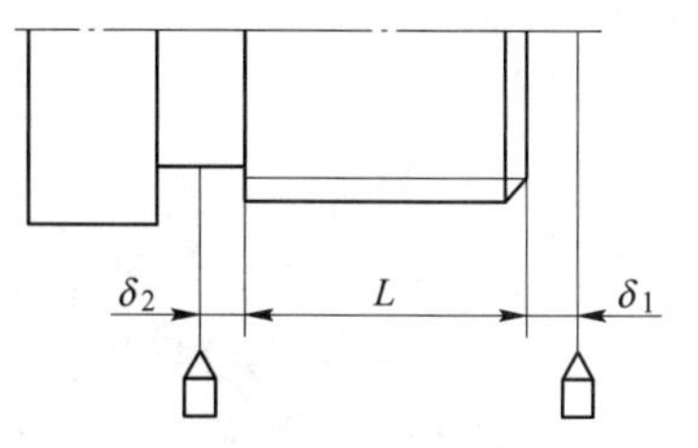

图 2-82　升速进刀段和减速退刀段

δ_1 和 δ_2 的数值与工件螺距和主轴转速有关，一般 δ_1 取 1~2P，δ_2 取 0.5P。实际生产中，δ_1 值一般取 2~5mm，对大螺距和高精度的螺纹取大值；δ_2 值一般为退刀槽宽度的一半左右，取 1~3mm。若螺纹收尾处没有退刀槽时，收尾处的形状与数控系统有关，一般按 45°退刀收尾。

（4）螺纹加工的切削用量。

1）主轴转速 n。

数控车床加工螺纹时，主轴转速受数控系统、螺纹导程、刀具、工件尺寸和材料等多种因素影响。不同的数控系统，有不同的推荐主轴转速范围，操作者仔细查阅说明书后，根据具体情况选用。大多数数控车床车削螺纹时，推荐主轴转速公式如下：

$$n \leqslant 1200/P - K$$

式中　P——螺纹的螺距（mm）；

K——保险系数，一般取 80；

n——主轴转速（r/min）。

2）切削深度或背吃刀量 ap，如表 2-31 所示。

3）进给量 F。

①单线螺纹的进给量等于螺距，即 $F=P$。

②多线螺纹的进给量等于导程，即 $F=L$。

2.5.2　编程指令

2.5.2.1　螺纹切削循环指令 G92

（1）功能。适用于循环加工小螺距螺纹零件，循环路线如图 2-83 所示。

（2）指令格式：

G92　X(U)__Z(W)__I(R)__F__；

式中　X，Z——螺纹终点的绝对坐标；

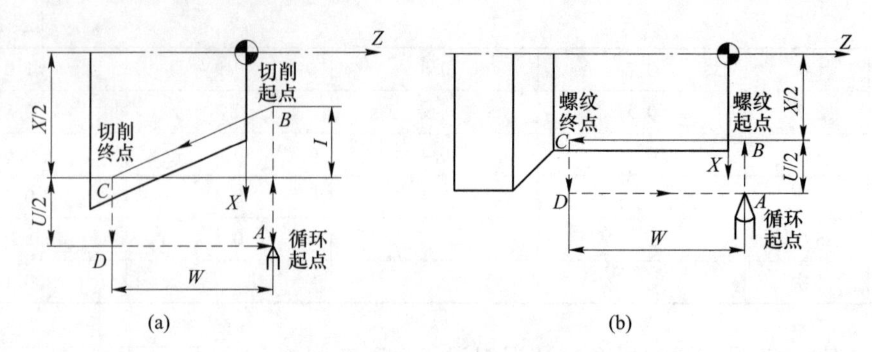

图 2-83　G92 指令循环路线

（a）圆柱螺纹循环路线；（b）圆锥螺纹循环路线

U，W——螺纹终点相对起点的坐标；

F——螺纹导程；

I(R)——圆锥螺纹起点半径与终点半径的差值。圆锥螺纹终点半径大于起点半径时 I(R) 为负值；圆锥螺纹终点半径小于起点半径时 I(R) 为正值。圆柱螺纹 I=0，可省略。

（3）注意事项。

螺纹车削时应注意以下事项。

1）车螺纹时不能使用恒线速度控制指令，要使用 G97 指令，粗车和精车主轴转速一样，否则会出现乱牙现象。

2）车螺纹时进给速度倍率、主轴速度倍率无效（固定为 100%）。

3）受机床结构及数控系统的影响，车螺纹时主轴转速有一定的限制。

4）圆锥螺纹，斜角在 45°以下时，螺纹导程以 Z 轴方向指定；斜角在 45°~90°时，螺纹导程以 X 轴方向指定。

【例 2-13】零件如图 2-84 所示，M24×2 螺纹外径已车至尺寸要求，4×2 的退刀槽已加工，用 G92 指令编制 M24×2 螺纹的加工程序。

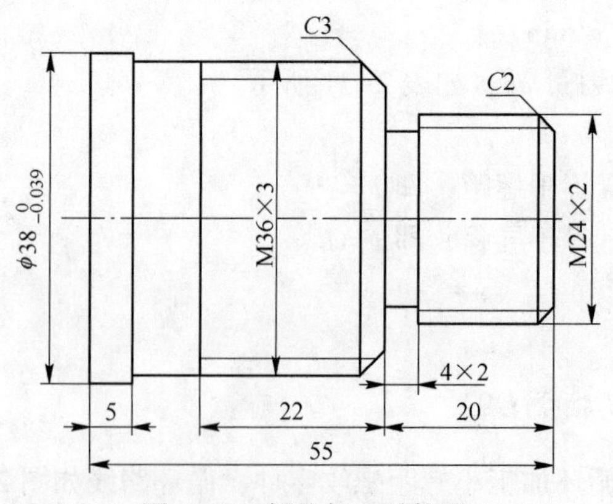

图 2-84　螺纹加工零件图

（1）尺寸计算：

实际车削时外圆柱面的直径 $d_{实}=d-0.1P=24-0.1\times2=23.8$mm

螺纹实际牙型高度 $d_{牙}=0.65P=1.3$mm

螺纹实际小径 $d_{小}=D-1.3P=21.4$mm。

升速进刀段 δ_1 取 5mm，减速退刀段 δ_2 取 2mm。

（2）切削用量：

主轴转速 n 取 400r/min，进给量 F 取 2mm。

由表 2-31 可知分五刀切削螺纹，被吃刀量分别为 0.9mm、0.7mm、0.6mm、0.3mm、0.1mm。

（3）加工程序：

M24×2 螺纹加工程序如表 2-32 所示。

表 2-32　M24×2 螺纹加工程序

O1411;	程序号
G40 G97 G99 M03 S400;	主轴正转，转速为 400r/min
T0404;	换 T04 号螺纹车刀，代入 04 号刀具补偿
G00 Z5.0;	刀具快速点定位至螺纹切削循环起点
G00 X25.0;	
G92 X23.1 Z-18.0 F2.0;	螺纹车削循环第一刀切深 0.9mm，进给量为 2mm
X22.4;	第二刀切深 0.7mm
X21.8;	第三刀切深 0.6mm
X21.5;	第四刀切深 0.3mm
X21.4;	第五刀切深 0.1mm
X21.4;	光车一刀，切深为 0mm
G00 X100.0;	快速退刀至换刀点
Z100.0;	
M30;	程序结束并返回起点

2.5.2.2　螺纹切削复合循环指令 G76

（1）功能。用于多次自动循环切削螺纹。它比 G92 指令更为简捷，只需要指定一次有关参数，螺纹即可加工。常用于加工不带退刀槽的螺纹和大螺距螺纹。G76 螺纹切削复合循环路线如图 2-85 所示。

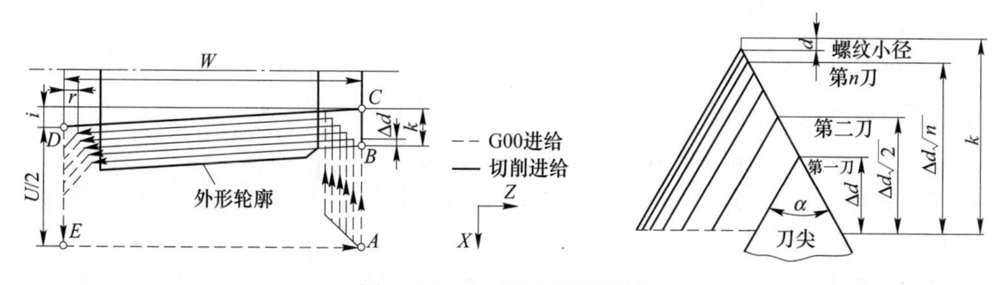

图 2-85　G76 指令循环路线

（2）指令格式：

$$G76\ P(m)(r)(\alpha)Q(\Delta d_{min})R(d);$$
$$G76\ X(U)Z(W)R(i)P(k)Q(\Delta d)F(L);$$

式中 m——精车重复次数（01~99 单位：次）；

r——螺纹尾部倒角量，用 00~99 之间的两位整数来表示（单位 0.1×L，L 为螺距）；

α——刀尖角度，即牙型角；

Δdmin——最小车削深度，用半径值指定（μm）；

d——精车余量，螺纹警车的切削深度，用半径值指定（mm）；

X(U)，Z（W）——螺纹终点绝对坐标或增量坐标（mm，X(U) 是直径值）；

i——螺纹锥度值，螺纹两端的半径差，圆柱螺纹，i 取 0；

k——螺纹高度，用半径值指定（μm）；

Δd——为第一次车削深度，用半径值指定（μm）；

L——导程，单头为螺距（mm）。

（3）注意事项。

编写螺纹切削复合循环指令应注意以下事项。

1）i 、k 和 Δd 数值以无小数点形式表示。

2）m 、r、α、Δd_{min} 和 d 是模态量。

3）外螺纹 X(U) 值为螺纹小径，内螺纹 X(U) 值为螺纹大径。

【例 2-14】零件如图 2-86 所示，M36×3 螺纹外径已车至尺寸要求，用 G76 编制 M36×3 螺纹部分的加工程序。

（1）尺寸计算：

实际车削时外圆柱面的直径：$d_{实}=d-0.1P=36-0.1\times3=35.7$mm

螺纹实际牙型高度：$d_{牙}=0.65P=0.65\times3=1.95$mm

螺纹实际小径：$d_{小}=D-1.3P=36-1.3\times3=32.1$mm

升速进刀段：δ_1 取 5mm

（2）螺纹参数：

精车重复 2 次，m 取 02；

螺纹尾部无倒角，r 取 00；

三角形螺纹刀尖角 60°，α 取 60；

最小车削深度 Δd_{min} 为 0.05mm，Q 取 50μm；

留 0.1mm 精车余量，R 取 0.1mm；

根据零件图计算螺纹终点坐标（X32.1，Z-42.0）

圆柱螺纹 i 为 0，可以省略；

螺纹高度 k 为 1.95mm，P 取 1950μm；

第一次车削深度 Δd 为 0.45mm，Q 取 450μm；

单头螺纹螺距为 3，F 取 3。

（3）加工程序：M36×3 螺纹加工程序见表 2-33。

表 2-33　M36×3 螺纹加工程序

O1412；	程序号
G40 G97 G99 M03 S400；	主轴正转，转速为 400r/min
T0404；	换 T04 号螺纹车刀，代入 04 号刀具补偿
G00 Z-15.0；	刀具快速点定位至螺纹切削复合循环起点
G00 X37.0；	
G76 P020060 Q50 R0.1；	G76 螺纹切削复合循环
G76 X32.1 Z-42.0 P1950 Q450 F3.0；	
G00 X100.0；	快速退刀至换刀点
Z100.0；	
M30；	程序结束并返回起点

2.5.3　加工操作

2.5.3.1　螺纹刀具安装

螺纹加工时，常用刀具有图 2-86 机夹外螺纹车刀和图 2-87 机夹可转位外螺纹车刀。

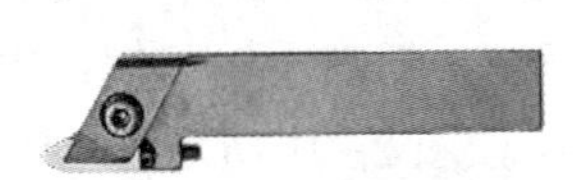

图 2-86　机夹外螺纹车刀

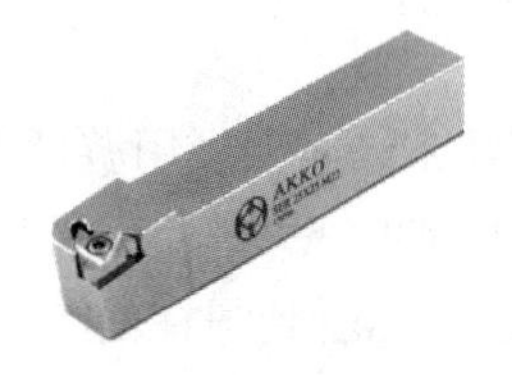

图 2-87　可转位外螺纹车刀

以可转位外螺纹刀为例，数控机夹可转位螺纹刀由于刀杆制造精度高，一般只要把刀杆靠紧刀架的侧边即可。注意：螺纹刀伸出过长刚性会降低，从而影响零件加工质量，如图 2-88 所示。

图 2-88　可转位外螺纹车刀安装

2.5.3.2　外螺纹刀的对刀操作与刀具补偿输入

（1）安装工件；

（2）安装螺纹刀；

（3）将机床工作方式切换到手轮方式，MDI 输入指令 M03 S400 循环启动使主轴正转；

（4）摇动手轮使外螺纹车刀靠近工件；

（5）螺纹刀 *Z* 轴对刀；

将螺纹刀尖与工件端面对齐，如图 2-89 所示。然后点击 进入【刀偏】中的【形状】功能，将“光标”移动到刀具所在“补偿”号位置，输入“Z0”，然后点击【测量】完成 *Z* 轴对刀。

图 2-89　外螺纹刀对 *Z* 轴

（6）螺纹刀 *X* 轴对刀。用游标卡尺对零件外圆柱面进行测量（如图 2-90 所示），记录下测量值（如 *X*39.32），然后操作面板点击主轴正转，摇动手轮将螺纹刀刀尖与工件外圆柱面轻微接触，当看到圆柱面有一条“亮线”后（如图 2-91 所示），再次进入【刀偏】中的【形状】功能，对应所在刀具补偿号，输入“X39.32”，然后点击【测量】。

图 2-90　测量直径

图 2-91　外螺纹刀对 *X* 轴

2.5.4　精度检验

2.5.4.1　螺纹测量工具

（1）螺距规。螺距规主要用于低精度螺纹工件的螺距和牙型角的检验，如图 2-92 所示。螺纹样板的各工作面均不应有锈蚀、碰伤、毛刺以及影响使用或外观质量的其他缺陷。样板与护板的连接应能使样板围绕轴心平滑地转动，不应有卡住或松动现象。

图 2-92　螺距规

（2）螺纹量规。螺纹量规有两种：螺纹塞规和螺纹环规，如图 2-93 所示。螺纹量规都有通端和止端，通端用字母“T”表示，止端用字母“Z”表示。螺纹通规具有完整的牙型，螺纹长度等于被测螺纹的旋合长度；螺纹止规具有截短牙型，螺纹长度比通规螺纹长度短。螺纹量规适用于批量零件生产检测使用。

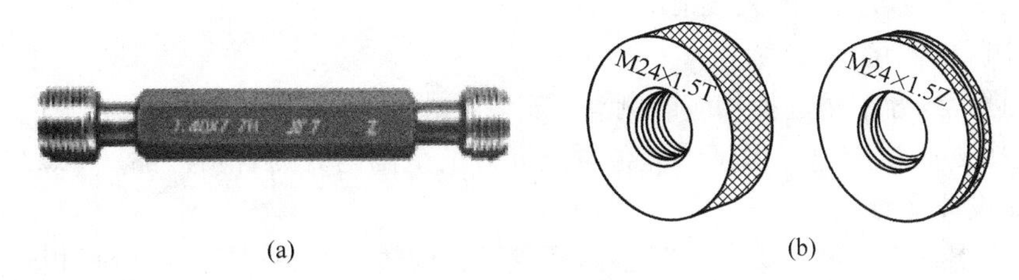

图 2-93　螺纹量规
（a）螺纹塞规；（b）螺纹环规

（3）螺纹千分尺。螺纹千分尺如图 2-94 所示，螺纹千分尺的刻度线及读数方式可参照外径千分尺，所不同的是螺纹千分尺附有两套测量头分别是 60°和 55°，适用于不同的牙型角和螺距，测量头可根据测量的需要进行更换，适用于单一螺纹的加工检测，通用性强。

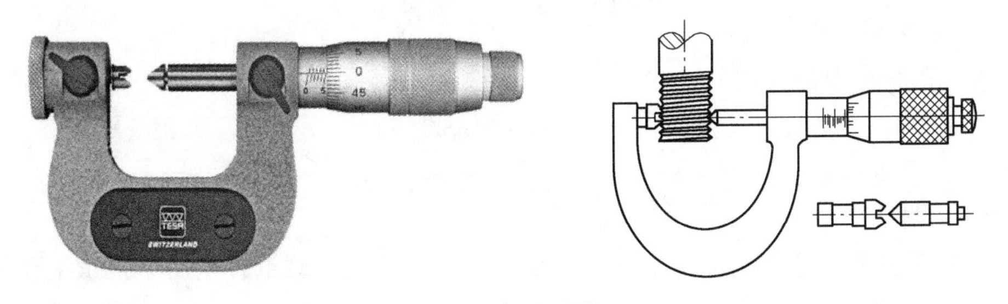

图 2-94　螺纹千分尺

2. 5. 4. 2　螺纹的测量方法

（1）螺距的测量。对于一般精度要求的螺纹，螺距常用游标卡尺和螺距规测量，测量方法如下：

1）测量螺距时，如图 2-95 所示，用螺距规作为样板，卡在被测螺纹零件上，如果不密合，换另外一片，直到密合为止，这时该螺距规上记录的数字即为被测螺纹零件的螺距。应尽可能利用螺纹工作部分长度，把螺距规卡在螺纹牙廓上，使测量结果较为准确。

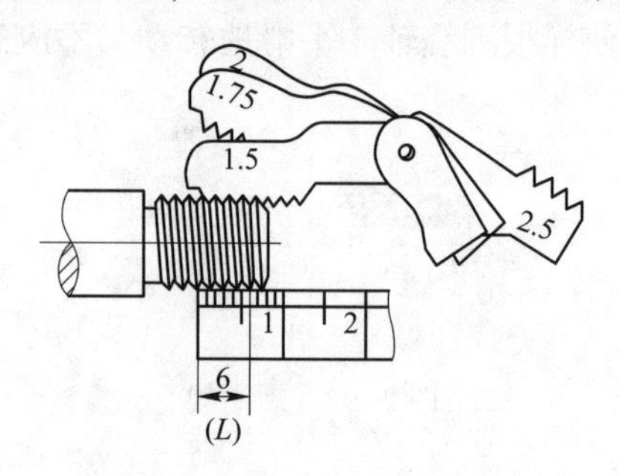

图 2-95　螺距规测量方法

2）测量牙型角时，把零件和与之相匹配的螺距规靠在一起，然后检查它们之间的接触情况，如果发现出现不均匀间隙且透光，说明被测螺纹的牙型角不准确。这种测量方法只能粗略判断牙型角误差的大概情况。

（2）中径的测量。

1）根据螺纹牙型角选择相应的侧头，安装在螺纹千分尺上，并校尺。

2）选择 2~3 次测量，保证误差值都在允许的范围内方为合格产品。（螺纹中径公差可查询机械设计手册）。

3）使用螺纹千分尺测量（如图 2-96 所示），读数的方法与外径千分尺相同。

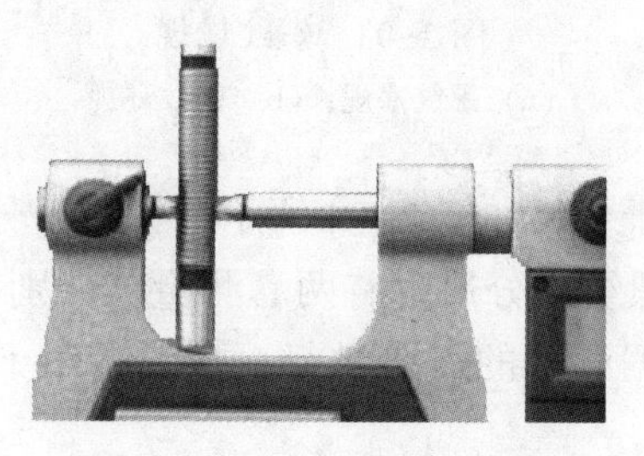

图 2-96　螺纹千分尺的测量方法

普通三角螺纹中径计算公式：

$$d_{中}=d-0.65P$$

式中　d——公称直径；

　　P——螺距。

（3）综合测量。综合测量主要是用螺纹量规对螺纹进行测量，适用于批量生产的螺纹检测，具有测量效率高的优点。

如图 2-97 所示，测量时，被测螺纹如果能够与螺纹通规自由旋合通过，与螺纹止规不能旋入或者不超过两个螺距，则表明被测螺纹的作用中径没有超出其最大实体牙型的中径。作用中径没有超出其最小实体牙型的中径，被测螺纹合格。

任务实施

（1）图样分析：

该零件加工表面有 M24×2 螺纹、$\phi30^{0}_{-0.025}$、$\phi38^{0}_{-0.039}$、$\phi30^{0}_{-0.033}$外圆面、R10 圆弧、R2

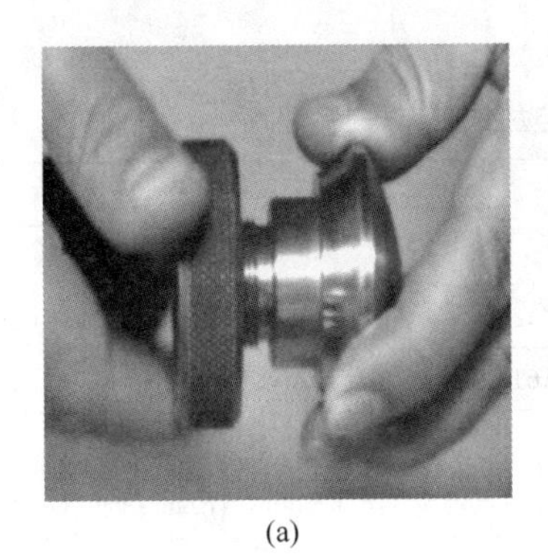
(a)

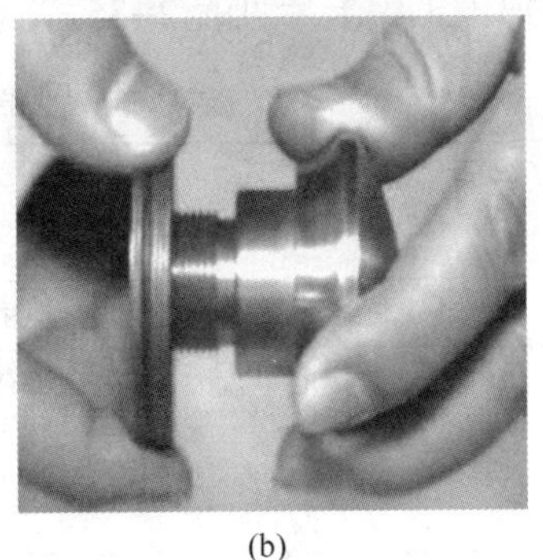
(b)

图 2-97　螺纹环规的测量方法

（a）通规测量；（b）止规测量

圆弧、5mm 螺纹退刀槽、$C2$ 倒角等，表面粗糙度 $Ra3.2$。

（2）加工工艺方案制订：

1）加工方案：

①使用三爪自定心卡盘装卡，零件伸出卡盘 58mm 左右，找正；

②对刀，设零件右端面中心为编程原点；

③粗车零件右端 $\phi30^{0}_{-0.025}$mm、$\phi38^{0}_{-0.039}$mm 外圆柱面及 M24×2 螺纹大径圆柱面；

④精车零件右端外轮廓至尺寸要求；

⑤车 5mm 退刀槽；

⑥车 M24×2 螺纹；

⑦掉头装卡零件右端 $\phi30^{0}_{-0.025}$mm 台阶轴，百分表找正，保证两端同轴度 0.04mm；

⑧粗车 $R10$ 球头、$\phi30^{0}_{-0.033}$mm 外圆柱面；

⑨精车零件左端至尺寸要求，保证零件长度尺寸。

2）刀具选用：

T01　90°硬质合金外圆偏刀（刀尖角 80°）

T02　切槽刀（刀刃宽度 3mm）

T03　数控机夹螺纹刀（60°刀尖角）

螺纹轴零件数控加工刀具卡片见表 2-34。

表 2-34　数控加工刀具卡片

零件名称		螺 纹 轴		零件图号		2-73		
序号	刀具号	刀具名称		数量	加工表面	刀尖半径 R/mm	刀尖方位 T	备注
1	T01	90°外圆偏刀		1	粗精车外轮廓	0.4	3	
2	T02	3mm 切槽刀		1	切槽			
3	T03	60°螺纹刀		1	车螺纹			
编制		审核		批准		日期	共 1 页	第 1 页

3）加工工序：

螺纹轴零件加工工序卡片见表2-35。

表2-35 数控加工工序卡片

<table>
<tr><td rowspan="2">单位名称</td><td rowspan="2" colspan="3">天津工业职业学院</td><td>零件名称</td><td colspan="2">零件图号</td></tr>
<tr><td>螺纹轴</td><td colspan="2">2-73</td></tr>
<tr><td>程序号</td><td>夹具名称</td><td colspan="2">使用设备</td><td>数控系统</td><td colspan="2">场地</td></tr>
<tr><td>O131</td><td>三爪自定心卡盘</td><td colspan="2">CKA6140</td><td>FANUC 0i-Mate</td><td colspan="2">数控实训车间</td></tr>
<tr><td>工序号</td><td>工序内容</td><td>刀具号</td><td>主轴转速 n/r·min^{-1}</td><td>进给量 F/mm·r^{-1}</td><td>背吃刀量 a_p/mm</td><td>备注</td></tr>
<tr><td>1</td><td>装夹零件外伸58mm，并找正</td><td></td><td></td><td></td><td></td><td rowspan="4">手动</td></tr>
<tr><td>2</td><td>对外圆偏刀</td><td>T01</td><td></td><td></td><td></td></tr>
<tr><td>3</td><td>对切槽刀</td><td>T02</td><td></td><td></td><td></td></tr>
<tr><td>4</td><td>对螺纹刀</td><td>T03</td><td></td><td></td><td></td></tr>
<tr><td>5</td><td>粗车右端外轮廓，留0. 5mm径向余量</td><td>T01</td><td>600</td><td>0. 2</td><td>1. 5</td><td rowspan="4">0251</td></tr>
<tr><td>6</td><td>精车右端外轮廓</td><td>T01</td><td>1200</td><td>0. 1</td><td>0. 25</td></tr>
<tr><td>7</td><td>切槽</td><td>T02</td><td>400</td><td>0. 05</td><td>3</td></tr>
<tr><td>8</td><td>车螺纹</td><td>T03</td><td>400</td><td>2</td><td>$0.05 < a_p < 0.4$</td></tr>
<tr><td>9</td><td>掉头装夹，找正</td><td></td><td></td><td></td><td></td><td rowspan="2">手动</td></tr>
<tr><td>10</td><td>对外圆偏刀</td><td>T01</td><td></td><td></td><td></td></tr>
<tr><td>11</td><td>粗车左端外轮廓，留0. 5mm径向余量</td><td>T01</td><td>600</td><td>0. 2</td><td>1. 5</td><td rowspan="2">0252</td></tr>
<tr><td>12</td><td>精车左端外轮廓</td><td>T01</td><td>1200</td><td>0. 1</td><td>0. 25</td></tr>
<tr><td>编制</td><td>审核</td><td>批准</td><td>日期</td><td></td><td>共1页</td><td>第1页</td></tr>
</table>

（3）程序编制

（4）工件加工：

1）根据图纸要求选择合理的毛坯，安装毛坯。

2）正确安装零件加工所需刀具。

3）进行对刀操作，建立工件坐标系。

4）参照表2-36在机床输入零件右端程序。

表2-36 螺纹轴零件右端程序

<table>
<tr><td>O251;</td><td>程序号</td></tr>
<tr><td>G40 G97 G99 M03 S600 F0. 2;</td><td>取消刀具半径补偿，取消主轴恒转速度，设每转进给量，主轴正转，转速为600r/min，设进给量为0. 2mm/r</td></tr>
<tr><td>T0101;</td><td>换T01外圆车刀，代入01号刀具补偿</td></tr>
<tr><td>M08;</td><td>打开切削液</td></tr>
<tr><td>G42 G00 Z5. 0;</td><td rowspan="2">设置刀尖半径右补偿，快速进刀至粗车循环起点</td></tr>
<tr><td>X41. 0;</td></tr>
<tr><td>G71 U1. 5 R0. 5;</td><td rowspan="2">设置外圆粗车循环</td></tr>
<tr><td>G71 P10 Q20 U0. 5 W0. 05;</td></tr>
</table>

续表 2-36

N10 G00 X0. 0；	精加工零件右端外轮廓
G01 Z0. 0；	
X23. 8 C2. 0；	
Z-20. 0；	
X29. 987 C0. 5；	
Z-42. 0；	
X37. 981 C0. 5；	
N20 Z-54. 0；	
G40 G00 X150. 0；	取消刀尖半径补偿，返回换刀点
Z150. 0；	
M05；	主轴停止、程序停止
M00；	
M03 S1200 F0. 1；	主轴正转，转速为 1200r/min，设置精车进给量 0. 1mm/r
T0101；	换 T01 外圆车刀，代入 01 号刀具补偿
G42 G00 X41. 0；	设置刀尖半径右补偿，快速进刀至精车循环起点
Z5. 0；	
G70 P10 Q20；	设置外圆精车循环
G40 G00 X150. 0；	取消刀尖半径补偿，返回换刀点
Z150. 0；	
M05；	主轴停止、程序停止
M00；	
M03 S400 F0. 05；	主轴正转，转速为 400r/min，设置切槽进给量 0. 05mm/r
T0202；	换 T02 切槽刀，代入 02 号刀具补偿
G00 X26. 0；	加工 5×3 的槽及 *C*2 左倒角
Z-20. 0；	
G01 X18. 2；	
X26. 0；	
W2. 0；	
X18. 2；	
X26. 0；	
W2. 0；	
X23. 8；	
U-4. 0 W-2. 0；	
X18. 0；	
Z-20. 0；	
U0. 5 W0. 25；	斜线退刀

续表 2-36

G00 X150.0；	返回换刀点
Z150.0；	
M05；	主轴停止、程序停止
M00；	
M03 S400 T0303；	主轴正转，转速为400r/min，调用T03螺纹车刀，代入03刀具补偿
G00 X26.0；	快速进刀至M24×2螺纹循环起点
Z4.0；	
G76 P020060 Q50 R0.1；	设置G76螺纹循环复合指令加工M24×2螺纹
G76 X21.4 Z-17.0 P1300 Q400 F2.0；	
G00 X150.0；	返回换刀点
Z150.0；	
M09；	切削液关
M30；	程序结束并返回程序头

5）按照图纸要求完成零件右端外圆轮廓、槽、螺纹的加工，如图2-98所示。

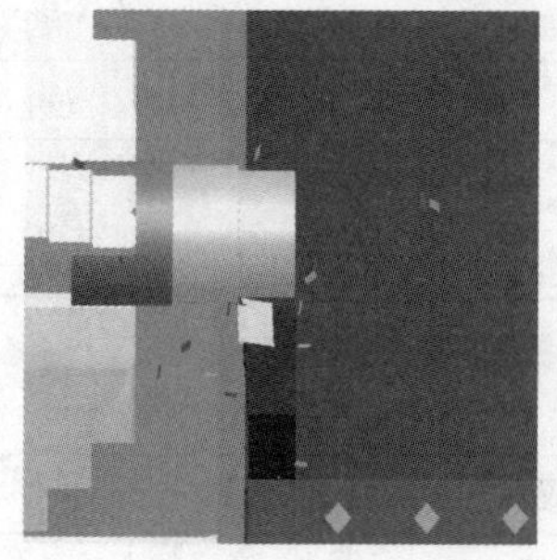
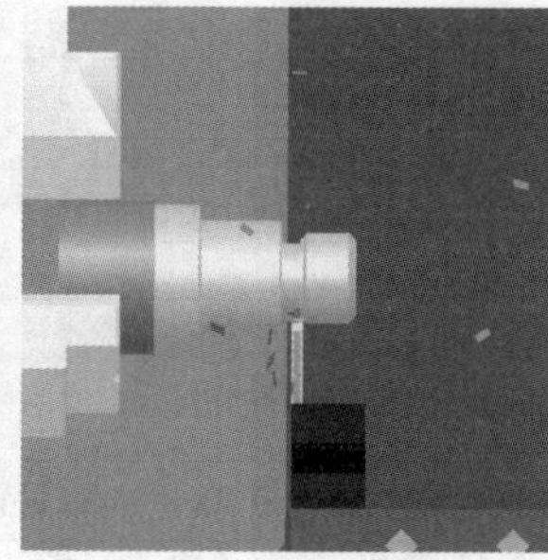
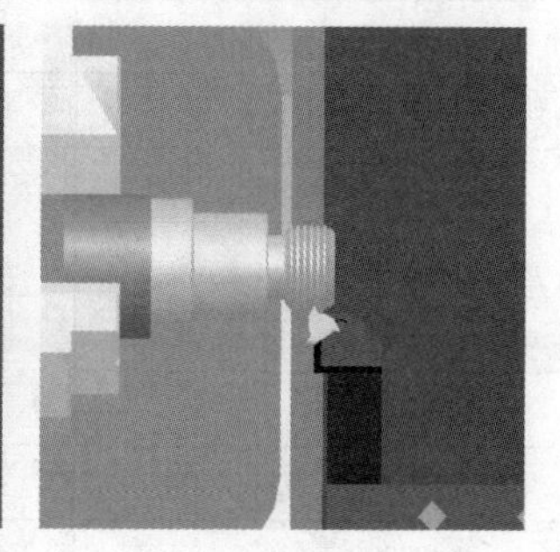

图2-98　零件右端加工

6）掉头装夹，找正。

7）外圆车刀对刀。

零件掉头后，*X*轴无需对刀，*Z*轴对刀，先车端面，延*X*正向退刀，用游标卡尺测量零件总长（或已加工完的某一端面与*Z*轴对刀所车削端面的长度方向尺寸）。计算：对刀*Z*轴测量值=被测零件实际长度-图纸该部位长度尺寸。输入测量值完成对刀。用此方法对刀，计算后*Z*轴测量值有效值为正。

8）参照表2-37在机床输入零件左端程序。

表2-37　螺纹轴零件左端程序

O252；	程序号
G40 G97 G99 M03 S600 F0.2；	取消刀具半径补偿，取消主轴恒转速度，设每转进给量，主轴正转，转速为600r/min，设进给量为0.2mm/r
T0105；	换T01外圆车刀，代入05号刀具补偿
M08；	打开切削液

续表 2-37

G42 G00 Z8. 0；	设置刀尖半径右补偿，快速进刀至粗车循环起点
X41. 0；	
G71 U1. 5 R0. 5；	设置外圆粗车循环
G71 P10 Q20 U0. 5 W0. 05；	
N10 G00 X00；	精加工零件左端外轮廓
G01 Z0. 0；	
G03 X20. 0 Z-10. 0 R10. 0；	
G01 X29. 983 C0. 5；	
X33. 983 R2. 0；	
X38. 983 C1. 0；	
N20 W-1. 0；	
G40 G00 X150. 0；	取消刀尖半径补偿，返回换刀点
Z150. 0；	
M05；	主轴停止、程序停止
M00；	
M03 S1200 F0. 1；	主轴正转，转速为 1200r/min，设置精车进给量 0. 1mm/r
T0105；	换 T01 外圆车刀，代入 05 刀具补偿
G42 G00 X41. 0；	设置刀尖半径右补偿，快速进刀至精车循环起点
Z5. 0；	
G70 P10 Q20；	设置外圆精车循环
G40 G00 X150. 0；	取消刀尖半径补偿，返回换刀点
Z150. 0；	
M09；	切削液关
M30；	程序结束并返回程序头

9）按照图纸要求完成零件左端外圆轮廓加工，如图 2-99 所示。

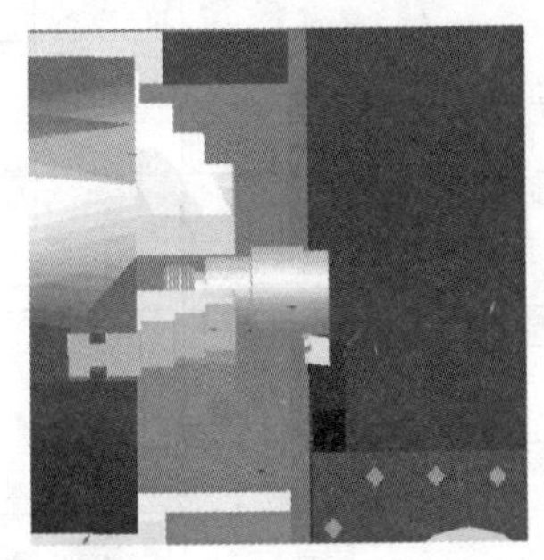

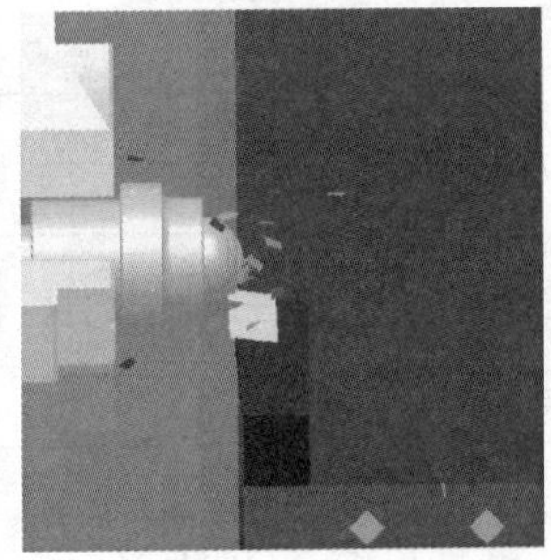

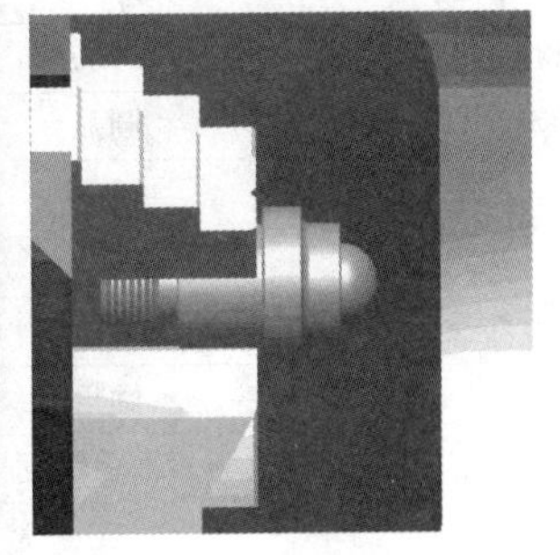

图 2-99　零件左端加工

（5）尺寸检测

1）使用 25-50 规格外径千分尺测量 $\phi30^{0}_{-0.025}$、$\phi38^{0}_{-0.039}$、$\phi30^{0}_{-0.033}$外径尺寸。

2）使用游标卡尺测量 10mm、72mm 长度尺寸。

3）使用游标卡尺测量 5mm 槽的尺寸。

4）使用螺纹量规测量 M24×2 螺纹尺寸。

5）使用 $R10$ 圆弧样板测量 $R10$ 圆弧。

6）使用粗糙度样板检测零件表面粗糙度。

任务评价

任务评价见表 2-38。

表 2-38　任务评价表

项目	序号	评分标准		配分	得分
加工工艺（10%）	1	加工步骤	符合数控车工工艺要求	3	
	2	加工部位尺寸	计算相关部位尺寸	3	
	3	刀具选择	刀具选择合理，符合加工要求	4	
程序编制（30%）	4	程序号	无程序号无分	2	
	5	程序段号	无程序段号无分	2	
	6	切削用量	切削用量选择不合理无分	4	
	7	原点及坐标	标明程序原点及坐标，否则无分	2	
	8	程序内容	不符合程序逻辑及格式要求每段扣 2 分	20	
			程序内容与工艺不对应扣 5 分		
			出现危险指令扣 5 分		
加工操作（20%）	9	程序输入与检索	每错一段扣 2 分	10	
	10	加工操作 60min	工件坐标系原点设定错误无分	10	
			误操作无分		
			超时无分		
尺寸检测（30%）	11	外圆	$\phi30^{0}_{-0.025}$ 超差 0. 01 扣 3 分	3	
	12		$\phi30^{0}_{-0.033}$ 超差 0. 01 扣 3 分	3	
	13		$\phi38^{0}_{-0.039}$ 超差 0. 01 扣 3 分	3	
	14		$\phi18$ 降级无分	1	
	15	螺纹	M24×2 超差无分	4	
	16	球头	$R10$ 超差无分	3	
	17	圆弧	$R2$ 超差无分	2	
	18	轴向长度	72±0. 05	3	
	19		10±0. 03	3	
	20	倒角	2×$C2$	2	
	21	倒角	4×$C0.5$	2	
	22	表面粗糙度	$Ra3.2$ 降级无分	1	
职业能力（10%）	23	学习能力		2	
	24	表达沟通能力		2	
	25	团队合作		2	
	26	安全操作与文明生产		4	
合计					

同步思考与训练

（1）思考题：

1）分析 G92、G76 指令在加工螺纹中的异同点。

2）加工导程较大的螺纹，主轴转速档位该如何选择？

3）如果所加工外螺纹用通规、止规均拧不进去，该如何修正？

4）如果加工的外螺纹出现用通规拧不进去，止规拧得进去的现象，是什么原因造成的？

（2）同步训练题：

1）毛坯：ϕ35，材料：铝合金，编写程序并加工图 2-100 所示工件。

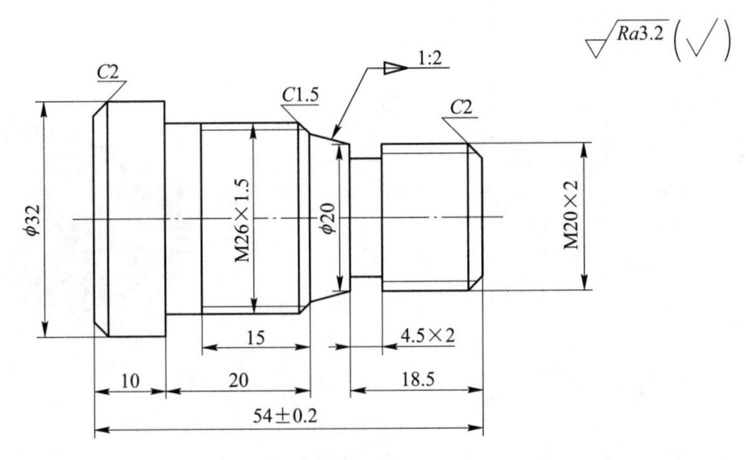

图 2-100　同步训练 1

2）毛坯：ϕ50×69，材料：铝合金，编写程序并加工如图 2-101 所示。

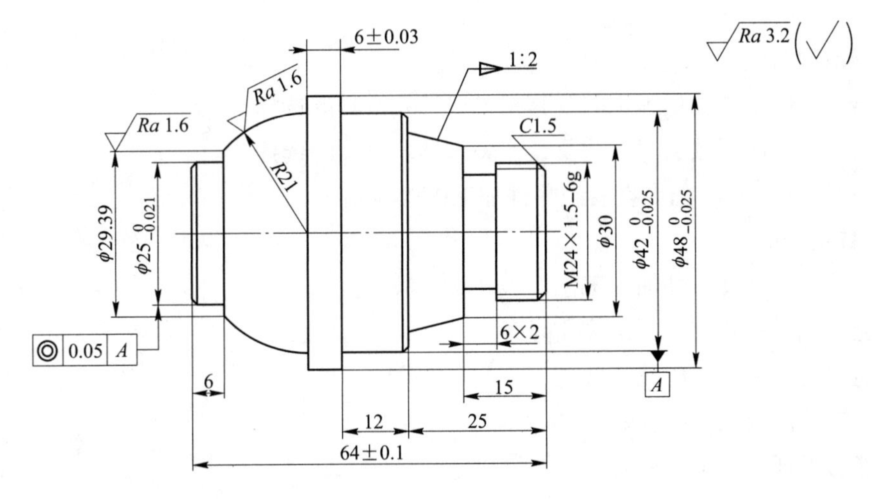

技术要求

1.未注倒角均为$C1$ 锐角倒钝。
2.未注线性尺寸公差应符合GB/T 1804—2000的要求。

图 2-101　同步训练 2

任务 2.6　套类零件的数控加工

任务描述

图 2-102 所示零件，材料硬铝合金 LY15，毛坯为 ϕ40×45mm 棒料，单件生产，与任务 5 零件配合，使用数控车床加工，编写加工程序，检测加工质量。

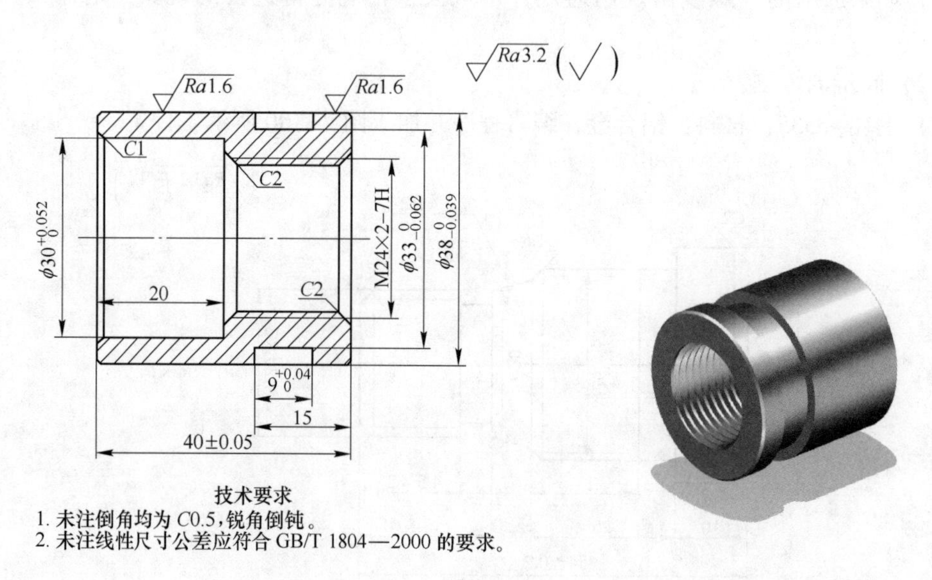

图 2-102　零件图

任务目标

（1）知识目标：

1）熟悉车床中镗孔加工工艺及其相关刃具、量具的选择。

2）熟悉车床内螺纹加工工艺及其相关刃具、量具的选择。

3）解释 G71、G76 指令在套类零件加工中的应用。

（2）技能目标：

1）能够分析套类零件加工工艺。

2）能够编写镗孔、车削螺纹孔程序。

3）能够正确使用车床尾座钻孔。

4）能够检测内孔、内孔螺纹。

5）能够保证零件配合良好。

（3）素质目标：

1）培养安全意识、规范操作意识。

2）针对配合工件，培养精益求精的工匠精神。

相关知识

2.6.1 工艺准备

2.6.1.1 套类零件加工常用刀具

套类零件的加工表面既有外表面也有内表面，外表面加工刀具与之前的任务一致，内表面加工刀具常用到镗孔刀、中心钻、麻花钻及相关附件，当内表面有螺纹时，还会用到内螺纹刀。

(1) 常用钻具。中心钻如图 2-103 (a) 所示，麻花钻头如 2-103 (b) 所示，安装用钻夹头如图 2-103 (c) 所示，安装锥柄的莫氏变径套如图 2-103 (d) 所示。

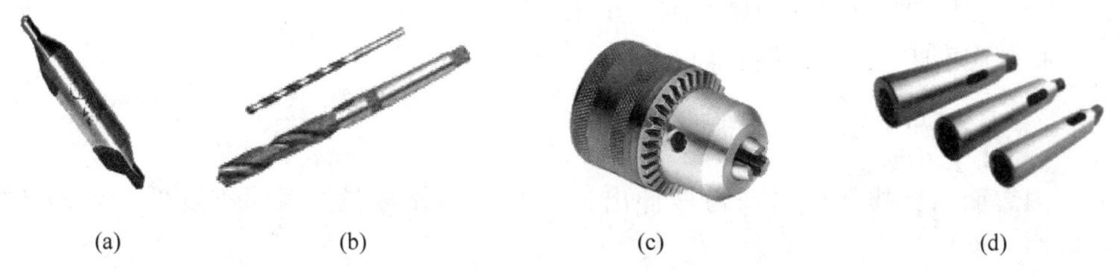

(a) (b) (c) (d)

图 2-103 常用钻具

(a) 中心钻；(b) 麻花钻；(c) 钻夹头；(d) 莫式尾座套筒

(2) 镗孔刀。镗孔刀是车床中最常用的内孔加工刀具，可以很好的保证加工的精度，常用镗孔刀如图 2-104 所示。

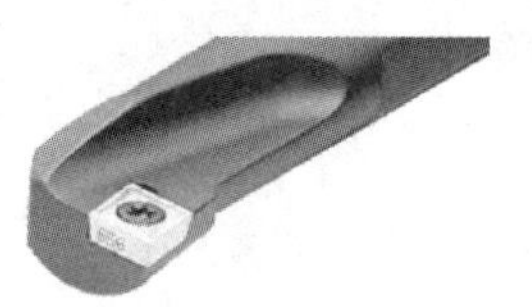

图 2-104 镗孔刀

(3) 内螺纹刀

与外螺纹刀加工原理相同，常用内孔螺纹刀如图 2-105 所示。

图 2-105 内孔螺纹刀

(4) 车床尾座。车床尾座如图 2-106 所示，摇动手轮 1 可实现套筒 Z 方向的前进和后退；锁紧尾座固定手柄 2 可以固定尾座；锁定主轴卡紧手柄 3 可以坐定尾座套筒的伸缩。车床尾座主要有以下三个作用。

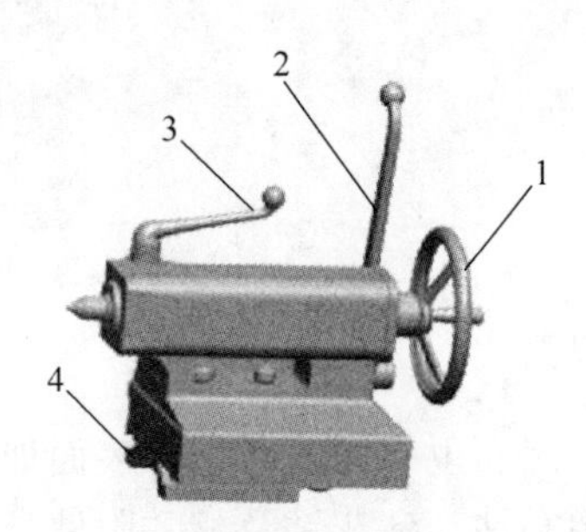

图 2-106　车床尾座

1—主轴手轮；2—尾座固定手柄；3—主轴卡紧手柄；4—尾座底板

1）利用尾座钻孔，打中心孔。

2）机床精度检测时，利用尾座调整车床精度。

3）利用尾座加工长轴、偏心轴。

2.6.1.2　内表面加工切削用量选择

加工内表面时，排屑困难、刀杆伸出长、刀头部分薄弱，因而刚度低，容易产生振动，所以内表面切削用量比外表面低些。

2.6.1.3　内表面加工工步安排

在实心材料上加工内表面，首先车削端面，然后钻中心孔（底孔精度不高时，可以不钻中心孔），再选用合适的钻头手动钻孔，选用合适的镗刀加工内轮廓表面，最后车内螺纹。

2.6.2　编程指令

2.6.2.1　刀具补偿

（1）刀尖方位 T 值。

内孔车刀刀尖方位 $T=2$。

（2）刀具补偿指令。前置刀架的数控车床，使用 G41 指令进行内孔刀具补偿。

2.6.2.2　粗加工复合循环指令 G71

（1）指令格式：

G71 U(Δd)R(e)；

G71 P(ns)Q(nf)U(Δu)W(Δw)；

（2）参数说明同前。

（3）注意：加工内表面时 Δu 为负值。

2.6.2.3　套类零件编程时进、退刀点的改变

为保证实际加工的安全性，套类零件编程时应注意刀具进、退刀点的改变，改变内容见表 2-39。

表 2-39　内外轮廓编程中，进、退刀点对照表

序　号	内　容	外轮廓编程	内轮廓编程
1	进刀点（G71/G70 前一点）X 值	工件毛坯直径	钻孔孔径
2	进刀编程，X、Z 先后顺序	先 Z 轴，后 X 轴	先 X 轴，后 Z 轴
3	退刀编程，X、Z 先后顺序	先 X 轴，后 Z 轴	先 Z 轴，后 X 轴

（1）外轮廓编程时，进刀点 X 值为工件毛坯直径，而内孔编程时，加工进刀点位置为钻孔孔径。

（2）外轮廓编程时，进刀点建议先运行 Z 轴，后运行 X 轴；内轮廓编程应首先运行 X 轴，后进行 Z 轴进刀。

（3）外轮廓编程时，退刀点建议先输入 X 值，后输入 Z 值；内轮廓编程应首先退 Z 方向，后进行 X 方向进刀。

注：G70 指令在内孔编程时，进、退刀方式同样参照 G71 编程改变。

【例 2-15】零件如图 2-107 所示，外圆轮廓已加工至尺寸要求，编写内轮廓加工程序。参考程序见表 2-40。

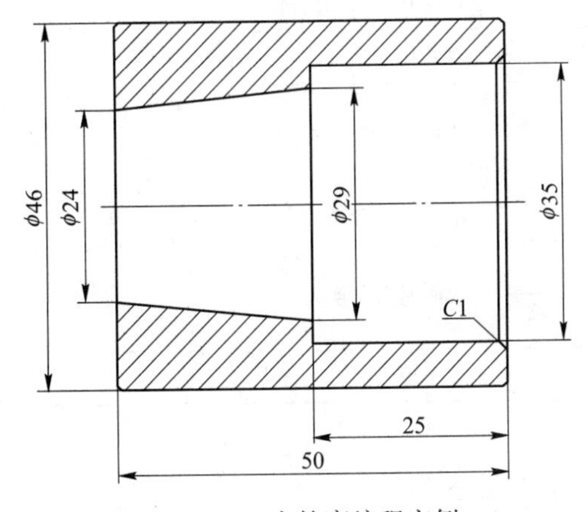

图 2-107　内轮廓编程实例

表 2-40　内轮廓加工程序

O262;	程序号
G40 G97 G99 M03 S400 F0.1;	取消刀具半径补偿，取消主轴恒转速度，设每转进给量，主轴正转，转速为 400r/min，设进给量为 0.1mm/r
T0202;	换 T02 镗孔刀，代入 02 号刀具补偿
M08;	打开切削液
G00 X20.;	快速进刀至 X 方向粗车循环起点（钻孔直径位置）
Z5.0;	快速进刀至 Z 方向粗车循环起点
G71 U1.5 R0.5;	设置内孔粗车循环，切削深度 1.5mm，退刀量为 0.5mm
G71 P30 Q40 U-0.5 W0.05;	精车路线由 N10～N20 指定，X 向精车余量 0.5mm，Z 向精车余量 0.05mm

续表 2-40

N30 G00 X37.0;	精车内轮廓
G01 Z0;	
X35.0 C1.5.;	
Z-25.0;	
X29.0;	
X24.0 Z-50.0;	
N40 X20.0;	
G00 Z150.0;	快速退刀至换刀点
X150.0;	
M09;	关闭切削液
M05;	主轴停止
M00;	程序暂停
M03 S1200 F0.08;	主轴正转，转速为 1200r/min，设进给量为 0.08mm/r
T0202;	代入 02 号刀具补偿
M08;	打开切削液
G00 X20.0;	快速进刀至精车循环起点
Z5.0;	
G70 P10 Q20;	精加工复合循环
G00 Z150.0;	快速退刀至换刀点
X150.0;	
M30;	程序结束同时光标返回程序头

2.6.2.4 螺纹切削复合循环指令 G76

（1）内螺纹加工优先使用 G76 指令。G76 指令采用斜进法切削螺纹，即螺纹刀始终用一个刀刃进行切削，减小了切削阻力，提高了刀具寿命和螺纹精车质量，此外，该指令加工效率较高，所以加工内螺纹优先使用 G76 指令。

（2）指令格式：

$$G76 \quad P(m)(r)(\alpha)Q(\Delta dmin)R(d);$$

$$G76 \quad X(U)Z(W)R(i)P(k)Q(\Delta d)F(L);$$

参数说明同前。

（3）内螺纹加工尺寸计算：

1）实际车削内圆柱面的直径：

塑性材料：$D_{实}=D-P$

脆性材料：$D_{实}=D-1.05P$

2）螺纹牙型高度：$h_{牙}=0.65P$

3）内螺纹 X 向的终点坐标（即底孔加工直径）为螺纹公称直径：$D_{底}=D$

【例 2-16】零件如图 2-108 所示，外轮廓和螺纹底孔已车至尺寸要求，编制 M28×1.5 内螺纹的加工程序。参考程序如表 2-41 所示。

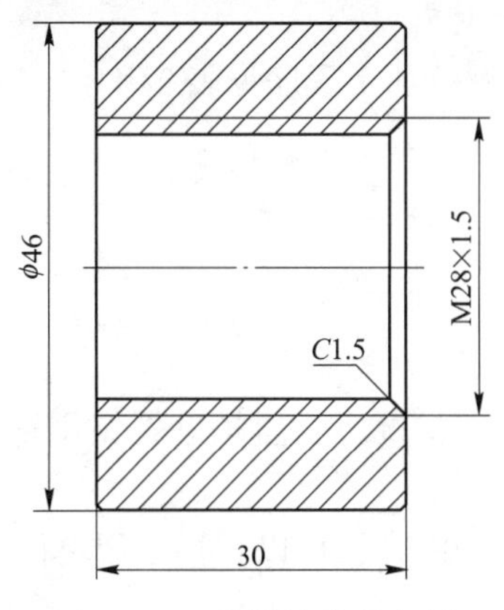

图 2-108　内螺纹编程实例

表 2-41　M28×1.5 螺纹加工程序

O263；	程序号
G40 G97 G99 M03 S400	主轴正转，转速为 400r/min
T0404；	换 T04 内孔螺纹刀，代入 04 号刀具补偿
M08；	打开切削液
G00 X25.0；	快速进刀至加工循环起点
Z5.0；	
G76 P020060 Q30 R0.05；	螺纹切削复合循环
G76 X28. Z-31. P975 Q400 F1.5.；	
G00 Z150.0；	快速退刀至换刀点
X150.0；	
M30；	程序结束同时光标返回程序头

（1）尺寸计算：

实际车削时内圆柱面的直径：$D_{实}=D-P=28-1.5=26.5$mm

螺纹实际牙型高度：$d_{牙}=0.65P=0.65\times1.5=0.95$mm

内螺纹 X 向的终点坐标：$D_{底}=D=28$mm

（2）螺纹参数：

精车重复 2 次，m 取 02；

螺纹尾部无倒角，r 取 00；

三角形螺纹刀尖角 60°，α 取 60；

最小车削深度 Δdmin 为 0.03mm，Q 取 30μm；

留 0.05mm 精车余量，R 取 0.05mm；

根据零件图计算螺纹终点坐标（X28.0 Z-31.0）

圆柱螺纹 i 为 0，可以省略；

螺纹高度 k 为 0.975mm，P 取 975μm；

第一次车削深度 Δd 为 0.4mm，Q 取 400μm；

单头螺纹螺距为 1.5，F 取 1.5。

（3）加工程序：M28×1.5 螺纹加工程序见表 2-41。

2.6.3 加工操作

2.6.3.1 钻头的安装与钻孔

（1）钻头的安装方法。

1）直柄麻花钻的安装。一般情况下，直柄麻花钻用钻夹头装夹，再将钻夹头的锥柄插入车床尾座锥孔内。

2）锥柄麻花钻的安装。锥柄麻花钻可以直接或用莫氏过渡锥套（变径套）插入车床尾座锥孔内。

（2）钻孔加工操作。

1）加工过程：首先平端面，然后钻中心孔，最后钻孔。精度不高时可以不钻中心孔。

2）钻孔加工时应注意如下事项：

①钻孔前必须将端面车平。

②找正使钻头轴线与工件回转轴线重合。

③当钻头接触工件端面和钻通孔快要钻透时，进给量要小，以防钻头折断。

④钻小而深的孔时，应先用中心钻钻中心孔，避免将孔钻歪。

⑤钻深孔时，切屑不易排出，要经常把钻头退出清除切屑。

⑥钻削钢料时，必须浇注充分的切削液，使钻头冷却；钻铸铁时可不用切削液。

2.6.3.2 镗孔刀的安装与对刀

（1）镗孔刀的安装。

1）应使刀杆与工件轴线基本平行。

2）镗刀刀杆的伸出长度应尽可能的短，一般比孔深长出一个刀头宽度即可。

3）刀尖等高或略高于主轴的回转中心，防止刀杆在切削力作用下弯曲产生“扎刀”现象。

（2）镗孔刀的对刀操作。

1）试切工件内孔，如图 2-109 所示。

图 2-109　试切工件内孔

2）沿 Z 方向退刀后，主轴完全停转，测量试切的内孔直径。

3）设置 X 向补正。

4）试切削端面。

5）设置 Z 向补正。

2.6.3.3 内螺纹刀的安装与对刀

（1）内螺纹刀的安装。

1）刀杆与工件轴线基本平行。

2）刀尖等高或略高于主轴的回转中心，防止刀杆在切削力作用下弯曲产生“扎刀”现象。

3）内螺纹刀刀杆的伸出长度应尽可能的短，以增加刀杆的刚性，防止产生震动。

（2）内螺纹刀的对刀操作。

1）移动刀具至内孔与端面的交接处，如图 2-110 所示。

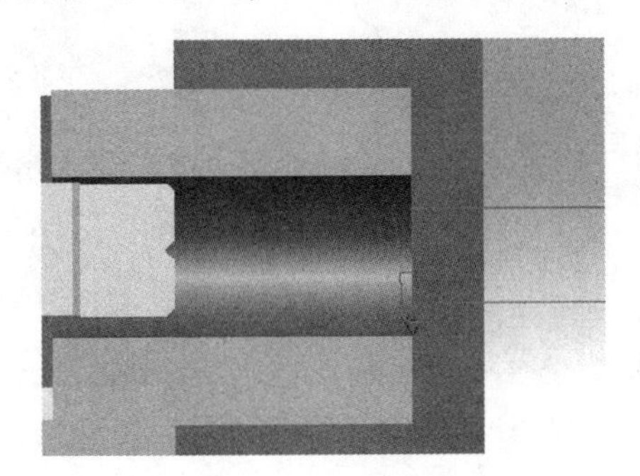

图 2-110 内螺纹刀对刀

2）设置 Z 向补正。

3）设置 X 向补正：在 X 向刀具参数处，输入镗孔刀车削后的孔径，完成内螺纹刀对刀。

2.6.4 精度检验

2.6.4.1 内径百分表

（1）结构。内径百分表为组合套件，由钟面式百分表和杠杆测量架组成，是对内孔进行相对测量的量具，如图 2-111 所示。

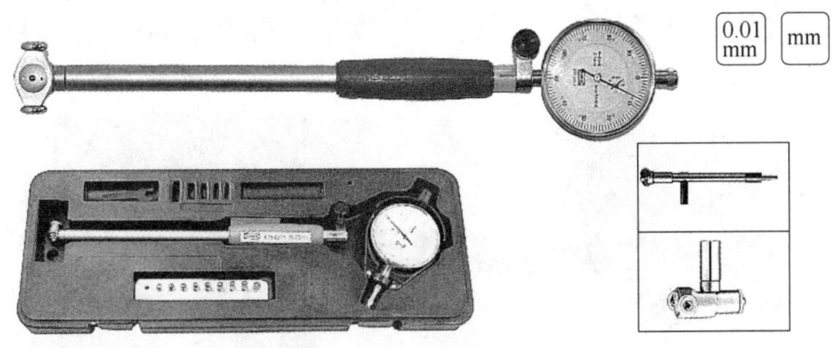

图 2-111 内径百分表

（2）读数。表盘上刻有100个等分格，其刻度值为0.01mm。当指针转一圈时，小指针即转动一小格，转数指示盘的刻度值为1mm。

用手转动表圈时，表盘也跟着转动，可使指针对准任一刻线。该功能可方便操作者进行校表。

（3）校表。

1）选择百分表并检查或修改计量标签。

2）用手轻轻按压百分表，观察指针转动是否灵活，如图2-112所示。

3）检查内径杠杆测量架内侧有无杂物。按压测量触头，触头可以灵活弹出。

4）安装百分表。把百分表插入测量架的轴孔中，压缩百分表小指针在大约0.5mm处，固定表头，如图2-113所示。

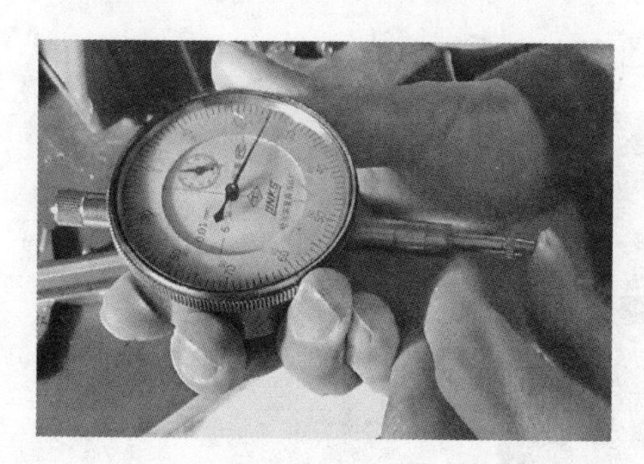

图2-112　检查百分表

图2-113　固定表头

5）按测量范围选取并安装调整可换测头。

6）用手按压活动测头，百分表指针转动应平稳灵活。

7）用游标卡尺粗略测量，可换测头长度应比校零长度长约0.5mm。

8）将校准后的外径千分尺调到要校零的尺寸。

9）将活动侧头抵住，千分尺（或环规）一端，轻压可换侧头放入其中，测杆与被测部位表面必须垂直。

10）轻轻摇动量表，使表针停到最小位置。

11）转动表盘使指针指到零处，拧紧螺钉。

（4）使用方法。

使用内径百分表测量孔径尺寸，如图2-114所示，测量时应该使量爪与被测的内孔充分接触。

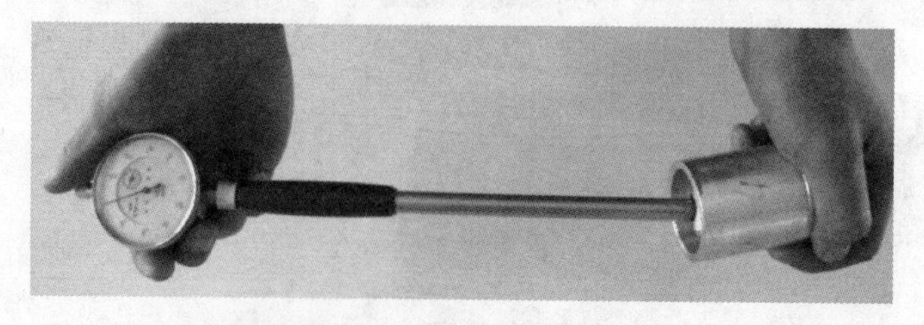

图2-114　内径百分表测量工件

2.6.4.2 内径千分尺

（1）应用。内径千分尺主要用于测量精度较高的孔径和槽的宽度等尺寸。

（2）结构。内径千分尺由活动量爪、导向管、固定量爪、微分筒、棘轮旋柄组成，如图 2-115 所示。

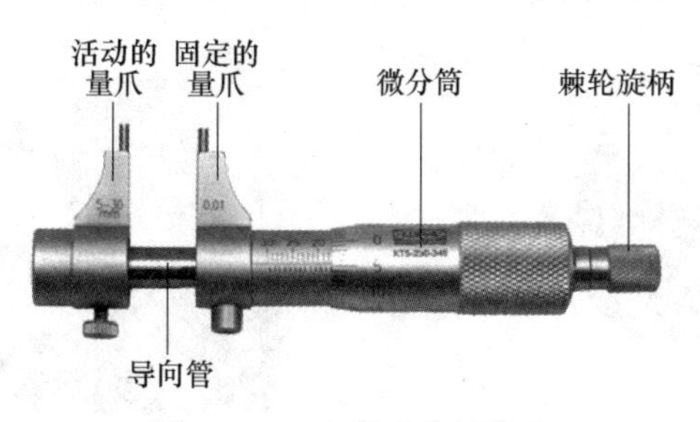

图 2-115　内径千分尺结构

（3）使用方法。测量时，先校准零位，然后将内径千分尺放入被测孔内，接触的松紧程度合适，读出直径的正确数值。

（4）读数与注意事项

与外径千分尺基本相同。

2.6.4.3 螺纹量规

测量内螺纹的量规一般称作螺纹塞规，如图 2-116 所示。用于通过的一端为通端，用字母“T”表示；用于限制通过的一端为止端，用字母“Z”表示。

图 2-116　螺纹塞规

螺纹塞规使用方法如下：

（1）用螺纹通规与被测螺纹旋合，如果能够通过，就表明被测螺纹的作用中径没有超过其最大实体牙型的中径；

（2）用螺纹止规与被测螺纹旋合，旋合量不超过两个螺距，即螺纹止规不完全旋合通过，表明单一中径没有超出其最小实体牙型的中径，被测螺纹中径合格。

任务实施

（1）图样分析：

该零件总长 40±0.05mm，是一典型的套类零件，既有外表面也有内表面，外表面 $\phi38^{0}_{-0.039}$外圆被底面直径 $\phi33^{0}_{-0.062}$、宽 9mm 的槽分为左右两段，内表面有 $\phi30^{+0.052}_{0}$ 内孔、M24×2-7H 内螺纹，表面粗糙度均为 $Ra3.2$。该零件外表面有宽 9mm 的槽，要求与任务 2.5 图纸相配合。

（2）加工工艺方案制订：

1）加工方案：

①使用三爪自定心卡盘装夹，零件伸出卡盘35mm左右。

②加工零件左端$\phi38_{-0.039}^{0}$外轮廓至尺寸要求。

③加工零件左端$\phi30_{0}^{+0.052}$内轮廓至尺寸要求，完成工件左端效果如图2-117所示。

④调头装夹零件，装夹零件长度约20mm，如图2-118所示。

图2-117　工件左端加工效果

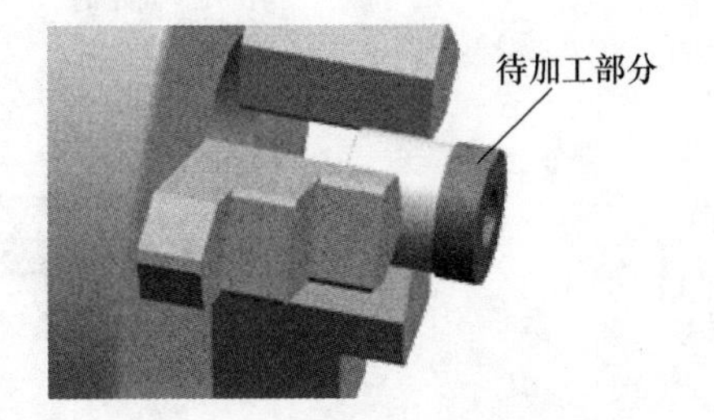

图2-118　工件装夹示意图

⑤找正工件。

⑥加工零件右端$\phi38_{-0.039}^{0}$外轮廓至尺寸要求，保证工件总长40±0.05mm。

⑦加工底面直径$\phi33_{-0.062}^{0}$、宽9mm的槽。

⑧加工零件右端内螺纹M24×2-7H底孔至尺寸要求。

⑨加工零件内螺纹。

2）刀具选用：

该零件为单件生产，为了节省换刀时间与降低加工成本，在轮廓加工时端面、外圆粗加工、精加工时使用同一把外圆车刀，内孔和内孔台阶粗加工与精加工时也使用同一把内孔车刀，数控加工刀具卡见表2-42。

表2-42　数控加工刀具卡片

零件名称		螺纹套		零件图号	2-102		
序号	刀具号	刀具名称	数量	加工表面	刀尖半径 R/mm	刀尖方位 T	备注
1	手动	中心钻头	1	钻中心孔			A3
2	手动	麻花钻头	1	钻孔			$\phi20$
3	T01	外圆车刀	1	粗精车外轮廓	0.4	3	
4	T02	镗孔刀	1	粗精车内轮廓	0.4	2	$\phi16$
5	T03	4mm切槽刀	1	切槽	0	0	
6	T04	内螺纹车刀	1	车削螺纹	0	0	$\phi16$
编制		审核	批准	日期		共1页	第1页

T01　硬质合金外圆偏刀，刀尖角55°；

T02　切槽刀，刀片宽度为4mm；

T03　镗孔刀，刀具直径16mm；

T04　内孔螺纹刀，刀具直径 16mm。

3）加工工序：

左端加工工序卡片见表 2-43。

表 2-43　左端加工工序卡片

单位名称	天津工业职业学院		零件名称		零件图号	
			螺纹套		2-102	
程序号	夹具名称	使用设备	数控系统		场地	
O61	三爪自定心卡盘	CKA6150	FANUC 0i-Mate		数控加工实训中心	
工序号	工序内容	刀具号	主轴转速 n/r·min^{-1}	进给量 F/mm·r^{-1}	背吃刀量 a_p/mm	备注
1	装夹零件外伸 35mm，并找正					手动
2	钻中心孔	手动	1200	0.05		A3
3	工件钻孔	手动	600	0.05		$\phi20$
4	粗加工零件左端外轮廓，留 0.5mm 精加工余量	T01	800	0.2	2	
5	精加工零件左端外轮廓，使工件左端外轮廓 $\phi38^{0}_{-0.039}$ 部分至尺寸要求	T01	1500	0.1	0.25	
6	粗加工零件左端内轮廓，留 0.5mm 精加工余量	T02	400	0.1	1.5	
7	精加工零件左端内轮廓，使工件左端外轮廓 $\phi30^{+0.052}_{0}$ 部分至尺寸要求	T02	1200	0.08	0.25	
编制	审核	批准	日期		共 1 页	第 1 页

右端加工工序卡片见表 2-44。

表 2-44　右端加工工序卡片

单位名称	天津工业职业学院		零件名称		零件图号	
			螺纹套		2-102	
程序号	夹具名称	使用设备	数控系统		场地	
O62	三爪自定心卡盘	CKA6150	FANUC 0i-Mate		数控加工实训中心	
工序号	工序内容	刀具号	主轴转速 n/r·min^{-1}	进给量 F/mm·r^{-1}	背吃刀量 a_p/mm	备注
1	掉头装夹工件，找正					手动
2	粗加工零件左端外轮廓，留 0.5mm 精加工余量	T01	800	0.2	2	
3	精加工零件左端外轮廓，使工件右端外轮廓 $\phi38^{0}_{-0.039}$ 部分至尺寸要求	T01	1500	0.1	0.25	
4	加工底径 $\phi33^{0}_{-0.062}$、宽 9mm 槽	T03	400	0.05	4	

续表 2-44

<table>
<tr><td rowspan="2">单位名称</td><td colspan="5" rowspan="2">天津工业职业学院</td><td colspan="2">零件名称</td><td colspan="2">零件图号</td></tr>
<tr><td colspan="2">螺纹套</td><td colspan="2">2-102</td></tr>
<tr><td colspan="2">程序号</td><td colspan="2">夹具名称</td><td colspan="2">使用设备</td><td colspan="2">数控系统</td><td colspan="2">场地</td></tr>
<tr><td colspan="2">O62</td><td colspan="2">三爪自定心卡盘</td><td colspan="2">CKA6150</td><td colspan="2">FANUC 0i-Mate</td><td colspan="2">数控加工实训中心</td></tr>
<tr><td>工序号</td><td colspan="4">工序内容</td><td>刀具号</td><td>主轴转速
$n/\mathrm{r \cdot min^{-1}}$</td><td>进给量
$F/\mathrm{mm \cdot r^{-1}}$</td><td>背吃刀量
a_p/mm</td><td>备注</td></tr>
<tr><td>5</td><td colspan="4">粗加工零件右端内轮廓，留 0.5mm 精加工余量</td><td>T02</td><td>400</td><td>0.1</td><td>1.5</td><td></td></tr>
<tr><td>6</td><td colspan="4">精加工零件右端内轮廓，保证螺纹底孔尺寸</td><td>T02</td><td>1200</td><td>0.08</td><td>0.25</td><td></td></tr>
<tr><td>7</td><td colspan="4">加工内孔螺纹，保证螺纹合格。</td><td>T04</td><td>400</td><td>2</td><td>0.3</td><td></td></tr>
<tr><td>编制</td><td></td><td>审核</td><td></td><td>批准</td><td></td><td>日期</td><td></td><td>共 1 页</td><td>第 1 页</td></tr>
</table>

（3）程序编制

左端加工程序见表 2-45。

表 2-45　左端加工程序

O61;	程序号
G40 G97 G99 M03 S800 F0.2;	取消刀具半径补偿，取消主轴恒转速度，设每转进给量，主轴正转，转速为 800r/min，设进给量为 0.2mm/r
T0101;	换 T01 外圆车刀，代入 01 号刀具补偿
M08;	切削液开
G00 Z5.0;	快速进刀至 Z 方向粗车循环起点
X42.0;	快速进刀至 X 方向粗车循环起点（毛坯直径加 2mm 位置）
G71 U1.5 R0.5;	设置外圆粗车循环，切削深度 1.5mm，退刀量为 0.5mm
G71 P10 Q20 U0.5 W0.05;	精车路线由 N10~N20 指定，X 向精车余量 0.5mm，Z 向精车余量 0.05mm
N10 G00 X36.981;	左端精车外轮廓
G01 Z0.0;	
X37.981 C0.5;	
Z-30.0;	
N20 X42.0;	
G00 X150.0;	快速退刀至换刀点
Z150.0;	
M09;	关闭切削液
M05;	主轴停止
M00;	程序暂停
M03 S1500 F0.1;	主轴正转，转速为 1500r/min，设进给量为 0.1mm/r
T0101;	代入 01 号刀具补偿

续表 2-45

M08;	打开切削液
G00 Z5. 0;	快速进刀至精车循环起点
X42. 0;	
G70 P10 Q20;	精加工复合循环
G00 X150. 0;	快速退刀至换刀点
Z150. 0;	
M09;	关闭切削液
M05;	主轴停止
M00;	程序暂停
M03 S400 F0. 1;	主轴正转，转速为 400r/min，设进给量为 0. 1mm/r
T0202;	换 T02 镗孔刀，代入 02 号刀具补偿
M08;	打开切削液
G00 X20. 0;	快速进刀至 *X* 方向粗车循环起点（钻孔直径位置）
Z5. 0;	快速进刀至 *Z* 方向粗车循环起点
G71 U1. 5 R0. 5;	设置内孔粗车循环，切削深度 1. 5mm，退刀量为 0. 5mm
G71 P30 Q40 U-0. 5 W0. 05;	精车路线由 N30~N40 指定，*X* 向精车余量 0. 5mm，*Z* 向精车余量 0. 05mm
N30 G00 X32. 026;	左端精车内轮廓
G01 Z0. 0;	
X30. 026 C1. 0;	
Z-20. 0;	
X26. 0;	
X21. 0 W-2. 5;	
N40 X20. 0;	
G00 Z150. 0;	快速退刀至换刀点
X150. 0;	
M09;	关闭切削液
M05;	主轴停止
M00;	程序暂停
M03 S1200 F0. 08;	主轴正转，转速为 1200r/min，设进给量为 0. 08mm/r
T0202;	代入 02 号刀具补偿
M08;	打开切削液
G00 X20. 0;	快速进刀至精车循环起点
Z5. 0;	
G70 P30 Q40;	精加工复合循环
G00 Z150. 0;	快速退刀至换刀点
X150. 0;	
M30;	程序结束同时光标返回程序头

零件右端加工程序见表 2-46。

表 2-46　零件右端加工程序

O62;	程序号
G40 G97 G99 M03 S800 F0.2;	取消刀具半径补偿，取消主轴恒转速度，设每转进给量，主轴正转，转速为 800r/min，设进给量为 0.2mm/r
T0101;	换 T01 外圆车刀，代入 01 号刀具补偿
M08;	切削液开
G00 Z5.0;	快速进刀至 *Z* 方向粗车循环起点
X42.0;	快速进刀至 *X* 方向粗车循环起点（毛坯直径加 2mm 位置）
G71 U1.5 R0.5;	设置外圆粗车循环，切削深度 1.5mm，退刀量为 0.5mm
G71 P10 Q20 U0.5 W0.05;	精车路线由 N10~N20 指定，*X* 向精车余量 0.5mm，*Z* 向精车余量 0.05mm
N10 G00 X36.981;	左端精车外轮廓
G01 Z0.0;	
X37.981 C0.5;	
Z-10.5;	
N20 X42.0;	
G00 X150.0;	快速退刀至换刀点
Z150.0;	
M09;	关闭切削液
M05;	主轴停止
M00;	程序暂停
M03 S1500 F0.1;	主轴正转，转速为 1500r/min，设进给量为 0.1mm/r
T0101;	代入 01 号刀具补偿
M08;	打开切削液
G00 Z5.0;	快速进刀至精车循环起点
X42.0;	
G70 P10 Q20;	精加工复合循环
G00 X150.0;	快速退刀至换刀点
Z150.0;	
M09;	关闭切削液
M05;	主轴停止
M00;	程序暂停
M03 S400 F0.05;	主轴正转，转速 400r/min ，进给量 0.05mm/r
T0303;	换 T03　4mm 刀宽切槽刀，代入 03 号刀具补偿
M08;	打开切削液
G00 Z5.0;	刀具快速定位至进刀点
X40.0;	

续表 2-46

G01 Z-15.0 F20;	刀具定位至切槽处
X33.4 F0.05;	切槽，直径留单边 0.2 加工余量
X38.5 F5;	退刀
W3.5;	刀具向右平移
X33.4 F0.05;	切槽，直径留单边 0.2 加工余量
X38.5 F5;	退刀
W1.52;	刀具向右平移，保证尺寸
X32.969 F0.05;	切槽至尺寸
Z-15.0;	刀具向左平移至起点
X38.5;	退刀
G00 X150.0;	快速退刀至换刀点
Z150.0;	
M09;	关闭切削液
M05;	主轴停止
M00;	程序暂停
M03 S400 F0.1;	主轴正转，转速为 400r/min，设进给量为 0.1mm/r
T0202;	换 T02 镗孔刀，代入 02 号刀具补偿
M08;	打开切削液
G00 X20.0;	快速进刀至 *X* 方向粗车循环起点（钻孔直径位置）
Z5.0;	快速进刀至 *Z* 方向粗车循环起点
G71 U1.5 R0.5;	设置内孔粗车循环，切削深度 1.5mm，退刀量为 0.5mm
G71 P30 Q40 U-0.5 W0.05;	精车路线由 N30～N40 指定，*X* 向精车余量 0.5mm，*Z* 向精车余量 0.05mm
N30 G00 X26.0;	左端精车内轮廓
G01 Z0.0;	
X24 C2.0;	
Z-21.0;	
N40 X20.0;	
G00 Z150.0;	快速退刀至换刀点
X150.0;	
M09;	关闭切削液
M05;	主轴停止
M00;	程序暂停
M03 S1200 F0.08;	主轴正转，转速为 1200r/min，设进给量为 0.08mm/r
T0202;	代入 02 号刀具补偿
M08;	打开切削液

续表 2-46

G00 X20.0;	快速进刀至精车循环起点
Z5.0;	
G70 P30 Q40;	精加工复合循环
G00 Z150.0;	快速退刀至换刀点
X150.0;	
M09;	关闭切削液
M05;	主轴停止
M00;	程序暂停
M03 S400;	主轴正转，转速为400r/min
T0404;	换T04内孔螺纹刀，代入04号刀具补偿
M08;	打开切削液
G00 X20.0;	快速进刀至加工循环起点
Z5.0;	
G76 P020060 Q30 R0.05;	螺纹切削复合循环
G76 X24.0 Z-21.0 P1300 Q400 F2.0;	
G00 Z150.0;	快速退刀至换刀点
X150.0;	
M30;	程序结束同时光标返回程序头

（4）工件加工

1）该零件为调头加工零件，因此在加工时首先应考虑调头装夹时是否可以装夹、是否拥有压表找正部位、是否方便测量及两个方向加工是否会出现明显痕迹等问题。因此在加工工件左端外圆时，选择加工至零件外圆槽中间部位，保证零件外圆加工尺寸要求即可。

2）加工内孔部分时，需要注意G71粗加工复合循环指令第二行中的U值设置为负值。同时如果需要在机床【刀偏/磨损】位置输入数值，进行半精加工保证零件尺寸精度，留余量同样需要输入负值；在编程时，同样需要注意进刀、退刀时的运行先后顺序，保证快速安全定位。

3）内孔测量时，由于内径百分表是相对测量，用来校核的外径千分尺的精度同样至关重要，有必要首先校准外径千分尺的精度，同时需要注意外径千分尺读数。测量时，应在轴向和径向上多次找点测量。

4）加工工件右端时，注意调头时如何确定工件坐标系 Z 值，此步骤是保证工件总长度的关键。

5）调头加工时，由于刀具相对于工件回转中心（主轴中心）的距离没有变化，所以刀具的 X 值坐标系参数不需要重新确定。

6）加工外圆槽时，应注意槽刀刀片宽度，仔细计算后编程。

7）加工内孔螺纹底孔时，螺纹是否合格主要取决于中径尺寸，所以底孔尺寸不是主要尺寸，可不进行半精加工，使用游标卡尺测量即可。

8）G76 指令采用斜进法切削螺纹，即螺纹刀始终用一个刀刃进行切削，减小了切削阻力，提高了刀具寿命和螺纹精车质量，此外，该指令加工效率较高，所以加工该内螺纹优先使用 G76 编程指令。零件完成效果如图 2-119 所示。

图 2-119　零件完成效果图

（5）尺寸检测：

1）本任务为最终任务，游标卡尺只允许刀具确定工件坐标系时使用，不用游标卡尺做最终工件的测量。

2）外径尺寸部分：采用 25-50 外径千分尺测量两端外径尺寸 $\phi38^{0}_{-0.039}$，槽直径 $\phi33^{0}_{-0.062}$。

3）内径尺寸部分：采用内径百分表测量内孔尺寸 $\phi30^{+0.052}_{0}$。

4）长度尺寸部分：采用 25～50 外径千分尺测量工件长度尺寸 40±0.05；采用 5～30 内测千分尺测量槽宽尺寸 $\phi9^{+0.04}_{0}$；采用 0～25 深度千分尺测量工件内孔深度尺寸 20。

5）内孔螺纹部分：采用 M24×2-7H 螺纹塞规检测内孔螺纹是否合格。

6）表面粗糙度：采用粗糙度比较样块比较工件整体粗糙度是否合格。

（6）零件装配：

任务 2.6 零件与任务 2.5 零件完成装配，装配图如图 2-120 所示。可见螺纹轴件的 $\phi30$ 外圆与螺纹轴件的 $\phi30^{+0.052}_{0}$ 内孔相配合；M24×2 外螺纹与内孔螺纹相配合；配合间隙要求 2±0.03，同时保证总长度 72±0.02。

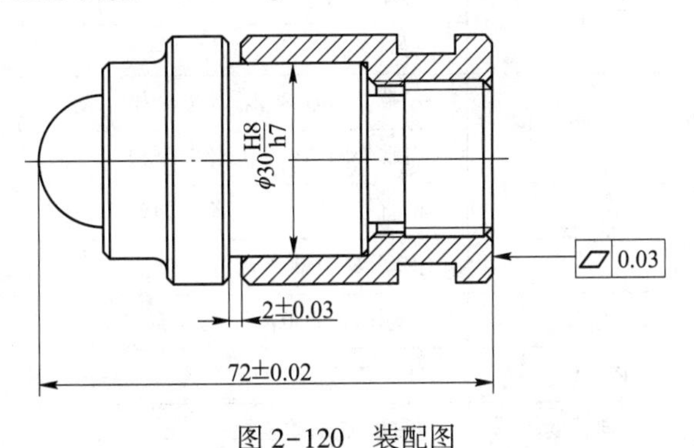

图 2-120　装配图

配合间隙保证方法：在保证螺纹轴尺寸的情况下（任务 2.5 零件完成），任务 2.6 零件螺纹套内孔深度尺寸 20 应在首先保证配合间隙 2±0.03 要求的前提下完成加工。

配做装配时，应首先清理内外螺纹上的切屑，防止切屑旋入螺纹，导致螺纹无法取下。

首次装配零件，应注意保持两工件轴线重合，一只手端稳工件，另一只手尝试平稳旋转工件，切记不可使用蛮力。

该装配组件存在螺纹及圆柱面两个配合部位，装配时出现无法旋入的情况，应首先分别检测两个零件的圆柱面及螺纹是否合格。尺寸精度修改完成后，再调整工件长度，保证零件配合间隙。

任务评价

任务评价见表 2-47。

表 2-47　任务评价表

项目	序号	评分标准		配分	得分
加工工艺（10%）	1	加工步骤	符合数控车工工艺要求	3	
	2	尺寸计算	计算相关部位尺寸	3	
	3	刀具选择	刀具选择安装合理，符合加工要求	4	
程序编制（30%）	4	程序号	无程序号无分	1	
	5	程序段号	无程序段号无分	1	
	6	切削用量	切削用量选择不合理无分	4	
	7	原点及坐标	标明程序原点及坐标，否则无分	2	
	8	程序内容	不符合程序逻辑及格式要求每段扣 2 分	22	
			程序内容与工艺不对应扣 5 分		
			出现危险指令扣 5 分		
加工操作（20%）	9	程序输入与检索	每错一段扣 2 分	10	
	10	加工操作 30min	工件坐标系原点设定错误无分	10	
			误操作无分		
			超时无分		
尺寸检测（30%）	11	外圆	左端外圆 $\phi38^{0}_{-0.039}$ 超差全扣	3	
	12		右端外圆 $\phi38^{0}_{-0.039}$ 超差全扣	2	
	13	外圆槽	槽直径 $\phi33^{0}_{-0.062}$ 超差全扣	2	
	14		槽宽尺寸 $9^{+0.04}_{0}$ 超差全扣	1	
	15	内孔	左端内孔 $\phi30^{+0.052}_{0}$ 超差全扣	5	
	16	内螺纹	M24×2-7H	5	
	17	轴向长度	40±0. 05 超差全扣	2. 5	
	18	倒角	5 处	2. 5	
	19	表面粗糙度 *Ra*3. 2	外轮廓及端面，每处不合格扣 0. 5 分	1. 5	
	20		内轮廓，降级无分	1. 5	
	21	配合尺寸	配合间隙 2±0. 03 超差全扣	4	
职业能力（10%）	22	学习能力		2	
	23	表达沟通能力		2	
	24	团队合作		2	
	25	安全操作与文明生产		4	
合计		100			

同步思考与训练

(1) 思考题：

1) 简述内孔部分编程与外轮廓编程区别。

2) 简述内测千分尺的使用方法。

3) 简述配合尺寸加工的方法。

(2) 同步训练题：

写出加工程序并加工图 2-121、图 2-122 所示工件，并完成装配，装配图如图 2-123、图 2-124 所示。

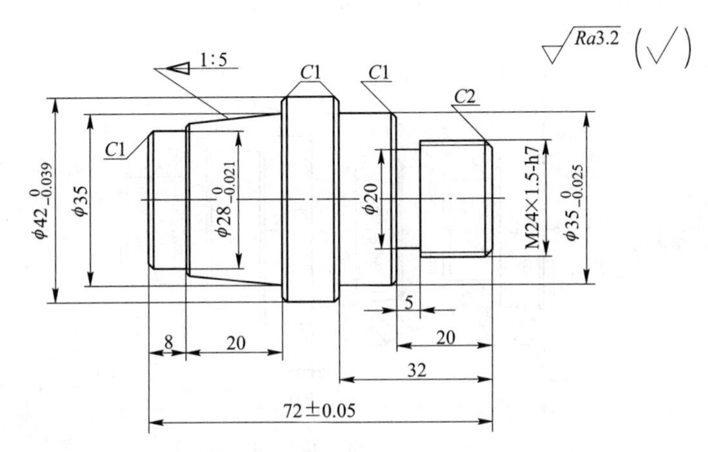

技术要求

1.未注倒角均为C0.5，锐角倒钝。

2.未注线性尺寸公差应符合GB/T 1804—2000的要求。

图 2-121　件 1

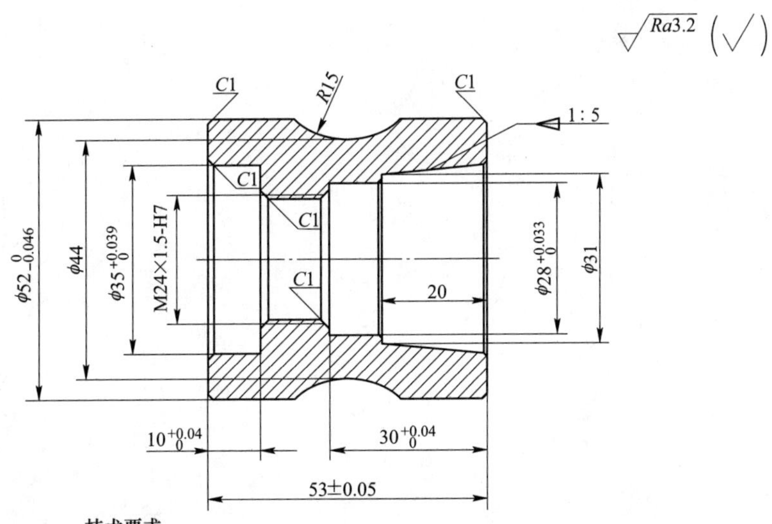

技术要求

1.未注倒角均为C0.5，锐角倒钝。

2.未注线性尺寸公差应符合GB/T 1804—2000的要求。

图 2-122　件 2

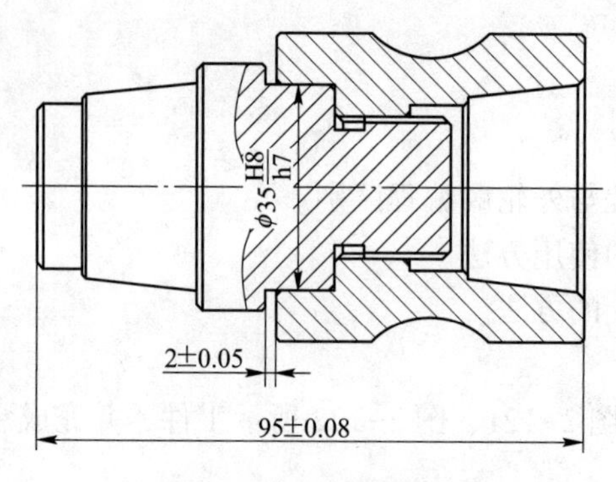

图 2-123　装配图 1

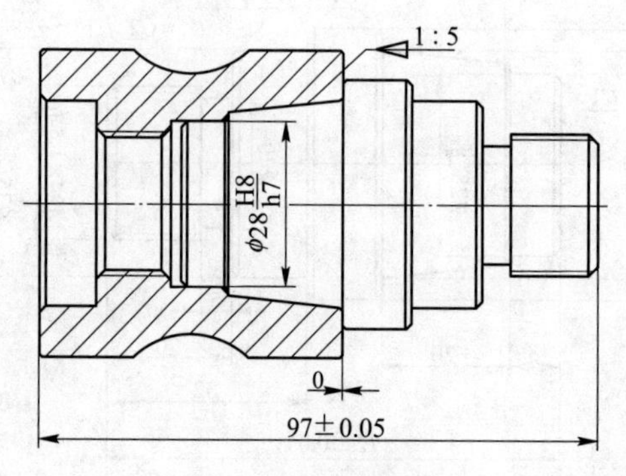

图 2-124　装配图 2

附　　录

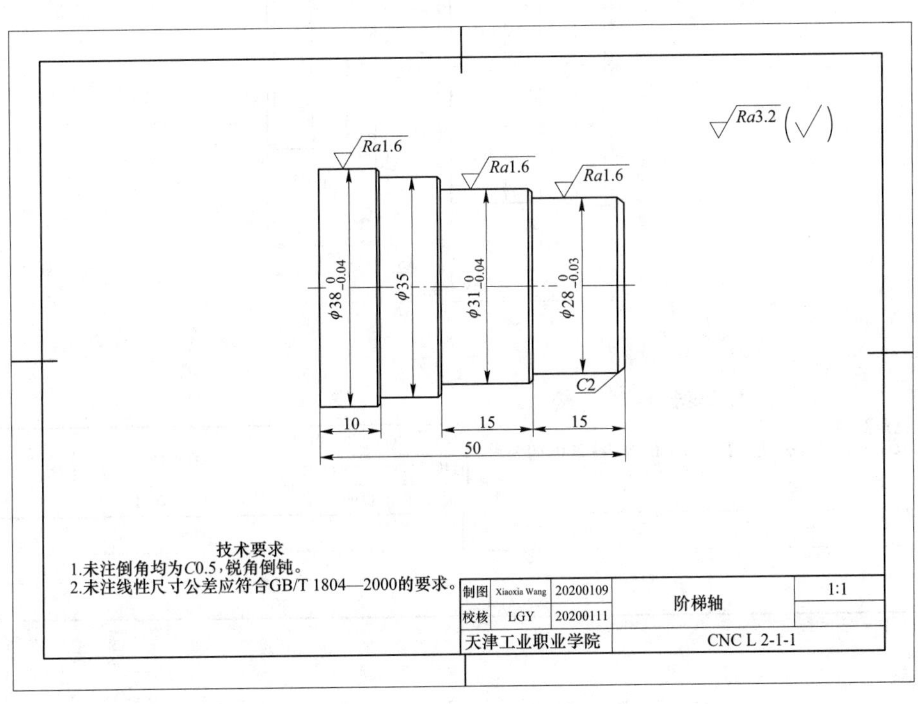

Ra3.2 (√)
Ra1.6
Ra1.6
Ra1.6
φ38 0 -0.04
φ35
φ31 0 -0.04
φ28 0 -0.03
C2
10
15
15
50
技术要求
1.未注倒角均为C0.5,锐角倒钝。
2.未注线性尺寸公差应符合GB/T 1804—2000的要求。
制图
Xiaoxia Wang
20200109
校核
LGY
20200111
阶梯轴
1:1
天津工业职业学院
CNC L 2-1-1

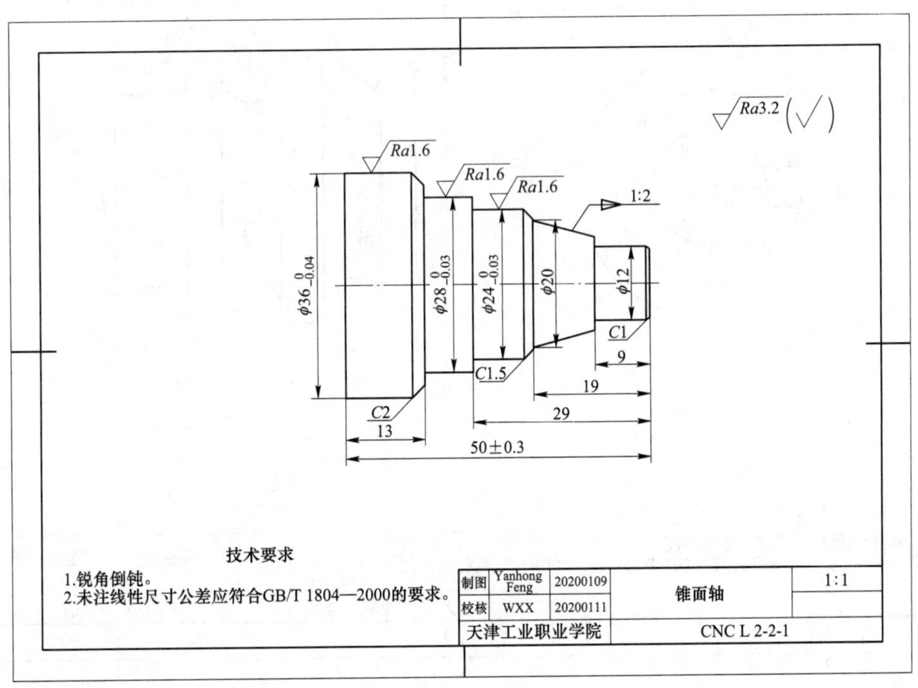

Ra3.2 (√)
Ra1.6
Ra1.6
Ra1.6
1:2
φ36 0 -0.04
φ28 0 -0.03
φ24 0 -0.03
φ20
φ12
C1
C1.5
C2
9
19
29
13
50±0.3
技术要求
1.锐角倒钝。
2.未注线性尺寸公差应符合GB/T 1804—2000的要求。
制图
Yanhong Feng
20200109
校核
WXX
20200111
锥面轴
1:1
天津工业职业学院
CNC L 2-2-1

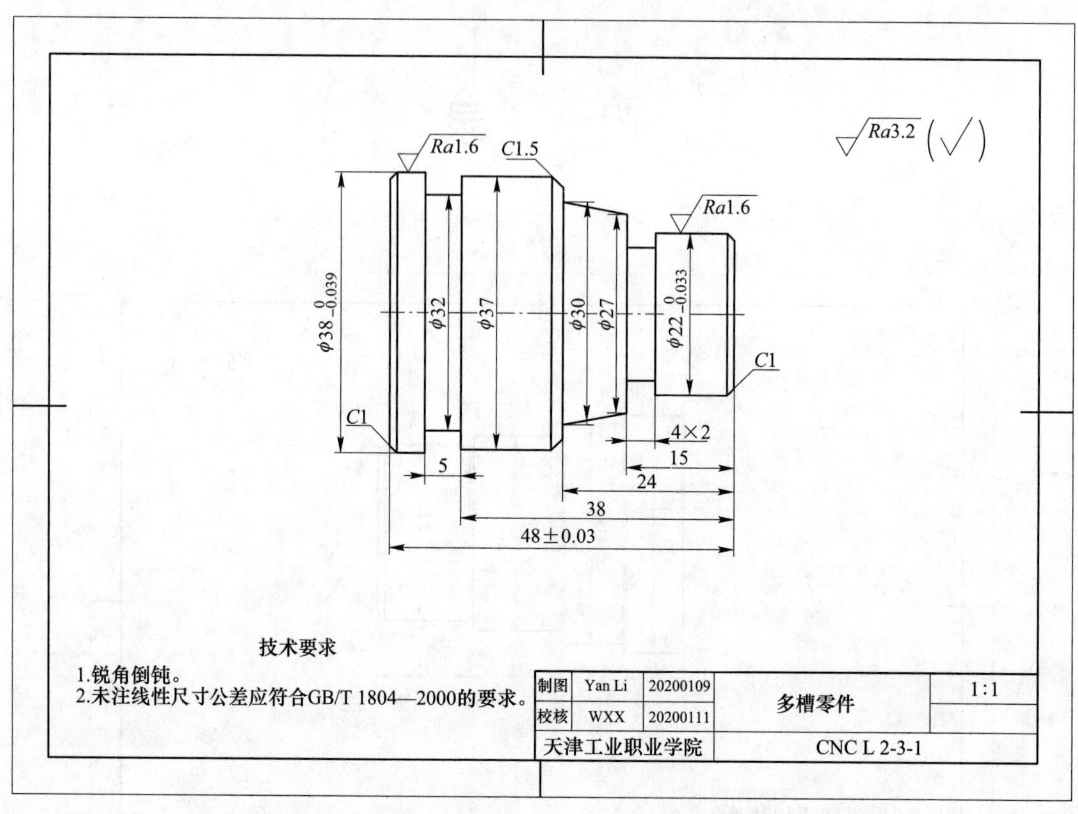

Ra1.6
C1.5
Ra3.2
Ra1.6
$\phi38_{-0.039}^{\ 0}$
$\phi32$
$\phi37$
$\phi30$
$\phi27$
$\phi22_{-0.033}^{\ 0}$
C1
C1
4×2
5
15
24
38
48±0.03
技术要求
1.锐角倒钝。
2.未注线性尺寸公差应符合GB/T 1804—2000的要求。
制图 Yan Li 20200109
校核 WXX 20200111
多槽零件
1:1
天津工业职业学院
CNC L 2-3-1

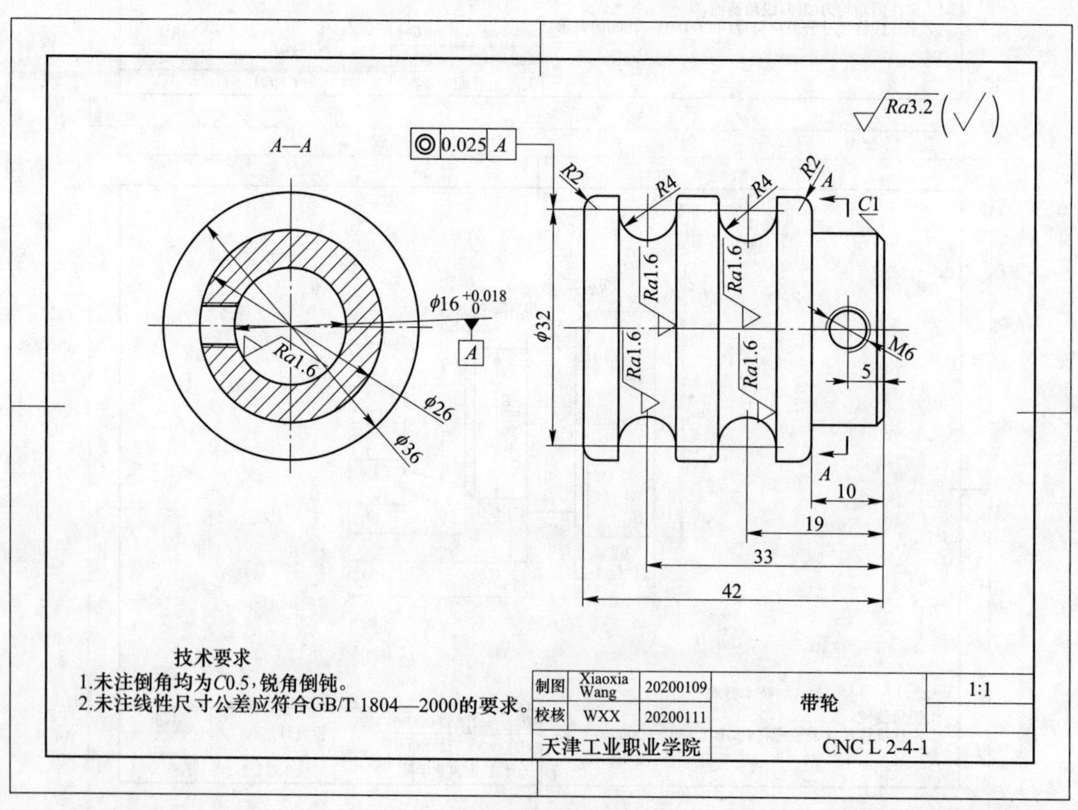

Ra3.2
A—A
0.025 A
R2
R4
R4
R2
A
C1
$\phi16_{\ 0}^{+0.018}$
A
Ra1.6
Ra1.6
Ra1.6
Ra1.6
Ra1.6
$\phi32$
$\phi26$
$\phi36$
M6
5
A
10
19
33
42
技术要求
1.未注倒角均为C0.5，锐角倒钝。
2.未注线性尺寸公差应符合GB/T 1804—2000的要求。
制图 Xiaoxia Wang 20200109
校核 WXX 20200111
带轮
1:1
天津工业职业学院
CNC L 2-4-1

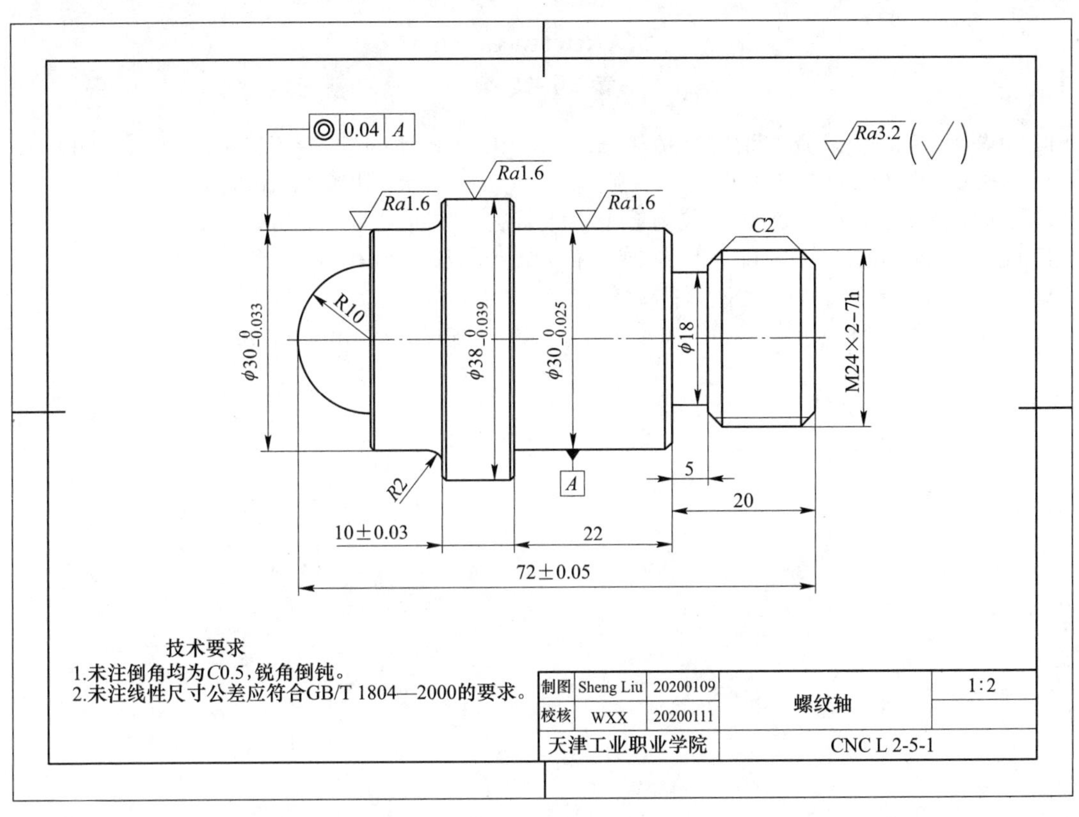
0.04 A
Ra3.2
Ra1.6
Ra1.6
Ra1.6
C2
R10
φ30 0 −0.033
φ38 0 −0.039
φ30 0 −0.025
φ18
M24×2−7h
R2
A
5
20
10±0.03
22
72±0.05
技术要求
1.未注倒角均为C0.5，锐角倒钝。
2.未注线性尺寸公差应符合GB/T 1804—2000的要求。
制图
Sheng Liu
20200109
校核
WXX
20200111
螺纹轴
1:2
天津工业职业学院
CNC L 2-5-1

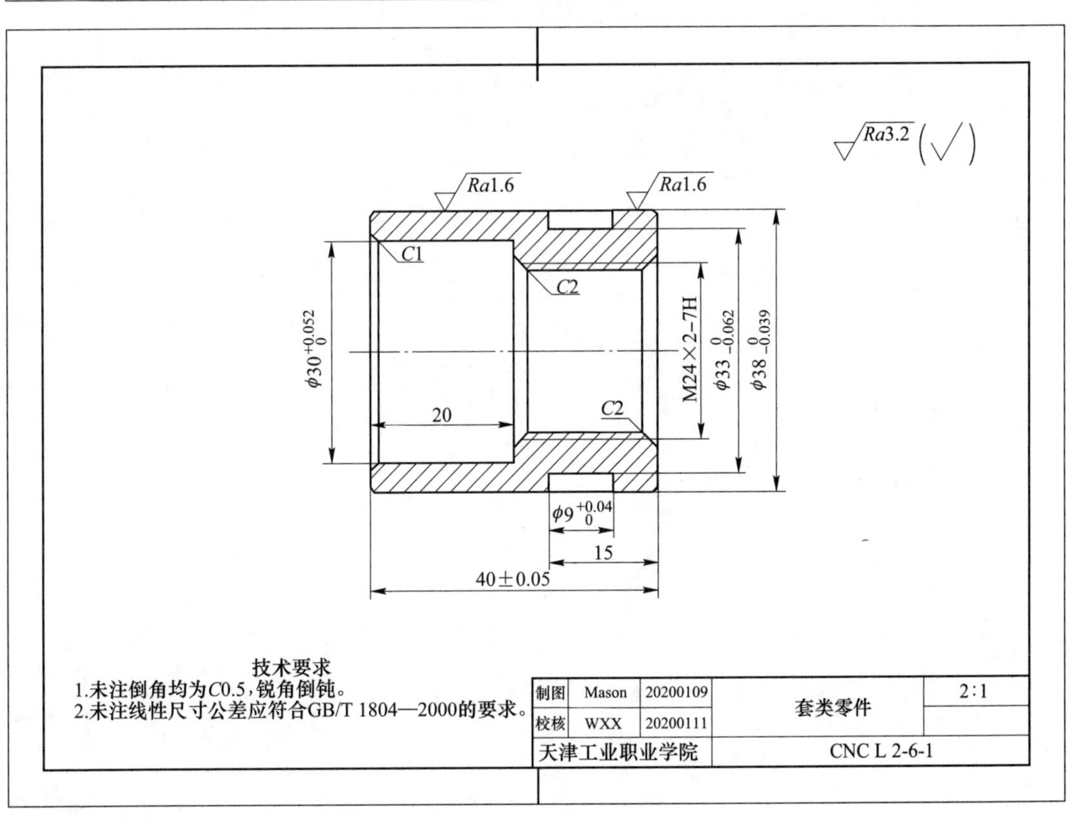
Ra3.2
Ra1.6
Ra1.6
C1
C2
C2
φ30 +0.052 0
20
M24×2−7H
φ33 0 −0.062
φ38 0 −0.039
φ9 +0.04 0
15
40±0.05
技术要求
1.未注倒角均为C0.5，锐角倒钝。
2.未注线性尺寸公差应符合GB/T 1804—2000的要求。
制图
Mason
20200109
校核
WXX
20200111
套类零件
2:1
天津工业职业学院
CNC L 2-6-1

References
参 考 文 献

［1］中华人民共和国人力资源和社会保障部. 6180101 国家职业技能标准——车工［S］. 北京，2019，1.
［2］李桂云 . 数控编程及加工技术［M］. 3 版 . 大连：大连理工大学出版社，2018，6.
［3］北京第一通用机械厂 . 机械工人切削手册［M］. 8 版 . 北京：机械工业出版社，2009.
［4］成大先 . 机械设计手册［M］. 6 版 . 北京：化学工业出版社，2016，5.
［5］李梅红 . 机械制图［M］. 北京：高等教育出版社 . 2019，8.
［6］李桂云 . 数控技术应用专业英语［M］. 4 版 . 大连：大连理工大学出版社，2018，6.